RAM Research in Applied Mathematics
Series Editors : P.G. CIARLET and J.-L. LIONS

FLUIDS AND PERIODIC STRUCTURES

C. CONCA

Professor at the University of Chile
Santiago, Chile

J. PLANCHARD

Research Engineer at Electricité de France
Paris, France

M. VANNINATHAN

Reader at Tata Institute of Fundamental Research
IISc-TIFR Mathematics Programme
Bangalore, India

JOHN WILEY & SONS
Chichester ● New York ● Brisbane ● Toronto ● Singapore

1995

MASSON
Paris ● Milan ● Barcelona

La collection **Recherches en Mathématiques Appliquées** a pour objectif de publier dans un délai très rapide des textes de haut niveau en Mathématiques Appliquées, notamment :

— des cours de troisième cycle,
— des séries de conférences sur un sujet donné,
— des comptes rendus de séminaires, congrès,
— des versions préliminaires d'ouvrages plus élaborés,
— des thèses, en partie ou en totalité.

Les manuscrits, qui doivent comprendre de 120 à 250 pages, seront reproduits directement par un procédé photographique. Ils devront être réalisés avec le plus grand soin, en observant les normes de présentation précisées par l'Éditeur.
Les manuscrits seront rédigés en français ou en anglais. Dans tous les cas, ils seront examinés par au moins un rapporteur. Ils seront soumis directement soit au

The aim of the **Recherches en Mathématiques Appliquées** series (Research in Applied Mathematics) is to publish high level texts in Applied Mathematics very rapidly :

— Post-graduate courses
— Lectures on particular topics
— Proceedings of congresses
— Preliminary versions of more complete works
— Theses (partially or as a whole)

Manuscripts which should contain between 120 or 250 pages will be printed directly by a photographic process. They have to be prepared carefully according to standards defined by the publisher.
Manuscripts may be written in English or in French and will be examined by at least one referee.
All manuscripts should be submitted to

Professor P.G. Ciarlet, Analyse numérique, T. 55,
Université Pierre et Marie Curie, 4, place Jussieu, 75005 Paris
soit au / or to
Professor J.-L. Lions, Collège de France,
11, place Marcelin-Berthelot, 75005 Paris

© Masson, Paris, 1995

ISBN Masson : 2-225-84821-1

ISBN Wiley : 0-471-96059-4

ISSN : 0298-3168

MASSON S.A.
JOHN WILEY AND SONS Ltd

120, bd Saint-Germain, 75280 Paris Cedex 06
Baffins Lane, Chichester, West Sussex PO 19 1UD, England

Contents

Chapter I
Elements of Spectral Theory with Examples

Chapter II
Spectral Problems in Fluid-Solid Structures

Chapter III
Asymptotic Methods in Fluid-Solid Structures

Preface

This book is addressed to mathematicians as well as engineers.

In the last few decades, many of them have concentrated their efforts in the understanding of various phenomena involved in fluid-solid structures. In particular, the problem of vibrations of a mechanical structure which is interacting with a fluid surrounding it has been of interest to many researchers. The purpose of this work is to show one mathematical approach to comprehend such structures.

To convince the reader of the significance of this subject, let us start by mentioning one practical situation in which vibrations of fluid-solid structures occur. It is well-known that in the process of producing steam in nuclear reactors, heat exchanges take place by placing tube bundles inside a moving fluid. This results in the movement of the tubes which modifies the pressure field of the fluid and thus acts on the neighbouring tubes. This is one aspect of the phenomenon of interaction between fluids and solids. Of course, there is a pleanty of other situations where such interactions take place. Let us cite:

- high factory chimnies, skyscrapers, electric cables and cableways which vibrate under the action of wind,
- ships and off-shore oil platforms vibrating under the action of ocean waves.

These vibrations which may be of small or large amplitudes cause damages to the very structure in the long run (wear and tear of materials through friction and shocks) and, of course, resonance can occur which may even destroy the structure quickly†. It is easy to understand the importance of these problems in the context of nuclear reactors. In fact, any possible deterioration of the mechanical structures in the combustion chamber may release radioactive products into the cooling circuit and contaminate the entire system. In the event of such an accident, the reactor has to be stopped and an expensive and long process of decontamination has to be undertaken.

It is worth pointing out that the various phenomena which may arise in fluid-solid interactions were somewhat underestimated in the past. However, the pioneering work of CONNORS [1970] triggered off an abundant experimental, numerical and theoretical research concerning these coupled structures

---

† A famous example of such a phenomenon was the destruction of the Tacoma suspension bridge in the State of Washington in the west coast of U.S.A.

in different countries. A measure of these activities is reflected by the ever-increasing number of articles, conference proceedings, books and the recent appearance of specific journals in this field. In the engineering side, remarkable progress has been made by different teams around the world and a special mention may be made of the groups headed by R.D. Blevins, S.S. Chen and M.P. Paidoussis. Applied mathematicians, realizing the importance of the subject, were not lagging behind. Groups have been formed around some individuals in various research laboratories and universities. As a sample, let us point out the several significant contributions made in this field by E. Sánchez-Palencia and his collaborators in France and by the team headed by S.S. Antman in U.S.A.

Among many ways of modelling the physical phenomena involved, our aim in this book is to provide a detailed mathematical analysis of the models introduced by J. Planchard in the early 80's. The results presented here are the outcome of collaborative research of the authors pursued for the last ten years or so. At the outset, we would like to remark that each way of modelling has its own virtues and drawbacks and that the very conception and the approach we follow towards this subject are not the only ones to understand the various phenomena arising in fluid-solid structures. In fact, there are several other points of view which differ from ours in both mathematical and physical contents. Not only the models treated are different but also the mathematical analysis used to study them are not the same. This is the case for instance with the models considered by S.S. Antman and his collaborators or those studied by H. Morand and R. Ohayon. If one wants to place our work among the existing books in the literature, the closest one that comes to our mind naturally is the book by SÁNCHEZ-HUBERT and SÁNCHEZ-PALENCIA [1989]. When we planned the book, our intention was all the time to remain within reasonable bounds and never to write a treatise. There are therefore several aspects of Vibration Theory not considered here for which the reader can consult classical works such as STOKER [1950], ROSEAU [1966], [1978].

We now outline the contents of the book which is divided into three chapters. Each of them has several sections and they are further divided into paragraphs. The principal tools we use to understand the fluid-solid structures are borrowed from Functional Analysis and more specifically Spectral Theory of Linear Operators. Keeping in mind to serve a wide and varied audience, we have decided to include sufficient mathematical background to make the book more self-contained. This is why, Chapter I is mainly devoted to the introduction of various concepts and terminologies from the above theory that we shall constantly apply in the sequel. In the same chapter, the reader is gradually introduced to the phenomena of vibrations through different examples taken from real life. The corresponding mathematical theory is presented in such a way that the results are first formally deduced through intuitive explanations and then rigourously justified. However, for a better understanding of Chapter I, it is desirable that the reader be familiar with Functional Analysis and Sobolev Spaces.

In Chapter II, we introduce a class of fluid-solid structures which constitutes the subject matter of the book. Here, we will be concerned with some aspects of modelling the vibrations of such structures, with the validation of such models by proving existence theorems as well as exhibiting qualitative properties by obtaining practical apriori bounds on the spectrum. In conclusion of this chapter some of the most favourite numerical methods employed by engineers to compute frequencies of vibrations of such structures are presented along with the mathematical justification.

Chapter III forms the crux of the subject because of its relevance in practice. We undertake to describe the asymptotic behaviour of the vibrations when the fluid-solid mixture gets finer and finer. To motivate the reader, let us go back to the example of the nuclear reactor given earlier. In order to exploit efficiently the heat produced, it is necessary to increase the area of the interface between the tubes and the fluid. One way of achieving this is to increase the number of tubes present. A mathematical model of this situation studied in this work assumes periodic distribution of the tubes and this justifies the title of the book. We will be therefore concerned here with a problem of homogenization. This is a task which we accomplish in two different ways. First, by following the *standard homogenization method* and secondly, by proposing an alternative which we will refer to as *non-standard homogenization procedure*. While the latter one is based on the so-called Bloch Wave Decomposition which is a technique that is often used in Solid State Physics, the standard homogenization method exploits the classical multiple scale expansions and a more recent tool known as two-scale convergence technique. The conclusions reached in each case are qualitatively different but not contradictory.

To understand the intricacies of this subject and enjoy its beauty, the reader is highly recommended to go through the masterpiece BENSOUSSAN, LIONS and PAPANICOLAOU [1978] in Homogenization Theory.

A good part of the different topics which constitute the book is illustrated by examples and supplemented by exercises. The interested students are encouraged to solve these exercises which are often supplied with hints and guidelines. The introduction of each chapter is very brief whereas a detailed one containing motivation and relevant bibliography may be found at the beginning of each section. Numbers of various formulae include that of the section to which they belong and in referring to them outside a chapter, explicit mention of the number of the chapter is made. We conclude the book with a list of Open Questions which arose naturally in the course of our discussions and for which we do not have complete answers.

To ease the presentation, we have started with simple cases and passed onto complicated ones step by step. Whenever we feel that the arguments are repetitive, we simply refer the reader to the simpler cases treated earlier without going into details. Therefore, a given section cannot be considered independent of earlier ones.

Acknowledgements

We wish to express our deep and sincere gratitude to many of our friends and collaborators who very generously assisted us during the period of association. Our thanks are especially due to F. Aguirre, M. Durán, M. Ibnou-Zahir, B. Thomas and the students of the Department of Mathematical Engineering of the University of Chile, L. Baffico and A. Osses. We are also deeply grateful to G. Allaire, S. Kesavan and E. Zuazua for going through the manuscript and for their comments and valuable remarks.

Part of this work was carried out while M. Vanninathan was visiting the Departamento de Ingeniería Matemática of the University of Chile at Santiago. He thanks the following organizations without whose generous support this book would not have seen the light of the day:
- Departamento de Extensión y Cooperación Internacional, Departamento de Post-Grado y Post-Título, Departamento Técnico de Investigación and Facultad de Ciencias Físicas y Matemáticas, all part of the University of Chile.
- Comisión Nacional de Investigación Científica y Tecnológica (CONICYT), Chile.

The text was entirely typed by Ms. Regina Mateluna and we are deeply grateful for her efficient and rapid work.

This introduction will be incomplete if we do not mention the moral support, constant encouragement and the help rendered by Professor Jacques-Louis Lions of Collège de France from the beginning of our research career.

Chapter I

Elements of Spectral Theory with Examples

This chapter is divided into four sections. The first one mainly presents the mathematical background which will be in constant use throughout the text. In section two, we illustrate the phenomena of vibrations with examples drawn from various situations in practical life and several mathematical models are presented. In this connection, we stress the difference between the vibrations which correspond to eigenvalues and eigenvectors of these models and those which do not. Section 3 is devoted to the classical spectral theory of self-adjoint operators. This is done in such a way that it provides in particular a mathematical justification of various phenomena observed in the earlier sections. We end this chapter with a section where we examine the effects on the spectrum under perturbations of operators.

§1. Some Function Spaces and their Properties

The goal of this section is to define certain function spaces which will be used in the formulation of various spectral problems that we shall consider in this book. We also recall some of the essential properties of these spaces. The reader is assumed to be familiar with the theory of distributions, Fourier transformation and Sobolev spaces. Since we do not intend to provide any details of the proofs, we suggest that the readers consult the existing literature on these topics. For example, concerning the theory of distributions and Fourier transform, we cite DAUTRAY and LIONS [1988-90] Volume II and SCHWARTZ [1965]. For Lebesgue and Sobolev spaces, we cite ADAMS [1975], BREZIS [1983] Chapters VIII and IX, GIRAULT and RAVIART [1986] Chapter I, GRISVARD [1985], MAZ'JA [1985], NEČAS [1967] and RAVIART and THOMAS [1983] Chapter I. In our analysis of coupling of fluid-solid structures, we will have occasion to use other types of function spaces apart from the usual Lebesgue and Sobolev spaces. These will be introduced as and when they are needed. We will provide proofs of the main results concerning these later spaces.

1.1. Preliminary Notations

Throughout this text, we will systematically deal with complex-valued functions u, v, w,..., ϕ, ψ,.... unless mentioned very explicitly to the contrary. The real part, the imaginary part, the conjugate and the absolute value of a function u are denoted respectively as $\mathfrak{Re}(u)$ or simply $\mathfrak{Re}\, u$, $\mathfrak{Im}(u)$ or $\mathfrak{Im}\, u$, $\bar{u}$ and $|u|$.

Let Ω be any open set in $\mathbb{R}^N$ where a generic point is denoted as $x = (x_1,...,x_N)$. We shall denote by Γ its boundary. The space of (class of) *square integrable functions* on Ω will be denoted by $L^2(\Omega)$, i.e., it is the space of Lebesgue measurable functions u such that

$$(1.1) \qquad \|u\|_{0,\Omega} \overset{\text{def}}{=} \left(\int_\Omega |u|^2 dx \right)^{1/2} < \infty,$$

where $dx = dx_1...dx_N$ stands for the Lebesgue measure in Ω. It is well-known that $L^2(\Omega)$ is a Hilbert space with the inner product

$$(u,v) \overset{\text{def}}{=} \int_\Omega u\bar{v}\,dx,$$

which gives rise to the norm (1.1). When no confusion can arise, we will continue to use same symbols for the norms and the inner products for vector-valued functions in $[L^2(\Omega)]^r$ where r is any integer ≥ 1.

The symbols

$$D^\alpha = \frac{\partial^{|\alpha|}}{\partial x_1^{\alpha_1} \cdots \partial x_N^{\alpha_N}} \quad \text{with} \quad \alpha = (\alpha_1,...,\alpha_N) \in (\mathbb{Z}^+)^N, \ |\alpha| = \alpha_1 + \cdots + \alpha_N$$

denote the standard operators of partial derivatives. For a non-negative integer m, let $\mathcal{C}^m(\Omega)$ be the space of all functions ϕ such that $D^\alpha \phi$, $|\alpha| \leq m$ are continuous in Ω. We abbreviate $\mathcal{C}^0(\Omega) = \mathcal{C}(\Omega)$.

The classical space of infinitely differentiable functions with compact support in Ω is written as usual $\mathcal{D}(\Omega)$ and its topological dual space $\mathcal{D}'(\Omega)$ is the well-known space of *distributions* on Ω. We define also the space $\mathcal{C}^m(\bar{\Omega})$ to consist of all functions ϕ such that $D^\alpha \phi$, $|\alpha| \leq m$ are bounded and uniformly continuous on Ω. Note that

$$\mathcal{C}^m(\overline{\mathbb{R}}^N) \neq \mathcal{C}^m(\mathbb{R}^N).$$

It is a classical fact that $\mathcal{C}^m(\bar{\Omega})$ is a Banach space for the norm

$$(1.2) \qquad \|\phi\|_{m,\infty,\Omega} = \max_{0 \leq |\alpha| \leq m} \sup_{x \in \Omega} |D^\alpha \phi(x)|$$

and that the functions ϕ in $C^m(\bar{\Omega})$ as well as their derivatives $D^\alpha\phi$, $|\alpha| \leq m$ possess unique, bounded and continuous extensions to $\bar{\Omega}$. We pose

$$C^\infty(\bar{\Omega}) = \bigcap_{m \geq 0} C^m(\bar{\Omega}).$$

The subspace of $C^\infty(\bar{\Omega})$ which consists of functions having compact support in $\bar{\Omega}$ is denoted by $\mathcal{D}(\bar{\Omega})$.

We will now introduce *Sobolev spaces*. Physically, elements of these spaces represent the configurations of mechanical systems with finite energy. The classical C^m spaces are not suitable for this purpose. For $m \in \mathbb{N}$, we define then

(1.3) $\qquad H^m(\Omega) = \{ u \in \mathcal{D}'(\Omega) \mid D^\alpha u \in L^2(\Omega) \ \forall \alpha, \ |\alpha| \leq m \},$

where the derivatives D^α are taken in the sense of distributions. This space is endowed with the norm

(1.4) $\qquad \|u\|_{m,\Omega} \overset{\text{def}}{=} \Big(\sum_{|\alpha| \leq m} \|D^\alpha u\|_{0,\Omega}^2 \Big)^{1/2},$

which makes $H^m(\Omega)$ a complete space. It becomes a Hilbert space with the following inner product:

$$((u,v)) \overset{\text{def}}{=} \sum_{|\alpha| \leq m} \int_\Omega D^\alpha u D^\alpha \bar{v} dx.$$

We use the notation $H_0^m(\Omega)$ for the space which is the closure of $\mathcal{D}(\Omega)$ in $H^m(\Omega)$.

1.2. Trace Results and Green's Formulae

We are now going to introduce some elements of *trace theory*, i.e., restriction of Sobolev class functions on the boundary Γ. For this, we need some regularity properties of the open set Ω. For our requirements, the following hypothesis is sufficient:

(1.5) $\qquad \Omega$ is bounded and its boundary Γ is Lipschitz,

i.e., the boundary Γ which is compact can be locally represented by the graph of a Lipschitz function. Under this condition, a $H^1(\Omega)$ function admits restriction on Γ in the following sense:

Theorem 1.1. *If Ω satisfies* (1.5), *then $C^\infty(\bar{\Omega})$ is dense in $H^1(\Omega)$ and there exists a unique linear continuous mapping γ_1 from $H^1(\Omega)$ into $L^2(\Gamma)$ such that for all $u \in H^1(\Omega) \cap C(\bar{\Omega})$ we have $\gamma_1 u = u_{|\Gamma}$. Moreover, the kernel of γ_1 is nothing but the space $H_0^1(\Omega)$.* ∎

When no confusion can arise, we simply put u in the place of $\gamma_1 u$ or $u_{|\Gamma}$. The trace map γ_1 is not onto $L^2(\Gamma)$. Its image is a subspace of $L^2(\Gamma)$ and is denoted as $H^{1/2}(\Gamma)$. This space is identified algebraically and topologically with the quotient space $H^1(\Omega)/H_0^1(\Omega)$ provided with the quotient norm

$$(1.6) \qquad \|u\|_{1/2,\Gamma} \stackrel{\text{def}}{=} \inf_{\substack{v \in H^1(\Omega) \\ \gamma_1 v = u}} \|v\|_{1,\Omega}.$$

A classical application of Theorem 1.1 is the Green's formulae given below:

$$(1.7) \qquad \int_\Omega u \frac{\partial \bar{v}}{\partial x_i} dx = -\int_\Omega \frac{\partial u}{\partial x_i} \bar{v} dx + \int_\Gamma u \bar{v} n_i ds$$

for all $i = 1, ..., N$, $u, v \in H^1(\Omega)$. In (1.7), $\mathbf{n} = (n_i)$ is the unit outward normal to the boundary Γ. It is well-defined almost everywhere with respect to the boundary measure ds because Γ is assumed to be Lipschitz. We remark that (1.7) is well-known in the class $C^\infty(\bar{\Omega})$ and Theorem 1.1 allows us to generalize it to the class $H^1(\Omega)$ by density arguments.

If $\Omega = \mathbb{R}^N$ then $\mathcal{D}(\mathbb{R}^N)$ is dense in $H^1(\mathbb{R}^N)$ (i.e. $H^1(\mathbb{R}^N) = H_0^1(\mathbb{R}^N)$) and (1.7) becomes

$$(1.8) \qquad \int_{\mathbb{R}^N} u \frac{\partial \bar{v}}{\partial x_i} dx = -\int_{\mathbb{R}^N} \frac{\partial u}{\partial x_i} \bar{v} dx$$

for all $i = 1, ..., N$, $u, v \in H^1(\mathbb{R}^N)$.

1.3. A Compactness Criterion

It is a standard practice to identify $L^2(\Omega)$ with its dual. This is consistent with the standard embedding of $L^2(\Omega)$ into $\mathcal{D}'(\Omega)$. As $\mathcal{D}(\Omega)$ is dense in $H_0^1(\Omega)$ and also in $L^2(\Omega)$, we can identify the dual of $H_0^1(\Omega)$ with a subspace of $\mathcal{D}'(\Omega)$. Denoting the dual by $H^{-1}(\Omega)$, we have the following dense and continuous injections:

$$\mathcal{D}(\Omega) \hookrightarrow H_0^1(\Omega) \hookrightarrow L^2(\Omega) \hookrightarrow H^{-1}(\Omega) \hookrightarrow \mathcal{D}'(\Omega).$$

The above injections are not only continuous but also compact as shown by the following more general result:

Theorem 1.2. (Rellich's Lemma)
  (i) *For bounded Ω, the canonical embedding $H_0^1(\Omega) \hookrightarrow L^2(\Omega)$ is compact, i.e., every bounded set in $H_0^1(\Omega)$ is relatively compact in $L^2(\Omega)$.*
 (ii) *If Ω satisfies the regularity condition (1.5), then the embedding of $H^1(\Omega)$ into $L^2(\Omega)$ and the trace map from $H^1(\Omega)$ into $L^2(\Gamma)$ are compact.* ∎

1.4. Poincaré Type Inequalities

Next, we proceed to recall the so-called Poincaré Inequality and some of its variants.

Theorem 1.3. (Poincaré Inequality) *If Ω is bounded, then there is a constant $c = c(\Omega) > 0$ such that*

$$\|u\|_{0,\Omega} \le c\|\nabla u\|_{0,\Omega} \quad \forall u \in H_0^1(\Omega),$$

where ∇u is simply the gradient of u. ∎

The above result implies that the semi-norm $\|\nabla u\|_{0,\Omega}$ is a norm on $H_0^1(\Omega)$ which is equivalent to the norm $\|u\|_{1,\Omega}$. A similar result holds on the space $H_{\Gamma_0}^1(\Omega)$ which consists of functions $u \in H^1(\Omega)$ which vanish on Γ_0, i.e.,

$$(1.9) \qquad H_{\Gamma_0}^1(\Omega) \stackrel{\text{def}}{=} \{u \in H^1(\Omega) \mid \gamma_1 u = 0 \quad \text{on} \quad \Gamma_0\},$$

which is a closed subspace of $H^1(\Omega)$. Here, Γ_0 is taken to be a non-empty open subset of Γ.

Let us also write down the inequalities of Poincaré-Wirtinger type: If Ω is connected then there exists a constant $c = c(\Omega) > 0$ such that

$$(1.10) \qquad \left\|u - \frac{1}{\text{meas}\,(\Omega)} \int_\Omega u\,dx\right\|_{0,\Omega} \le c\|\nabla u\|_{0,\Omega},$$

$$(1.11) \qquad \left\|u - \frac{1}{\text{meas}\,(\Gamma_0)} \int_{\Gamma_0} u\,ds\right\|_{0,\Omega} \le c\|\nabla u\|_{0,\Omega}$$

for all $u \in H^1(\Omega)$.

1.5. Properties of the Spaces $H^m(\Omega)$

Most of what we have said above in the case of $H^1(\Omega)$ can be generalized to $H^m(\Omega)$ spaces. In particular, the trace map γ_m is a continuous and linear map from $H^m(\Omega)$ into $(L^2(\Gamma))^m$ which extends the classical restriction map $u \longmapsto (u, \partial u/\partial n, ..., \partial^{m-1}u/\partial n^{m-1})_{|\Gamma}$ for $u \in C^m(\bar{\Omega})$. Here, $\partial^k u/\partial n^k$ stands for the kth derivative of u in the exterior normal direction $\mathbf{n}$ to the boundary Γ. As in Theorem 1.1, we can also characterize $H_0^m(\Omega)$ to be the kernel of the trace operator γ_m and the image of γ_m is equal to $H^{m-1/2}(\Gamma) \times H^{m-3/2}(\Gamma) \times \cdots \times H^{1/2}(\Gamma)$ defined in a manner analogous to (1.6) via quotient norm. Finally, the topological dual of $H_0^m(\Omega)$ is called $H^{-m}(\Omega)$ and that of $H^{m-1/2}(\Gamma)$ is $H^{-m+1/2}(\Gamma)$ for all $m \geq 1$. If $u \in H^2(\Omega)$ and $v \in H^1(\Omega)$, then we have the following Green's formula which is obtained as an easy consequence of (1.7):

$$(1.12) \qquad \int_{\Omega} \nabla u \cdot \nabla \bar{v} dx = - \int_{\Omega} \Delta u \bar{v} dx + \int_{\Gamma} \frac{\partial u}{\partial n} \bar{v} ds.$$

If $\Omega = \mathbb{R}^N$, there is an alternative definition of $H^m(\mathbb{R}^N)$ via Fourier transformation. To present it, let $v \in \mathcal{S}'(\mathbb{R}^N)$ be a *tempered distribution* and $\hat{v}$ be its Fourier transform. If $v \in L^1(\mathbb{R}^N)$, we recall the expression of $\hat{v}$:

$$(1.13) \qquad \hat{v}(\xi) = \frac{1}{(2\pi)^{N/2}} \int_{\mathbb{R}^N} v(x) e^{-ix \cdot \xi} dx \quad \forall \xi \in \mathbb{R}^N,$$

where $i = \sqrt{-1}$. We can now define the space $H^m(\mathbb{R}^N)$ as follows:

$$(1.14) \qquad H^m(\mathbb{R}^N) = \{v \in \mathcal{S}'(\mathbb{R}^N) \mid (1 + |\xi|^2)^{m/2} \hat{v}(\xi) \in L^2(\mathbb{R}^N)\}.$$

This is endowed with the norm

$$v \longmapsto \left\| (1 + |\xi|^2)^{m/2} \hat{v}(\xi) \right\|_{0, \mathbb{R}^N}.$$

Let us observe that the above definition makes sense for all $m \in \mathbb{R}$. When m is an integer these spaces coincide with the earlier ones with equivalent norms. This can be easily seen using the relations

$$(1.15) \qquad (\widehat{\frac{\partial v}{\partial x_k}})(\xi) = i\xi_k \hat{v}(\xi) \quad \forall k = 1, ..., N$$

and *Plancherel's Theorem* which states that $v \longmapsto \hat{v}$ is a unitary isomorphism of $L^2(\mathbb{R}^N)$:

$$(1.16) \qquad \int_{\mathbb{R}^N} |v(x)|^2 dx = \int_{\mathbb{R}^N} |\hat{v}(\xi)|^2 d\xi.$$

Also, $\mathcal{D}(\mathbf{R}^N)$ is dense in $H^m(\mathbf{R}^N)$ for all $m \in \mathbf{R}$ (i.e., $H_0^m(\mathbf{R}^N) = H^m(\mathbf{R}^N)$) and the Green's formula (1.12) reads as follows:

$$(1.17) \qquad \int_{\mathbf{R}^N} \nabla u \cdot \nabla \bar{v} dx = - \int_{\mathbf{R}^N} \Delta u \bar{v} dx$$

for all $u \in H^2(\mathbf{R}^N), v \in H^1(\mathbf{R}^N)$.

1.6. The Space $H(\Omega; \Delta)$

In the sequel, we will also be interested in another class of function spaces which involves Laplace operator in $\mathbf{R}^N$:

$$\Delta \overset{\text{def}}{=} \sum_{i=1}^{N} \frac{\partial^2}{\partial x_i^2}.$$

Such spaces were introduced and studied in the book by LIONS and MAGENES [1968] pp. 187–193, pp. 228–229. We will use only one of them which is defined as follows:

$$(1.18) \qquad H(\Omega; \Delta) = \{ u \in L^2(\Omega) \mid \Delta u \in L^2(\Omega) \}.$$

This space is a Hilbert space with the norm given by

$$\|u\|_{H(\Omega;\Delta)} \overset{\text{def}}{=} \left(\|u\|_{0,\Omega}^2 + \|\Delta u\|_{0,\Omega}^2 \right)^{1/2}.$$

We remark that $H^2(\Omega)$ is a proper subspace of $H(\Omega; \Delta)$ with a stronger norm. Nevertheless, the trace map $\gamma_2: u \longmapsto (u, \frac{\partial u}{\partial n})_{|\Gamma}$ extends from $H^2(\Omega)$ to $H(\Omega; \Delta)$ in the following sense:

Theorem 1.4. *Assume that Ω satisfies (1.5). Then there exists a unique linear continuous mapping $\tilde{\gamma}_2$ from $H(\Omega; \Delta)$ onto $H^{-1/2}(\Gamma) \times H^{-3/2}(\Gamma)$ such that for all $u \in H^2(\Omega)$ we have $\tilde{\gamma}_2 u = \gamma_2 u$. Moreover, the following generalized Green's formula holds: For all $u \in H^2(\Omega)$, $v \in H(\Omega; \Delta)$*

$$(1.19) \qquad \int_{\Omega} \bar{v} \Delta u dx - \int_{\Omega} u \Delta \bar{v} dx = \int_{\Gamma} \frac{\partial u}{\partial n} \bar{v} ds - \int_{\Gamma} \frac{\partial \bar{v}}{\partial n} u ds,$$

where the first integral on Γ on the right hand side is interpreted as a duality pairing between $H^{1/2}(\Gamma)$ and $H^{-1/2}(\Gamma)$ and the second one between $H^{3/2}(\Gamma)$ and $H^{-3/2}(\Gamma)$. Both extend the standard inner product $(u, v)_{L^2(\Gamma)} = \int_{\Gamma} u \bar{v} ds$ in $L^2(\Gamma)$. ∎

We remark that in case $\Omega = \mathbf{R}^N$ then $H(\mathbf{R}^N; \Delta)$ coincides with $H^2(\mathbf{R}^N)$ by Fourier transform.

1.7. Lax-Milgram Lemma

In this paragraph, we shall consider a classical mathematical framework which allows us to formulate and solve a class of linear elliptic boundary-value problems. Let V be a Hilbert space with norm $\|\cdot\|$ and inner product $(\cdot,\cdot)$. Many elliptic problems can be written in a weak form as follows: Find $u \in V$, such that

$$(1.20) \qquad a(u,v) = L(v) \quad \forall v \in V,$$

where $a(\cdot,\cdot)$ is a continuous sesquilinear form on V and $L \in V'$, where V' denotes the topological dual of V. The sesquilinear form $a(\cdot,\cdot)$ is said to be *V-elliptic* if there exists $\alpha > 0$ such that

$$(1.21) \qquad \mathfrak{Re}\, a(v,v) \geq \alpha \|v\|_V^2 \quad \forall v \in V.$$

Under this hypothesis, *Lax-Milgram Lemma* assures the existence and uniqueness of a solution to problem (1.20). Moreover, if $a(\cdot,\cdot)$ is Hermitian, i.e., $a(u,v) = \overline{a(v,u)}$, then the solution u of (1.20) can be characterized as the unique solution of the following minimization problem:

$$(1.22) \qquad \text{Find } u \in V, \; J(u) = \min_{v \in V} J(v),$$

where $J(v) = \frac{1}{2}a(v,v) - L(v)$.

◇ **Example 1.1.** (Homogeneous Dirichlet problem) If

$$V = H_0^1(\Omega), \quad a(u,v) = \int_\Omega \nabla u \cdot \nabla \bar{v}\, dx$$

and the linear form L is represented by an element $f \in H^{-1}(\Omega)$ then by virtue of Poincaré Inequality, $a(\cdot,\cdot)$ is V-elliptic and it can be seen that the unique solution of (1.20) solves the following boundary-value problem:

$$(1.23) \qquad \begin{cases} -\Delta u = f & \text{in } \Omega, \\ u = 0 & \text{on } \Gamma. \end{cases} \blacksquare$$

◇ **Example 1.2.** (Homogeneous Neumann problem) Take

$$V = H^1(\Omega), \quad a(u,v) = \int_\Omega (\nabla u \cdot \nabla \bar{v} + u\bar{v})dx \quad \text{and} \quad L(v) = \int_\Omega f\bar{v}\, dx$$

with $f \in L^2(\Omega)$. It is an interesting exercise to use the Green's formulae (1.12) and (1.19) to establish that the unique solution of (1.20) solves the boundary-value problem

$$(1.24) \qquad \begin{cases} -\Delta u + u = f & \text{in } \Omega, \\ \dfrac{\partial u}{\partial n} = 0 & \text{on } \Gamma. \end{cases} \blacksquare$$

◇ **Example 1.3.** (Homogeneous Robin problem) This time, we take

$$V = H^1(\Omega), \quad a(u,v) = \int_\Omega (\nabla u \cdot \nabla \bar{v} + u\bar{v})dx + \int_\Gamma bu\bar{v}ds, \quad L(v) = \int_\Omega f\bar{v}dx,$$

where $f \in L^2(\Omega)$ and $b \in L^\infty(\Gamma)$, $b \geq 0$ on Γ. Following the method of the previous example, it can be shown that the unique solution of (1.20) solves the boundary-value problem

(1.25)
$$\begin{cases} -\Delta u + u = f & \text{in } \Omega, \\ \dfrac{\partial u}{\partial n} + bu = 0 & \text{on } \Gamma. \end{cases} \blacksquare$$

◇ **Example 1.4.** (Homogeneous mixed problem) Let Γ_0 be a non-empty open subset of Γ. We make the following choice:

$$V = H^1_{\Gamma_0}(\Omega), \quad a(u,v) = \int_\Omega \nabla u \cdot \nabla \bar{v}dx \quad \text{and} \quad L(v) = \int_\Omega f\bar{v}dx,$$

where $f \in L^2(\Omega)$. Using the Poincaré Inequality in the space $H^1_{\Gamma_0}(\Omega)$, we see that there is a unique solution to (1.20) and this can be interpreted as the solution of

(1.26)
$$\begin{cases} -\Delta u = f & \text{in } \Omega, \\ u = 0 & \text{on } \Gamma_0, \\ \dfrac{\partial u}{\partial n} = 0 & \text{on } \Gamma \backslash \Gamma_0. \end{cases} \blacksquare$$

1.8. Hilbert Space-Valued Distributions

In the perturbation analysis of spectra of operators, we will use some Sobolev spaces which consist of distributions with values in a Hilbert space. So we shall, in this paragraph, briefly recall their definitions and main properties. For a more detailed study, see BARBU [1976], DIESTEL and UHL [1977], LIONS [1968], SCHWARTZ [1957] and YOSIDA [1980].

In what follows, V denotes a separable Hilbert space (with norm $\| \cdot \|$ and inner product $(\cdot, \cdot)$) and $I =]a, b[$ is an interval in $\mathbb{R}$, not necessarily bounded. We call *space of distributions on I with values in V*, the space of linear continuous maps from $\mathcal{D}(I)$ into V:

$$\mathcal{D}'(I;V) = \mathcal{L}(\mathcal{D}(I), V).$$

We say that a sequence $\{u_j\}$ of elements in $\mathcal{D}'(I;V)$ *converges strongly* (respectively *weakly*) to an element u in $\mathcal{D}'(I;V)$ if for all $\phi \in \mathcal{D}(I)$,

$$\langle u_j, \phi \rangle \longrightarrow \langle u, \phi \rangle \quad \text{in } V \quad \text{(respectively weakly in } V),$$

where $\langle \cdot, \cdot \rangle$ denotes the action of a distribution on the test function. The *distribution derivative* of $u \in \mathcal{D}'(I; V)$ is defined, as usual, by the formula

$$\langle \frac{du}{dt}, \phi \rangle = -\langle u, \frac{d\phi}{dt} \rangle \quad \forall \phi \in \mathcal{D}(I).$$

Definition. A function $u \colon I \longrightarrow V$ is said to be *measurable* if for all $L \in V'$, the scalar function $t \longmapsto L(u(t))$ is measurable in the usual sense. ∎

It is a consequence that if $u \colon I \longrightarrow V$ is measurable then the scalar function $t \longmapsto \|u(t)\|$ is measurable (see YOSIDA [1980] p.131). As in the scalar case, $L^p(I; V)$ $1 \le p \le \infty$ will be the space of (class of) measurable functions $u \colon I \longrightarrow V$ such that

(1.27)
$$\begin{cases} \|u\|_{L^p(L;V)} \overset{\text{def}}{=} \left(\int_I \|u(t)\|^p \, dt \right)^{1/p} < \infty & \text{if} \quad 1 \le p < \infty, \\[2mm] \|u\|_{L^\infty(I;V)} \overset{\text{def}}{=} \operatorname*{ess\,sup}_{t \in I} \|u(t)\| < \infty & \text{if} \quad p = \infty. \end{cases}$$

The space $L^p(I; V)$ is a Banach space for the norm (1.27). If $1 \le p < \infty$, it is separable. Indeed $\mathcal{C}(I; V)$ is dense in $L^p(I; V)$ for these values of p. $L^2(I; V)$ is a Hilbert space with the inner product

$$(u, v)_{L^2(I;V)} \overset{\text{def}}{=} \int_I (u(t), v(t)) \, dt.$$

For $1 \le p < \infty$, the dual space of $L^p(I; V)$ is $L^{p'}(I; V')$, where $(1/p) + (1/p') = 1$. As a consequence, $L^p(I; V)$ is reflexive for $1 < p < \infty$. Finally $L^\infty(I; V)$ being the dual of a separable space, its unit ball is weak* sequentially compact.

If $u \in L^1(I; V)$ and $E \subseteq I$ measurable then, as in the scalar case, we can approximate u by a sequence of step functions with values in V and one can define

$$\int_E u(t) \, dt$$

as an element of V. This is the so-called *Bochner integral* of u over E. The following three properties are classical:

(i)
$$\| \int_E u(t) \, dt \| \le \int_E \|u(t)\| \, dt \quad \forall u \in L^1(I; V).$$

(ii) If $A \in \mathcal{L}(V, W)$, where W is another separable Hilbert space, then for all $u \in L^1(I; V)$, $Au \in L^1(I; W)$ and

$$\int_E Au(t) \, dt = A \int_E u(t) \, dt.$$

(iii) If $u: I \longrightarrow V$ is *absolutely continuous* on a bounded subinterval $[c, d]$ of I, then u is almost everywhere differentiable in the classical sense, its derivative $u'(t)$ belongs to $L^1([c, d]; V)$ and the following formula holds:

$$u(d) - u(c) = \int_c^d u'(t)dt.$$

The standard local L^p spaces can be generalized to the case of Hilbert space-valued L^p spaces. Indeed, $L^p_{loc}(I; V)$ is the space of measurable functions $u: I \longrightarrow V$ such that $\phi u \in L^p(I; V)$ for all $\phi \in \mathcal{D}(I)$. By means of Bochner integral, each element $u \in L^p_{loc}(I; V)$ gives rise to a unique distribution in $\mathcal{D}'(I; V)$ defined by

$$\langle u, \phi \rangle \stackrel{\text{def}}{=} \int_I \phi(t)u(t)dt \quad \forall \phi \in \mathcal{D}(I).$$

The embedding $L^p_{loc}(I; V) \hookrightarrow \mathcal{D}'(I; V)$ defined above is strongly and weakly sequentially continuous.

We can now introduce *Sobolev spaces* of functions with values in V. For $m \in \mathbb{N}$, we define

$$H^m(I; V) \stackrel{\text{def}}{=} \left\{ u \in L^2(I; V) \ \Big| \ \frac{d^k u}{dt^k} \in L^2(I; V) \quad \forall k \leq m \right\},$$

where the derivatives are, of course, taken in the sense of distributions. $H^m(I; V)$ is a Hilbert space where the inner product and the norm are defined by

$$(u, v)_{H^m(I;V)} \stackrel{\text{def}}{=} \sum_{k=0}^m \int_I \left(\frac{d^k u}{dt^k}, \frac{d^k v}{dt^k} \right) dt,$$

$$\|u\|_{H^m(I;V)} \stackrel{\text{def}}{=} \left(\sum_{k=0}^m \int_I \left\| \frac{d^k u}{dt^k} \right\|^2 dt \right)^{1/2}.$$

We will use in the sequel the following properties of the space $H^1(I; V)$:

Theorem 1.5. *The space $H^1(I; V)$ is embedded continuously into the Hölder space $C^{0,1/2}(I; V)$ with exponent $1/2$. More precisely, each $u \in H^1(I; V)$ satisfies*

$$u(d) - u(c) = \int_c^d \frac{du}{dt}(t)dt$$

and

$$\|u(d) - u(c)\| \leq \|u\|_{H^1(I;V)} |d - c|^{1/2}$$

for all bounded subintervals $[c, d]$ of I. Moreover, u is absolutely continuous and its classical and distributional derivatives coincide. ∎

The following corollary results immediately from the classical *Arzelà-Ascoli Theorem*:

Corollary 1.6. *Let I be a bounded interval of $\mathbb{R}$. Then the embedding $H^1(I; V)$ into $C(I; V)$ is compact.* ∎

1.9. Analytic Functions with Values in a Banach Space

In several situations throughout the book, we will encounter functions defined on subsets of Banach spaces and which take values in another Banach space. In this paragraph, we introduce the notion of *analyticity* for such functions, give various equivalent forms of it and announce the classical *Implicit Function Theorem* in this category. A statement of the *Weierstrass Preparation Theorem* is also included for our later use. As is usual in this section, we will not present proofs of these results since they are well covered in other books; see for instance, BERGER [1977], KATO [1966], KRANTZ [1982], KRANTZ and PARKS [1992], LOJACIEWICZ [1991] and RUDIN [1979]. The reader is supposed to be familiar with the standard notions of differential calculus for Banach space-valued functions.

There are at least two very natural definitions of analyticity available in this general setting: a "weak" one and a "strong" one. They turn out to be one and the same if the function takes values in a Banach space. Indeed, surprisingly weaker conditions are equivalent to "strong" analyticity as shown by one of the results below.

Definition. Let X and Y be two Banach spaces over $\mathbb{C}$ and U be a connected open subset of X. A function $f: U \longrightarrow Y$ is said to be *complex-analytic* at a point $x \in U$ if for each $L \in Y'$ and $h \in X$ the complex-valued function $z \longmapsto L(f(x + zh))$ is analytic in the complex variable z at $z = 0$. f is *complex-analytic* in U if it is complex-analytic at each point $x \in U$. ∎

An immediate consequence of this definition which, in fact, corresponds to "weak" analyticity is the fact that for $|z|$ sufficiently small and $x \in U$ we have the expansion

$$(1.28) \qquad L(f(x + zh)) = \sum_{n=0}^{\infty} a_n(x, h)\frac{z^n}{n!},$$

where

$$(1.29) \qquad a_n(x, h) = \frac{d^n}{dz^n} L(f(x + zh))|_{z=0} \quad \forall n \geq 0.$$

In general, the radius of convergence of the series in (1.28) depends on h and L. But the result below, especially point (vi), shows that one can find a positive lower bound for this radius of convergence which is independent of h and L.

Theorem 1.7. *Let* $f: U \longrightarrow Y$ *be a function which is locally bounded. Then the following properties are equivalent:*
   (i) f *is complex-analytic in* U.
  (ii) f *is Gateaux differentiable in* U.
 (iii) f *is Fréchet differentiable in* U.
 (iv) f *has infinitely many Gateaux derivatives in* U.
  (v) f *has infinitely many Fréchet derivatives in* U.
 (vi) *For each* $x \in U$ *there exists* $r = r(x) > 0$ *such that the series*

$$(1.30) \qquad \sum_{n=0}^{\infty} \frac{f^{(n)}(x)h^n}{n!}$$

converges in Y *uniformly for* $\|h\| < r$ *and the sum is nothing but* $f(x + h)$, *where* $f^n(x)$ *denotes the* n^{th} *Fréchet derivative of* f *at the point* $x \in U$. ∎

The fact that "weak" analyticity is equivalent to "strong" one is exhibited by the equivalence between (i) and (vi) of the above result.

Theorem 1.8. (Analytic Implicit Function Theorem) *Suppose* U *and* V *are connected open subsets of Banach spaces* X *and* Y *respectively, and let* $F: U \times V \longrightarrow Z$ *be an analytic function taking values in another Banach space* Z. *Let* (x_0, y_0) *be a point in* $U \times V$ *such that* $F(x_0, y_0) = 0$. *Assume that the bounded linear operator defined by the Fréchet derivative* $F_y(x_0, y_0)$ *is invertible. Then there is one and only one solution* $y = y(x)$ *of the equation* $F(x, y) = 0$ *near* (x_0, y_0) *that is an analytic function of* x *near* x_0 *such that* $y(x_0) = y_0$. ∎

Theorem 1.9. (Weierstrass Preparation Theorem) *Let* $f(z)$ *be a complex-analytic function in a neighbourhood of the origin in* $\mathbb{C}^N$. *Assume, without loss of generality, that* $f(0, 0, ..., z_n) \not\equiv 0$. *Then* f *may be written as* $f(z) = H(z)V(z)$ *where* $V(z)$ *is complex-analytic, which vanishes nowhere and* $H(z)$ *is a polynomial of certain degree, say* m, *in the* z_N *variable and with coefficients* $a_j(z')$ *which are complex-analytic in* $z' = (z_1, ..., z_{N-1})$ *variables:*

$$H(z', z_N) = z_N^m + a_1(z')z_N^{m-1} + \cdots + a_m(z'). ∎$$

Let us now consider the case $X = \mathbb{R}^N$ and define the concept of analyticity for functions of real variables with values in a Banach space Y over $\mathbb{C}$ or $\mathbb{R}$.

Definition. Let U be a connected open subset of $\mathbb{R}^N$. A function $f: U \longrightarrow Y$ is said to be *real-analytic* at a point $x \in U$ if f admits a Taylor expansion at x, i.e., f is infinitely Fréchet differentiable at x and there exists $r = r(x) > 0$ such that the series

$$(1.31) \qquad \sum_{n=0}^{\infty} \frac{f^{(n)}(x)h^n}{n!}$$

converges in Y uniformly for all $h \in \mathbb{R}^N$ such that $|h| < r$ and the sum is $f(x+h)$. If f is real-analytic at each point of U, we say that f is *real-analytic* in U. ∎

The statements of Implicit Function Theorem and Weierstrass Preparation Theorem are valid in the class of real-analytic functions as well. For the proofs, the reader is referred to the literature cited above.

It is classically known that if $f: U \longrightarrow Y$ is real-analytic then f is the restriction of a complex-analytic function $\tilde{f}$ defined in a neighbourhood $\tilde{U}$ in $\mathbb{C}^N$ of U. This can be deduced from the observation that the series (1.31) converges indeed uniformly for all $h \in \mathbb{C}^N$ such that $|h| < r$.

§2. Some Classical Examples of Vibrating Systems

In this section, we consider some classical examples of vibrations of mechanical systems most of which we experience in our daily life. Intuitively we feel that the sound vibrations emitted by a body depend on its shape, its size, the constituting material and its interactions with the neighbourhood. Fundamental contributions in this area have been made by Lord RAYLEIGH in his famous book *The Theory of Sound* in the year 1877 (see RAYLEIGH [1945]). One standard way of modelling vibration phenomena is by means of evolution equations. A simplified model would be an eigenvalue problem which arises when one tries to separate the space and time variables in the above equations. We obtain such simplified models in the case of vibrating drums and variants, acoustic resonators, acoustic waves in free space and a model of church organ. In each of these cases, we examine the corresponding eigenvalue problem and obtain solutions, if they exist. We use these solutions to explain the phenomenon of resonance. There are situations where solutions do not exist in the classical sense. We then introduce "approximate" eigenvectors which will lead us naturally to the mathematical notion of *spectrum* of a linear operator. The basic difficulty is that these solutions do not have finite energy, but we try to convince the reader how to overcome this difficulty and prove in what sense these "eigenvectors" form a basis in the space of all functions of finite energy.

2.1. Vibrations of a Drum

Let us consider an elastic membrane which occupies a two-dimensional bounded connected domain Ω with smooth boundary Γ (see Figure 2.1). The membrane is stretched and fixed on its rigid boundary Γ. Let $u(x,t)$ denote the transverse displacement of the membrane at the point $x \in \Omega$ and at time t under a force (acting perpendicularly to the plane of the membrane) with density $f(x,t)$. Assuming that u and f are "small", we can neglect the deformations in the x-plane and we have the following wave equation connecting u and f (see SOBOLEV [1964]†):

$$(2.1) \qquad \frac{\partial^2 u}{\partial t^2} - c^2 \Delta u = f \quad \text{in} \quad \Omega \times \mathbb{R},$$

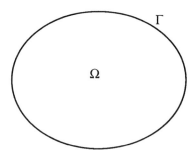

Figure 2.1.

where c^2 is a constant which is proportional to the tension of the membrane. Equation (2.1) is the classical wave equation with characteristic speeds c and $-c$. Since the membrane is supposed to be fixed on its boundary, we add the following Dirichlet boundary condition:

$$(2.2) \qquad u = 0 \quad \text{on} \quad \Gamma \times \mathbb{R}.$$

To arrive at the eigenvalue problem which describes free vibrations of the membrane, we take $f \equiv 0$ and look for solutions where the variables are separated, i.e., $u(x,t) = v(x)g(t)$, where $v = 0$ on Γ, v and g are taken to be not identically zero in order to eliminate the trivial solution $u \equiv 0$. Thus v and g satisfy the relation

$$g''(t)\frac{1}{g(t)} = c^2 \Delta v(x)\frac{1}{v(x)}$$

---

† Indeed, the content of this section is very well covered in the literature and to present it we have chosen to follow, mainly, the book of COURANT and HILBERT [1953].

whenever $g(t)$ and $v(x)$ are not zero. The left side of the above relation is dependent only on t and the right side depends only on x and therefore they are necessarily constant, which we denote by $-\lambda$. Hence we have

(2.3)
$$\begin{cases} -c^2 \Delta v(x) = \lambda v(x), \\ -g''(t) = \lambda g(t). \end{cases}$$

By using Green's formula, it is worth to note from (2.2) and (2.3) that

(2.4)
$$c^2 \int_\Omega |\nabla v(x)|^2 dx = -c^2 \int_\Omega \Delta v(x) \bar{v}(x) dx = \lambda \int_\Omega |v(x)|^2 dx$$

which shows that λ is non-negative. If $\lambda = 0$, then $v \equiv 0$ and hence $u \equiv 0$ and this possibility has already been eliminated. We conclude therefore that $\lambda > 0$. This number λ, if it exists, is called *eigenvalue* and $v(x) \not\equiv 0$ is called the corresponding *eigenvector*. They satisfy the following *eigenvalue problem*:

(2.5)
$$\begin{cases} -c^2 \Delta v = \lambda v \quad \text{in} \quad \Omega, \\ v = 0 \quad \text{on} \quad \Gamma. \end{cases}$$

We now define the *frequency* parameter

(2.6)
$$\omega = +\sqrt{\lambda}.$$

Going back to the equation (2.3) for g and integrating it, we get the general form of the solution as follows:

$$g(t) = a \cos(\omega t + \varphi),$$

where a and φ are constants. This gives rise to the sinusoidal solution

$$u(x, t) = a \cos(\omega t + \varphi) v(x)$$

of equation (2.1) with $f \equiv 0$.

Having defined formally the eigenvalue problem, we now give a proper formulation of it. We mention that it is reasonable to search for eigenvectors v in the space $H_0^1(\Omega)$ (see (2.4)). Therefore, taking test functions $\phi \in H_0^1(\Omega)$ and multiplying equation (2.5) by ϕ and integrating by parts, we obtain

(2.7)
$$\begin{cases} c^2 \int_\Omega \nabla v \cdot \nabla \bar{\phi} dx = \lambda \int_\Omega v \bar{\phi} dx \quad \forall \phi \in H_0^1(\Omega), \\ v \in H_0^1(\Omega), \ v \not\equiv 0. \end{cases}$$

This is a *weak formulation* of problem (2.5). To resolve this, it is a classical technique to introduce the so-called Green's operator $G\colon L^2(\Omega) \longrightarrow L^2(\Omega)$ which is defined as $Gh = z$, where $z \in H_0^1(\Omega)$ is solution of

$$(2.8) \qquad c^2 \int_\Omega \nabla z \cdot \nabla \bar\phi \, dx = \int_\Omega h\bar\phi \, dx \quad \forall \phi \in H_0^1(\Omega).$$

The existence and uniqueness of the solution z follow easily from Lax-Milgram Lemma (see Paragraph 1.7). Furthermore, applying Rellich's Lemma (Theorem 1.2), it is straightforward to verify that G is a compact operator. It is also easy to see that G is self-adjoint and that (λ, v) satisfies (2.7) iff

$$(2.9) \qquad \lambda Gv = v, \ v \not\equiv 0.$$

This means that λ is an eigenvalue of problem (2.5) with corresponding eigenvector v iff λ^{-1} is an eigenvalue of G with eigenvector v. To solve (2.9), we will merely need to apply the spectral theory of compact self-adjoint operators which we develop in §3 (see Examples 3.4 and 3.12). This will prove the existence of a non-decreasing sequence of eigenvalues $\{\lambda_n\}$ which tends to infinity. The corresponding eigenvectors $\{v_n\}$ can be chosen to be an orthonormal basis of $L^2(\Omega)$ and an orthogonal basis of $H_0^1(\Omega)$.

Before concluding this example, we would like to point out one interesting phenomenon called *resonance* associated with the wave equation (2.1). To this end, we seek general solutions of (2.1) and (2.2), where now $f \not\equiv 0$. Expanding u and f in terms of the basis $\{v_n\}$ of eigenvectors, we have

$$u(x,t) = \sum_{n=1}^\infty u_n(t)v_n(x),$$

$$f(x,t) = \sum_{n=1}^\infty f_n(t)v_n(x).$$

Introducing the above expressions into (2.1), we get the following equations for the unknown coefficients:

$$(2.10) \qquad u_n''(t) + \lambda_n u_n(t) = f_n(t) \quad \forall n \ge 1$$

whose general solution can be written as

$$(2.11) \qquad \begin{cases} u_n(t) = a_n \cos(\omega_n t + \varphi_n) + \dfrac{1}{\omega_n} \displaystyle\int_0^t \cos \omega_n(t - s) f_n(s)\, ds, \\[2mm] \omega_n = +\sqrt{\lambda_n}, \end{cases}$$

where a_n and φ_n are constants which can be determined from the initial conditions on u which can, for instance, be

$$(2.12) \qquad\qquad u(x,0), \ \frac{\partial u}{\partial t}(x,0) \quad \text{given}.$$

Let us now suppose that $f_n(t)$ is a sinusoidal function of the time variable, say $f_n(t) = \cos\omega t$. Then two cases can arise.

• Case (i): $\omega^2 \neq \omega_n^2$ for any n.

In this case, the convolution integral in (2.11) remains bounded as a function of t.

• Case (ii): $\omega^2 = \omega_n^2$ for some n.

In this case, the convolution integral in (2.11) gives rise to a term $t\cos\omega t$ which implies that the solution $u_n(t)$ oscillates with the frequency ω_n and whose amplitude increases linearly with the time t. This is the so-called *phenomenon of resonance*. Thus when a mechanical structure is excited by an external force containing a frequency ω such that ω^2 is an eigenvalue of the structure, then the amplitude of the vibrations of the structure increases linearly with time which could even destroy it. This is illustrated by the example of the Tacoma bridge which was cited in the Preface of this book†.

2.2. Spectral Problems with Neumann and Robin Boundary Conditions

We are going to present two simple variants of the eigenvalue problem (2.5) or (2.7) considered in the previous paragraph. One involves the so-called Neumann boundary condition and the other the Robin condition. While Dirichlet condition signifies the fact that the drum is fixed along its boundary, Neumann condition is imposed when the boundary is free. Robin condition can mathematically be interpreted as an intermediate condition between these two. We add to say that in modelling other physical situations such as heat and mass transfer across boundaries and interfaces or charge transfer across electrodes etc., Neumann and Robin conditions appear in a more natural manner.

The Neumann eigenvalue problem consists in finding λ for which there is a function v which is not identically zero in Ω such that

$$(2.13) \qquad\qquad \begin{cases} -c^2\Delta v = \lambda v \quad \text{in} \quad \Omega, \\ \dfrac{\partial v}{\partial n} = 0 \quad \text{on} \quad \Gamma. \end{cases}$$

---

† Let us also cite the catastrophic destruction of the suspension bridge at Angers in France (around 1850) under the effect of rhythmic marching of troops. After this tragic accident, a military order was issued to soldiers to walk in somewhat a disordered fashion while crossing bridges.

Here as usual Ω is a connected open set in $\mathbb{R}^N$ satisfying the regularity condition (1.5). To solve (2.13), we introduce a change of variable $\mu = \lambda + 1$ and replace the operator $-c^2 \Delta$ by $(-c^2 \Delta + I)$. Then (2.13) becomes

(2.14)
$$\begin{cases} -c^2 \Delta v + v = \mu v & \text{in} \quad \Omega, \\ \dfrac{\partial v}{\partial n} = 0 & \text{on} \quad \Gamma. \end{cases}$$

Next we use the Green's operator technique introduced in the previous paragraph. To this end, let $G_1 : L^2(\Omega) \longrightarrow L^2(\Omega)$ be defined as $G_1 f = u$, where $u \in H^1(\Omega)$ is solution of

(2.15)
$$\begin{cases} -c^2 \Delta u + u = f & \text{in} \quad \Omega, \\ \dfrac{\partial u}{\partial n} = 0 & \text{on} \quad \Gamma. \end{cases}$$

which we know, admits a unique weak solution (see Example 1.2). As in Paragraph 2.1, it follows that G_1 is a self-adjoint and compact operator for which the general theory of §3 applies and yields the result (see Example 3.5 and Exercise 3.13): Problem (2.14) possesses a countable number of solutions $\{(\mu_n, v_n)\}$, where $\{\mu_n\}$ is a non-decreasing sequence of numbers tending to infinity, $\mu_n \geq 1$ for all n and $\{v_n\}$ forms an orthonormal basis for $L^2(\Omega)$ and an orthogonal basis for $H^1(\Omega)$. Obviously $\mu_1 = 1$ (i.e., $\lambda_1 = 0$) and $v_1 \equiv 1$.

The eigenvalue problem corresponding to Robin boundary condition consists in finding λ for which there is a v (not identically zero) such that

(2.16)
$$\begin{cases} -c^2 \Delta v = \lambda v & \text{in} \quad \Omega, \\ c^2 \dfrac{\partial v}{\partial n} + bv = 0 & \text{on} \quad \Gamma, \end{cases}$$

where $b = b(x)$ is a given function defined on Γ. It is assumed that b is non-negative. Note that $b \equiv 0$ corresponds to the Neumann boundary condition and that $b \equiv \infty$ corresponds formally to the Dirichlet boundary condition. To solve (2.16), we make as before a change of variable $\mu = \lambda + 1$ and consider

(2.17)
$$\begin{cases} -c^2 \Delta v + v = \mu v & \text{in} \quad \Omega, \\ c^2 \dfrac{\partial v}{\partial n} + bv = 0 & \text{on} \quad \Gamma. \end{cases}$$

Following the Green's operator technique, we define $G_2 : L^2(\Omega) \longrightarrow L^2(\Omega)$ as $G_2 f = u$, where $u \in H^1(\Omega)$ is solution of

(2.18)
$$\begin{cases} -c^2 \Delta u + u = f & \text{in} \quad \Omega, \\ c^2 \dfrac{\partial u}{\partial n} + bu = 0 & \text{on} \quad \Gamma. \end{cases}$$

Once again, it can be proved that G_2 is self-adjoint and compact and so by using the general theory of §3 (see Example 3.5 and Exercise 3.13), we conclude that problem (2.17) possesses a non-decreasing sequence of eigenvalues $\{\mu_n\}$ which tend to infinity and the corresponding eigenvectors $\{v_n\}$ form a basis of $L^2(\Omega)$ and $H^1(\Omega)$.

2.3. Steklov Eigenvalue Problem

In Fluid Mechanics, we encounter sometimes eigenvalue problems of the following type wherein the eigenvalue parameter λ appears in the boundary condition instead of the equation in Ω:

$$(2.19) \qquad \begin{cases} \Delta v = 0 \quad \text{in} \quad \Omega, \\[2mm] \dfrac{\partial v}{\partial n} = \lambda v \quad \text{on} \quad \Gamma. \end{cases}$$

This model arises, for instance, in the study of surface waves (see BERGMANN and SCHIFFER [1953]). Also, one comes across this type of problems in the study of stability of mechanical oscillators immersed in a viscous fluid (see CONCA, PLANCHARD, THOMAS and VANNINATHAN [1994] Chapter VII or PLANCHARD and THOMAS [1993]).

It is not surprising that in this case the associated Green's operator G is defined on $L^2(\Gamma)$ instead of $L^2(\Omega)$. More precisely, it is not difficult to convince oneself that here G is defined as $Gg = u_{|\Gamma}$, where u is solution of

$$(2.20) \qquad \begin{cases} \Delta u = 0 \quad \text{in} \quad \Omega, \\[2mm] \dfrac{\partial u}{\partial n} + u = g \quad \text{on} \quad \Gamma. \end{cases}$$

A weak formulation of the above problem can, as usual, be obtained by multiplying (2.20) by a test function $\phi \in H^1(\Omega)$ and integrating by parts. This gives

$$(2.21) \qquad \begin{cases} \displaystyle\int_\Omega \nabla u \cdot \nabla \bar\phi \, dx + \int_\Gamma u \bar\phi \, ds = \int_\Gamma g \bar\phi \, ds \quad \forall \phi \in H^1(\Omega), \\[4mm] u \in H^1(\Omega). \end{cases}$$

Regarding the V-ellipticity of the bilinear form corresponding to the above problem, we remark that it suffices to use Poincaré-Wirtinger Inequality (1.10) and the triangle inequality. Lax-Milgram Lemma (see Paragraph 1.7) then ensures the existence of a unique solution of problem (2.21). The operator G is thus well-defined and it is self-adjoint. That it is compact follows from Theorem 1.2. Applying the general theory of §3 (see Example 3.6 and Exercise 3.13), we reach the same conclusions as in the previous examples except that now the eigenvectors of (2.19) form an orthogonal basis of $L^2(\Gamma)$.

2.4. The Acoustic Resonator

We start by describing the physical situation involved in this model. An *acoustic resonator* is represented by a bounded cavity Ω in $\mathbb{R}^3$ with boundary Γ which contains a perfect and compressible fluid. In the acoustic approximation, the pressure p of the fluid and its velocity potential ϕ satisfy the wave equation (see LANDAU and LIFCHITZ [1989] §64, p.356):

$$(2.22) \qquad \frac{\partial^2 u}{\partial t^2} - c^2 \Delta u = 0 \quad \text{in} \quad \Omega \times \mathbb{R},$$

where c denotes the speed of sound in the fluid. In general, the acoustic resonator is excited from the exterior by sending waves through a small opening Γ_0 in its boundary (see Figure 2.2). The boundary Γ of Ω is thus divided into two non-empty pieces: the hole Γ_0 and its complement Γ_1 which is assumed to be a rigid wall. This leads us to impose the condition

$$(2.23) \qquad \frac{\partial u}{\partial n} = 0 \quad \text{on} \quad \Gamma_1 \times \mathbb{R}.$$

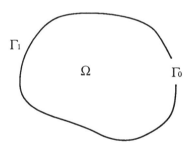

Figure 2.2.

The excitation from the exterior through the boundary Γ_0 is modelled by

$$(2.24) \qquad u = h \quad \text{on} \quad \Gamma_0 \times \mathbb{R},$$

where $h = h(x,t)$ is a given function. This signifies, for example, that the pressure is prescribed on Γ_0.

The first step to describe the vibrations of the acoustic resonator is to consider free vibrations, i.e., we take $h \equiv 0$ in (2.24) and look for solutions u by separation of variables:

$$u(x,t) = v(x)g(t).$$

Using the calculation done in Paragraph 2.1, we are led to the following eigenvalue problem for v which is a variant of (2.5):

(2.25)
$$\begin{cases} -c^2 \Delta v = \lambda v & \text{in } \Omega, \\ \dfrac{\partial v}{\partial n} = 0 & \text{on } \Gamma_1, \\ v = 0 & \text{on } \Gamma_0. \end{cases}$$

This is an example of an eigenvalue problem with *mixed boundary conditions*. It is easily seen that the Green's operator G associated with problem (2.25) is well-defined (see Example 1.4) and it is self-adjoint and compact. Consequently, there is a non-decreasing sequence $\{\lambda_n\}$ of eigenvalues for problem (2.25) with corresponding eigenvectors $\{v_n\}$ which form orthogonal basis in $L^2(\Omega)$ and in $H^1_{\Gamma_0}(\Omega)$.

The second step consists of developing u in terms of the eigenfunctions v_n:

$$u(x,t) = \sum_{n \geq 1} u_n(t) v_n(x).$$

We remark that the convergence of the above series is in the sense of $L^2(\Omega)$ for each fixed $t \in \mathbb{R}$. Even though $u \in H^1(\Omega)$, the convergence of the series is not strong enough and it does not permit us to restrict it on Γ_0; observe that $u = h$ on Γ_0 whereas v_n's all vanish there.

In order to compute the unknown Fourier coefficients $u_n(t)$, we multiply equation (2.22) by v_n and integrate by parts in Ω. Using the boundary conditions satisfied by u and v_n, we obtain

(2.26)
$$u_n''(t) + \lambda_n u_n(t) = -c^2 \int_{\Gamma_0} h \frac{\partial \bar{v}_n}{\partial n} ds.$$

Let $h_n(t)$ denote the right side of the above equation. Comparing (2.26) and (2.10), we see that there will be resonance if $h_n(t)$ is a sinusoidal function of t with frequency $\sqrt{\lambda_n}$. This justifies the terminology *acoustic resonator*. Thus, we are able to give conditions on the external excitation h so that it gives rise to resonance.

Remark. The above analysis can be readily extended to other boundary conditions on Γ_0. For example, the normal velocity of the fluid can be prescribed on the hole Γ_0:

(2.27)
$$c^2 \frac{\partial u}{\partial n} = h \quad \text{on } \Gamma_0 \times \mathbb{R}.$$

This is a good model if we imagine that the small fluid mass situated at the hole Γ_0 acts like a piston on the fluid. This hypothesis is frequently used in Acoustics literature. The eigenfunctions must then satisfy the condition

(2.28)
$$\frac{\partial v_n}{\partial n} = 0 \quad \text{on } \Gamma_0 \times \mathbb{R}.$$

In this case, we obtain the following equations for the Fourier coefficients $u_n(t)$ of the solution u:

$$(2.29) \qquad u_n''(t) + \lambda_n u_n(t) = c^2 \int_{\Gamma_0} h\bar{v}_n ds$$

and once again there will be resonance if the right hand side of (2.29) is a sinusoidal function of time with frequency $\sqrt{\lambda_n}$. ∎

2.5. Propagation of Waves in Free Space

In the preceding examples, in particular the one treated in Paragraph 2.1, we studied the propagation of waves in a bounded medium $\Omega \subset \mathbb{R}^N$. Another physically interesting example is the case of waves in the entire space $\mathbb{R}^N$. This arises for instance when one is looking for the local behaviour of waves around a fixed point in the medium. It is then intuitively clear that one can neglect the influence of boundary conditions in the various models studied in the earlier paragraphs and reduce the problem to $\mathbb{R}^N$. As we shall see, this situation is quite different from the previous examples; in particular, the nature of the vibrations is different in the sense that we will not have eigenvalues because Rellich's Lemma is no more applicable and moreover the usual Green's operator technique fails.

As in Paragraph 2.1, we consider the waves governed by the wave equation in $\mathbb{R}^N$ without source terms:

$$(2.30) \qquad \frac{\partial^2 u}{\partial t^2} - \Delta u = 0 \quad \text{in} \quad \mathbb{R}^N \times \mathbb{R}.$$

Here $u = u(x,t)$ denotes the amplitude of the waves and we have taken the speed of propagation to be unity. By physical considerations, we assume that the waves have finite energy at each instant; for instance, we suppose that $u(\cdot, t)$ satisfies

$$\int_{\mathbb{R}^N} |u(x,t)|^2 dx < +\infty \quad \forall t \in \mathbb{R}.$$

We seek solutions by separation of the variables x and t: $u(x,t) = v(x)g(t)$ and we obtain formally, as in Paragraph 2.1, the following eigenvalue problem for v: Find λ for which there is a function v not identically zero satisfying

$$(2.31) \qquad \begin{cases} -\Delta v = \lambda v \quad \text{in} \quad \mathbb{R}^N, \\ \quad v \in L^2(\mathbb{R}^N). \end{cases}$$

We will now easily see that there is no solution to the above problem. First, we remark that v, if it exists, belongs to $H(\mathbb{R}^N; \Delta)$ and hence to $H^2(\mathbb{R}^N)$.

Iterating this argument, we see that $v \in H^m(\mathbb{R}^N)$ for all m. Multiplying the equation (2.31) by $\bar{v}$ and applying formula (1.17), we get

$$\int_{\mathbb{R}^N} |\nabla v|^2 dx = \lambda \int_{\mathbb{R}^N} |v|^2 dx$$

which obviously implies $\lambda \geq 0$. Next, by considering the Fourier transform $\hat{v}(\xi)$ we see, using the relations (1.15) and (1.16), that (2.31) is transformed to

(2.32)
$$\begin{cases} (|\xi|^2 - \lambda)\hat{v}(\xi) = 0 \quad \text{in} \quad \mathbb{R}^N, \\ \hat{v} \in L^2(\mathbb{R}^N) \end{cases}$$

and therefore the support of $\hat{v}$ is contained in the sphere $\{\xi \in \mathbb{R}^N \mid |\xi|^2 = \lambda\}$ whose N-dimensional measure is zero. Thus $\hat{v} = 0$ and hence $v = 0$.

Therefore, we have to understand problem (2.31) in a generalized sense and this leads us to the notion of *spectrum*. We say that a complex number λ is in the spectrum σ of $(-\Delta)$ if one of the following condition holds: (a) $(-\Delta - \lambda I)$ is not injective in $L^2(\mathbb{R}^N)$, (b) $(-\Delta - \lambda I)$ is injective but not onto $L^2(\Omega)$. We have just seen above that the possibility (a) which corresponds to the case where λ is an eigenvalue has been eliminated. Thus, we shall interpret (2.31) as to finding λ such that $(-\Delta - \lambda I)$ is injective but not onto. Our next goal is to prove that $\sigma = [0, \infty)$.

To this end, we consider $(-\Delta)$ to be an unbounded operator with domain in $L^2(\mathbb{R}^N)$†. More exactly, let us define $D(-\Delta) = H^2(\mathbb{R}^N)$ which is dense in $L^2(\mathbb{R}^N)$. Thus $(-\Delta)$ is a well-defined operator from $D(-\Delta)$ into $L^2(\mathbb{R}^N)$. The following result characterizes the spectrum of this operator:

Proposition 2.1. $\sigma(-\Delta) = [0, \infty)$.

Proof. *First Step*: Here we prove that $(-\Delta - \lambda I)$ is onto if $\lambda \notin [0, \infty)$. Let $f \in L^2(\mathbb{R}^N)$ be arbitrary. We have to uniquely solve the following problem:

(2.33)
$$\begin{cases} -\Delta v - \lambda v = f \quad \text{in} \quad \mathbb{R}^N, \\ v \in L^2(\mathbb{R}^N). \end{cases}$$

By Fourier transform, this is equivalent to finding $\hat{v} \in L^2(\mathbb{R}^N)$ such that

$$\hat{v}(\xi) = \frac{\hat{f}(\xi)}{|\xi|^2 - \lambda} \quad \forall \xi \in \mathbb{R}^N.$$

Since $\hat{f} \in L^2(\mathbb{R}^N)$ and $(|\xi|^2 - \lambda)^{-1} \in L^\infty(\mathbb{R}^N)$, we are done by virtue of *Plancherel's Theorem*.

---

† This way of tackling (2.31) is different from the Green's operator technique considered in the earlier examples.

Second Step: Let us now show that if $\lambda \in [0, \infty)$, then $(-\Delta - \lambda I)$ is not onto. We suppose the contrary and denote by $R_\lambda = (-\Delta - \lambda I)^{-1}$ which exists by our assumption. Observe that the operator R_λ has a closed graph and so it is continuous by the classical *Closed Graph Theorem*.

We shall now contradict this statement by exhibiting a sequence $\{v_n\}$ in $L^2(\mathbb{R}^N)$ such that

$$(2.34) \qquad \begin{cases} \|v_n\|_{0,\mathbb{R}^N} = 1, \\ (-\Delta v_n - \lambda v_n) \longrightarrow 0 \quad \text{in} \quad L^2(\mathbb{R}^N). \end{cases}$$

This will complete the proof of Proposition 2.1. To construct such a sequence, we use the *plane waves* $e^{i\xi \cdot x}$ where $\xi \in \mathbb{R}^N$ and $|\xi|^2 = \lambda$ after localizing them by means of a sequence of cut-off functions ϕ_n with the following properties:

$$(2.35) \qquad \begin{cases} \phi_n \in \mathcal{D}(\mathbb{R}^N), \quad \phi_n = 1 \quad \text{in} \quad B(0, n), \quad \mathrm{supp}\, \phi_n \subseteq B(0, n+1), \\ \|\phi_n\|_{2,\infty,\mathbb{R}^N} \le c \quad \text{(independent of } n). \end{cases}$$

We define next

$$v_n(x) = \frac{\phi_n(x)}{\|\phi_n\|_{0,\mathbb{R}^N}} e^{i\xi \cdot x}.$$

Using the fact that the plane waves $e^{i\xi \cdot x}$ are exact solutions of $(-\Delta v - \lambda v) = 0$ and $\|\phi_n\|_{0,\mathbb{R}^N}$ is of the order of $n^{N/2}$, elementary calculations show that (2.34) holds. ∎

Although $(-\Delta - \lambda I)$ is injective for all $\lambda \in \mathbb{C}$, we see that for λ in the spectrum of $(-\Delta)$ the *plane waves* $e^{i\xi \cdot x}$ with $|\xi|^2 = \lambda$ can be regarded as "approximate" eigenvectors of $(-\Delta)$ with "eigenvalue" λ. These functions do not belong to $L^2(\mathbb{R}^N)$ but they span all of $L^2(\mathbb{R}^N)$ in the sense of *Fourier inversion*:

$$(2.36) \qquad u(x) = \lim_{M \to \infty} \frac{1}{(2\pi)^{N/2}} \int_{\{\xi \mid |\xi| \le M\}} \hat{u}(\xi) e^{i\xi \cdot x} d\xi \quad \forall u \in L^2(\mathbb{R}^N).$$

We conclude this paragraph by mentioning that the situation outlined above constitutes a prototype example of what is called *continuous spectrum*, a concept which will be defined in §3.

2.6. Vibrations of Air Trapped in many Narrow Channels

Continuing our efforts to illustrate the various aspects of vibrations of mechanical systems, we present in this paragraph another example which contains a new phenomenon, namely, the presence of eigenvalues of infinite multiplicity. This is again due to lack of compactness. While in the previous example we could not apply Rellich's Lemma because the domain was unbounded, in the present case, the origin of non-compactness is the degeneracy of the operator with regard to some space variables. Even though the domain is bounded in this case, Rellich's Lemma is not applicable.

Let us consider air (or any other perfect fluid) in a medium made up of many narrow channels, in the x_1-direction. Its vibrations in such a medium can be modelled via the *theory of homogenization*. This has been done by FLEURY and SÁNCHEZ-PALENCIA [1986] (see also SÁNCHEZ-HUBERT and SÁNCHEZ-PALENCIA [1989]) and they obtained a model which can, in two dimensions, be written as follows: Let Ω be the domain sketched in Figure 2.3, i.e.,

$$(2.37) \qquad \Omega = \{(x_1, x_2) \in \mathbb{R}^2 \mid 0 < x_1 < \ell(x_2),\ 0 < x_2 < 1\},$$

where the function $\ell(x_2)$ is either identically equal to a positive constant or a smooth strictly monotone positive function. It represents the average length of the channel at x_2. Let $\Gamma_0 = \{0\} \times [0, 1]$ and $\Gamma_\ell = \{(\ell(x_2), x_2) \mid 0 \leq x_2 \leq 1\}$ (see Figure 2.3 for details). The region Ω represents the *homogenized medium* and we have the following system of equations:

$$(2.38) \qquad \begin{cases} \dfrac{\partial^2 u}{\partial t^2} - \dfrac{\partial^2 u}{\partial x_1^2} = 0 & \text{in} \quad \Omega \times \mathbb{R}, \\[2ex] \dfrac{\partial u}{\partial x_1} = 0 & \text{on} \quad (\Gamma_0 \times \mathbb{R}) \cup (\Gamma_\ell \times \mathbb{R}). \end{cases}$$

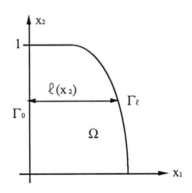

Figure 2.3.

Here $u = u(x,t)$ represents the amplitude of the vibrations. Observe the absence of the derivative with respect to x_2-variable in (2.38); x_2 plays the role of a parameter. This is the degeneracy alluded to above. Note also the absence of boundary condition on the part lying on $\{x_2 = 0\}$.

Following our practice of finding solutions via separation of variables of the form $u(x,t) = v(x)g(t)$, we obtain formally the following problem for v: Find λ such that there is a function v not identically zero satisfying

$$(2.39) \qquad \begin{cases} -\dfrac{\partial^2 v}{\partial x_1^2} = \lambda v \quad \text{in} \quad \Omega, \\[2mm] \dfrac{\partial v}{\partial x_1}(0, x_2) = \dfrac{\partial v}{\partial x_1}(\ell(x_2), x_2) = 0, \quad 0 < x_2 < 1. \end{cases}$$

As in the previous section, we interpret this problem as the definition of the spectrum associated with the operator $A = -\partial^2/\partial x_1^2$ in $L^2(\Omega)$. This is an unbounded operator whose domain is

$$(2.40) \qquad D(A) = \left\{ v \in L^2(\Omega) \mid \frac{\partial^2 v}{\partial x_1^2} \in L^2(\Omega), \ \frac{\partial v}{\partial x_1} = 0 \quad \text{on} \quad \Gamma_0 \cup \Gamma_\ell \right\}$$

We remark that the class of functions $v \in L^2(\Omega)$ such that $Av \in L^2(\Omega)$ is a space of the same type as $H(\Omega; \Delta)$ introduced in Paragraph 1.6. Following LIONS and MAGENES [1968], we see that we can give a meaning to $\partial v/\partial x_1$ as an element of $H^{-3/2}(\Gamma_0 \cup \Gamma_\ell)$. Hence the space $D(A)$ is well-defined.

Since (2.39) is a boundary-value problem involving an ordinary differential equation, we can explicitly write down its solutions:

$$(2.41) \qquad \begin{cases} v_n = \cos\left[\dfrac{n\pi}{\ell(x_2)} x_1\right] F_n(x_2), \\[3mm] \lambda_n = \left[\dfrac{n\pi}{\ell(x_2)}\right]^2, \quad n \geq 0, \end{cases}$$

where $F_n = F_n(x_2) \in L^2(0,1)$ is arbitrary. In case $\ell \equiv$ constant, (2.41) gives an infinite number of eigenvalues each one with infinite multiplicity since F_n is arbitrary. If $\ell(x_2)$ is not identically constant then taking $n = 0$, we see that 0 is an eigenvalue of infinite multiplicity. However for $n > 0$, the expression for λ_n in (2.41) does not define an eigenvalue of the operator A since there is dependence on x_2. Hence, there is a need in such a case to understand the solution given by (2.41) in a generalized sense. Thus we fix x_2 at a value γ and replace $F_n(x_2)$ by the Dirac mass $\delta(x_2 - \gamma)$. So, we may think that the generalized solutions are

$$(2.42) \qquad \begin{cases} u_n = \cos\left[\dfrac{n\pi}{\ell(\gamma)} x_1\right] \delta(x_2 - \gamma), \\[3mm] \lambda_n = \left[\dfrac{n\pi}{\ell(\gamma)}\right]^2, \quad 0 < \gamma < 1. \end{cases}$$

Having defined these, it is our aim to see that λ_n belongs to the spectrum $\sigma(A)$ of A defined in Paragraph 2.5. Indeed, we have the following result (see also Figure 2.4):

Proposition 2.2. $\sigma(A) = \{0\} \cup \{\lambda_n \mid n \geq 1, \, 0 < \gamma < 1\}$.

Proof. First, for $\lambda \neq 0$ and $\lambda \notin \{\lambda_n \mid n \geq 1, \, 0 < \gamma < 1\}$ the following family indexed by γ of second order differential equations admits unique solution v in $L^2(\Omega)$ for all $f \in L^2(\Omega)$:

(2.43)
$$\begin{cases} -\dfrac{\partial^2 v}{\partial x_1^2}(x_1, \gamma) - \lambda v(x_1, \gamma) = f(x_1, \gamma) \quad 0 < x_1 < \ell(\gamma), \\[2mm] \dfrac{\partial v}{\partial x_1}(0, \gamma) = \dfrac{\partial v}{\partial x_1}(\ell(\gamma), \gamma) = 0. \end{cases}$$

Thus $\lambda \notin \sigma(A)$ and this proves that $\sigma(A) \subseteq \{0\} \cup \{\lambda_n \mid n \geq 1, \, 0 < \gamma < 1\}$.

To prove the other inclusion, we first remark that 0 is already shown to be an eigenvalue of A with infinite multiplicity. We conclude by showing that $(A - \lambda_n I)$ is not onto $L^2(\Omega)$ for all $n \geq 1$ and $0 < \gamma < 1$. To this end, we *regularize* u_n given by (2.42) in contrast to the *localization technique* adapted in the proof of Proposition 2.1.

Let us therefore fix $n > 0$ and $\gamma \in]0, 1[$ and introduce mollifiers

$$\phi_k(y) = \phi(ky) \quad k \geq 1,$$

where ϕ is chosen in $\mathcal{D}(\mathbb{R})$ such that $\int_{\mathbb{R}} \phi \, dy = 1$. If we put $c = \int_{\mathbb{R}} \phi^2 dy$ then it is well-known that

(2.44)
$$\begin{cases} \displaystyle\int_{\mathbb{R}} \phi_k^2 \, dy = \dfrac{c}{k}, \\[2mm] k\phi_k(y - \gamma) \longrightarrow \delta(y - \gamma) \quad \text{in} \quad \mathcal{D}'(\mathbb{R}), \text{ as } k \to \infty. \end{cases}$$

We now construct the sequence of functions $v_k \in D(A)$ which regularize u_n:

(2.45)
$$v_k(x_1, x_2) = \sqrt{\dfrac{2k}{cl(\gamma)}} \, \phi_k(x_2 - \gamma) \cos\left[\dfrac{n\pi}{\ell(\gamma)} x_1\right].$$

Simple computations show that the sequence $\{v_k\}$ enjoys the properties

(2.46)
$$\begin{cases} \|v_k\|_{0,\Omega} \longrightarrow 1 \quad \text{as } k \to \infty, \\[2mm] -\dfrac{\partial^2 v_k}{\partial x_1^2} - \lambda_n v_k = 0. \end{cases}$$

This construction along with the *Closed Graph Theorem* argument presented in Proposition 2.1 terminates the proof. ∎

As in Paragraph 2.5, the elements $u_n = u_n(\gamma; x_1, x_2)$ defined by (2.42) with $n \geq 1$ and $0 < \gamma < 1$ can be regarded as "approximate" eigenvectors of A with "eigenvalues" $\lambda_n = \lambda_n(\gamma)$. These elements do not belong to $L^2(\Omega)$ but they, along with the eigenvectors corresponding to the eigenvalue zero, span $L^2(\Omega)$. To see this at least formally, let us introduce an orthonormal basis $\{e_m(x_2)\}_m$ of $L^2(0, 1)$. For every $f = f(x_1, x_2) \in L^2(\Omega)$, we can decompose

$$(2.47) \qquad f(x_1, x_2) = g(x_1, x_2) + \mathcal{M}(f)(x_2),$$

where $\mathcal{M}(f)(x_2)$ is the average of f with respect to x_1, i.e.,

$$(2.48) \qquad \mathcal{M}(f)(x_2) = \frac{1}{\ell(x_2)} \int_0^{\ell(x_2)} f(x_1, x_2) dx_1.$$

On the other hand, g can formally be written as

$$(2.49) \qquad g(x_1, x_2) = \int_0^1 g(x_1, \gamma) \delta(x_2 - \gamma) d\gamma.$$

For fixed $\gamma \in]0, 1[$, $g(x_1, \gamma)$ has zero average with respect to x_1 and so we can expand

$$(2.50) \qquad g(x_1, \gamma) = \sum_{n \geq 1} \alpha_n(\gamma) \cos\left[\frac{n\pi}{\ell(\lambda)} x_1\right],$$

where the Fourier coefficients $\alpha_n(\gamma)$ are given by

$$(2.51) \qquad \alpha_n(\gamma) = \int_0^{\ell(\gamma)} g(x_1, \gamma) \cos\left[\frac{n\pi}{\ell(\lambda)} x_1\right] dx_1.$$

But $\mathcal{M}(f)(x_2) \in L^2(0, 1)$ and so it can be expressed as a linear combination of e_m's, i.e.,

$$(2.52) \qquad \mathcal{M}(f)(x_2) = \sum_{m \geq 1} \beta_m e_m(x_2).$$

Combining (2.52) with (2.50), we get

$$(2.53) \qquad f = \int_0^1 \left[\sum_{n \geq 1} \alpha_n(\gamma) u_n(\gamma)\right] d\gamma + \sum_{m \geq 1} \beta_m e_m$$

which provides the formal decomposition of f in terms of "eigenvectors" of
A. As we shall see in §3, this is a general phenomenon which is a consequence
of the *spectral decomposition* of the identity.

We conclude by stating that the situation described above provides an
example of *continuous spectrum* together with eigenvalues of *infinite multiplicity*. These two form part of what is called *essential spectrum* a concept to
be defined in §3. It is to be noted that the presence of continuous spectrum is
due to the fact that the channels have different lengths. If the length remains
constant then there is no continuous spectrum but all the eigenvalues are of
infinite multiplicity†.

Figure 2.4.

§3. Spectral Theory of Linear Operators

As illustrated in the case of many physical examples of §2, the spectrum
of linear operators is a concept of basic physical interest as far as vibration problems are concerned. This assertion is confirmed further by many
examples in other areas also. In this section, we plan to develop a general
spectral theory for linear self-adjoint operators in Hilbert spaces. In the case
of arbitrary linear operators, it is often hard to obtain information about the
spectrum even of qualitative nature. The main goal of this section is to establish the so-called *Spectral Decomposition Theorem* for self-adjoint operators.
This will give a mathematical explanation for all the phenomena described
in the case of examples of §2. Another important aim here is to derive variational characterizations of eigenvalues in terms of *Rayleigh quotient*. This is
a very useful tool in applications because it provides good estimates of the
elements in the spectrum.

The material of this section is classical and it can be found together
with historical remarks in many books. Here we give a sample of references
which is clearly not exhaustive: BREZIS [1983] Chapter VI, DUNFORD and
SCHWARTZ [1964], HUET [1976], HUTSON and PYM [1980], KATO [1966], REED
and SIMON [1972-78] and RIESZ and NAGY [1955].

---

† A practical example which one can reflect upon is the organ found in churches. We observe
that the organ pipes have different lengths. Why?

3.1. Generalities on Unbounded Linear Operators

We begin by recalling certain definitions concerning unbounded operators. Let H be a Hilbert space over $\mathbb{C}$ with scalar product $(\cdot,\cdot)$ and the associated norm $\|\cdot\|$. A linear map $A: D(A) \subseteq H \longrightarrow H$ defined on a subspace $D(A)$ of H is called *unbounded linear operator* with *domain* $D(A)$. If there is a constant $c \geq 0$ such that

$$(3.1) \qquad \|Au\| \leq c\|u\| \quad \forall u \in D(A),$$

then A is said to be a *bounded* operator. The set of all bounded operators A with $D(A) = H$ is denoted as $\mathcal{L}(H)$. It is well-known that $\mathcal{L}(H)$ is a Banach space for the norm

$$A \longmapsto \|A\| \overset{\text{def}}{=} \sup_{\substack{u \in D(A) \\ u \neq 0}} \frac{\|Au\|}{\|u\|}.$$

If $D(A)$ is dense in H, i.e., if $\overline{D(A)} = H$ then A is said to be *densely defined*. By the *graph* of A, we understand the following subspace of $H \times H$:

$$\mathcal{G}(A) = \{[u, Au] \mid u \in D(A)\}$$

and A is said to be *closed* if $\mathcal{G}(A)$ is closed in $H \times H$. More precisely, A is closed if and only if for all sequences $\{u_n\}$ in $D(A)$ such that $u_n \longrightarrow u$ in H and $Au_n \longrightarrow f$ in H we have $u \in D(A)$ and $Au = f$.

In all our examples seen in §2 as well as the ones to be studied later, operators considered will always be closed and densely defined. We shall use the notations:

$$\mathcal{R}(A) = \text{ image of } A,$$
$$\mathcal{N}(A) = \text{ null space of } A,$$
$$A_\lambda = A - \lambda I \quad \lambda \in \mathbb{C},$$

where I is the identity operator in H.

Definitions. Let A be a closed unbounded operator which is densely defined. The *resolvent set* $\rho(A)$ of A is defined as the set of $\lambda \in \mathbb{C}$ such that $A_\lambda: D(A) \longrightarrow H$ is bijective and A_λ^{-1} is bounded. For $\lambda \in \rho(A)$, the operator A_λ^{-1} is usually called *resolvent operator* of A. ∎

Remark. We have introduced the above notions in the case of a closed operator A with dense domain. However, these can be extended to cover arbitrary operators; see for instance DAUTRAY and LIONS [1988-90] Chapter VIII. ∎

Definition. The complement of $\rho(A)$ in $\mathbb{C}$ is called the *spectrum* of A. It is denoted by $\sigma(A)$. ∎

Some elementary properties of the spectrum are summarized in the next result.

Proposition 3.1. *The resolvent set $\rho(A)$ is an open subset of* $\mathbb{C}$. *The resolvent operator $\lambda \in \rho(A) \longmapsto A_\lambda^{-1} \in \mathcal{L}(H)$ is a complex-analytic operator-valued map.*

Proof. If $\lambda_0 \in \rho(A)$, we will prove that $\lambda \in \rho(A)$ for all λ such that $|\lambda - \lambda_0| < (1/\|A_{\lambda_0}^{-1}\|)$. For each λ verifying the above condition, we introduce the *Neumann series* in $\mathcal{L}(H)$:

$$S(\lambda) = A_{\lambda_0}^{-1} \sum_{k \geq 0} (\lambda - \lambda_0)^k A_{\lambda_0}^{-k}.$$

We observe that this series is uniformly convergent because the general term is dominated by a convergent real geometric series:

$$\|(\lambda - \lambda_0)^k A_{\lambda_0}^{-k}\| \leq |\lambda - \lambda_0|^k \|A_{\lambda_0}^{-1}\|^k \leq \theta^k,$$

where $\theta = |\lambda - \lambda_0| \|A_{\lambda_0}^{-1}\| < 1$. Once the above convergence is established, it is easily seen that the operator $S(\lambda)$ provides the inverse of A_λ as shown below. We have

$$A_{\lambda_0} S(\lambda) = \sum_{k \geq 0} (\lambda - \lambda_0)^k A_{\lambda_0}^{-k},$$

$$(\lambda - \lambda_0) S(\lambda) = \sum_{k \geq 0} (\lambda - \lambda_0)^{k+1} A_{\lambda_0}^{-(k+1)}.$$

By subtraction, we obtain $A_\lambda S(\lambda) = I$. An analogous computation shows that $S(\lambda) A_\lambda = I$. Hence A_λ is invertible and its inverse is given by $A_\lambda^{-1} = S(\lambda) \in \mathcal{L}(H)$. The above proof also provides a power series expansion of A_λ^{-1} in terms of λ and so the resolvent operator is analytic. ∎

The above result provides the optimum properties of $\sigma(A)$ as a subset of $\mathbb{C}$ in the general case. Without additional assumptions on A, one cannot say anything more on the structure of $\sigma(A)$. If, for instance, A is bounded, then $\sigma(A)$ is compact and non-empty (see Proposition 3.2 below). The following examples illustrate the pathologies:

◇ **Example 3.1.** With $H = L^2(\mathbb{C})$, $D(A) = \{u \in H \mid zu(z) \in L^2(\mathbb{C})\}$ and $Au(z) = zu(z)$, we see easily that $\sigma(A) = \mathbb{C}$. If we take $H = L^2(0,1)$, $D(A) = \{u \in H^1(0,1) \mid u(0) = 0\}$ and $Au(x) = u'(x)$, then one can prove that $\sigma(A)$ is empty. ∎

Proposition 3.2. *If $A \in \mathcal{L}(H)$ then its spectrum $\sigma(A)$ is a non-empty compact subset of* $\mathbb{C}$.

Proof. We will show that $\sigma(A)$ is contained in the ball of radius $\|A\|$ and centered at the origin. This will prove that $\sigma(A)$ is compact. Let $\lambda \in \mathbb{C}$ be then such that $|\lambda| > \|A\|$. Then the Neumann series

$$S(\lambda) = \sum_{k \geq 0} \lambda^{-(k+1)} A^k$$

is uniformly convergent because $(\|A\|/|\lambda|) < 1$. Following the calculations of the proof of Proposition 3.1, it is easily seen that $S(\lambda)$ indeed provides an inverse of A_λ.

To prove that $\sigma(A)$ is non-empty, we suppose the contrary. Then the resolvent map $\lambda \longrightarrow A_\lambda^{-1}$ is analytic on $\mathbb{C}$. The Neumann series which defines A_λ^{-1} shows clearly that the resolvent map should then remain bounded on $\mathbb{C}$. By classical *Liouville's Theorem*, it follows that the resolvent map is constant which is impossible. ∎

Definitions. An element $\lambda \in \mathbb{C}$ is called an *eigenvalue* of A if there exists $v \in D(A)$, $v \neq 0$ such that $Av = \lambda v$. The corresponding non-zero vectors v are called *eigenvectors*. We define the associated *eigenspace* as follows:

$$\mathcal{N}(A_\lambda) = \{v \in D(A) \mid Av = \lambda v\}.$$

The dimension of the space $\mathcal{N}(A_\lambda)$ is known as (geometric) *multiplicity* of λ. The set $\sigma_p(A)$ consisting of eigenvalues is called the *point spectrum* of A. ∎

It is well-known that in finite dimensions the spectrum coincides with the point spectrum. This is due to the fact that in finite dimensions, all linear operators are continuous and they are injective iff they are bijective. The crucial difference in infinite dimensions is that the spectrum can, and usually does, contain elements other than eigenvalues; see for instance, the examples presented in Paragraphs 2.5 and 2.6. The next definition covers the other possibilities in infinite dimensions.

Definition. The set of $\lambda \in \sigma(A)$ for which A_λ is injective and $\mathcal{R}(A_\lambda)$ is dense (respectively not dense) in H will be called *continuous spectrum* (respectively *residual spectrum*). We will denote them by $\sigma_c(A)$ and $\sigma_r(A)$ respectively. ∎

Clearly, $\sigma(A)$ is decomposed into the three disjoint sets

$$(3.2) \qquad \sigma(A) = \sigma_p(A) \cup \sigma_c(A) \cup \sigma_r(A).$$

This decomposition of the spectrum is not very well adapted to the classification of self-adjoint operators which are of interest to us. This is because, as we shall see, $\sigma_r(A)$ is always empty for such operators. A different decomposition will be considered later for the spectrum of self-adjoint operators.

3.2. Self-Adjoint Operators

Definition. (Adjoint operator) Let $A: D(A) \longrightarrow H$ be an unbounded opera-tor which is densely defined. Identifying H with its dual, we will now define an unbounded operator $A^*: D(A^*) \subseteq H \longrightarrow H$ as follows: First, let

(3.3) $D(A^*) = \{v \in H \mid \exists c \geq 0 \text{ such that } |(v, Au)| \leq c\|u\| \quad \forall u \in D(A)\}.$

More precisely, $D(A^*)$ consists of those elements v in H for which the linear form $u \in D(A) \longmapsto (v, Au)$ is continuous for the norm in H. Since $D(A)$ is dense in H, the above linear form can be uniquely extended to a continuous linear form on H. Hence, by *Riesz Representation Theorem*, there is a unique element $g \in H$ such that

$$(v, Au) = (g, u) \quad \forall u \in D(A).$$

Next, for every v in $D(A^*)$, we define

(3.4) $A^*v = g.$

This operator A^* is clearly linear. It is called the *adjoint* of A. By the very definition, we have the following fundamental relation:

(3.5) $(v, Au) = (A^*v, u) \quad \forall u \in D(A), \forall v \in D(A^*).$ ∎

It is worth to note that if A is bounded then $D(A^*) = H$ and A^* is bounded too. In the general case, one can only say that A^* is closed. To prove this, we need to introduce the concept of *inverse graph* $\mathcal{G}'(A)$ which is the following subset of $H \times H$:

(3.6) $\mathcal{G}'(A) = \{[Au, u] \mid u \in D(A)\}.$

The role played by the graphs in the analysis of adjoint operators is illustrated by

Proposition 3.3. *Let* $A: D(A) \longrightarrow H$ *be an unbounded densely defined operator. Then* $\mathcal{G}'(-A^*) = \mathcal{G}(A)^\perp$, *the orthogonal complement of* $\mathcal{G}(A)$ *in* $H \times H$.

Proof. Let $[v, f] \in H \times H$. Then

$$[f, v] \in \mathcal{G}'(-A^*) \Leftrightarrow -A^*v = f$$
$$\Leftrightarrow -(f, u) = (v, Au) \quad \forall u \in D(A)$$
$$\Leftrightarrow [f, v] \in \mathcal{G}(A)^\perp. \blacksquare$$

Since $\mathcal{G}(A)^\perp$ is always closed in $H \times H$, an immediate consequence of the above result is

Corollary 3.4. A^* *is closed.* ∎

Concerning $D(A^*)$, we have

Proposition 3.5. *Let* $A: D(A) \longrightarrow H$ *be an unbounded densely defined operator which is closed. Then* $D(A^*)$ *is dense in* H.

Proof. Let $g \in D(A^*)^\perp$ be arbitrary. Then $(g, v) = 0 \; \forall v \in D(A^*)$. This can be written as $[0, g] \perp [-A^*v, v] \; \forall v \in D(A^*)$ and hence $[0, g] \in \mathcal{G}'(-A^*)^\perp$. Now using Proposition 3.3, we conclude that

$$[0, g] \in (\mathcal{G}(A)^\perp)^\perp = \overline{\mathcal{G}(A)} = \mathcal{G}(A),$$

since A is closed. It follows immediately that $g = 0$. Thus we reach the conclusion $D(A^*)^\perp = \{0\}$ and a standard application of the *Hahn Banach Theorem* terminates the proof. ∎

We are now in a position to prove a very useful relation between the range of A and the null space of A^*. This will enable us to analyze the residual spectrum of self-adjoint operators.

Theorem 3.6. *Let* $A: D(A) \longrightarrow H$ *be an unbounded densely defined operator. Then* $\overline{\mathcal{R}(A)} = \mathcal{N}(A^*)^\perp$.

Proof. First, if $Au \in \mathcal{R}(A)$ with $u \in D(A)$, then for all $v \in \mathcal{N}(A^*)$ we have $(Au, v) = (u, A^*v) = 0$. This shows that $\overline{\mathcal{R}(A)} \subset \mathcal{N}(A^*)^\perp$ because $\mathcal{N}(A^*)^\perp$ is closed. Secondly, to prove the other inclusion, it is enough to show that $\mathcal{R}(A)^\perp \subset \mathcal{N}(A^*)$. Let then w be such that $(w, Au) = 0 \; \forall u \in D(A)$. From (3.5), it follows that $w \in D(A^*)$ and $(A^*w, u) = 0 \; \forall u \in D(A^*)$. Since $D(A)$ is dense, $A^*w = 0$, i.e., $w \in \mathcal{N}(A^*)$. ∎

Definitions. A densely defined linear operator A is called *self-adjoint* if $A = A^*$ (in particular, $D(A) = D(A^*)$). A is said to be *unitary* if A is invertible and $A^{-1} = A^*$. ∎

Corollary 3.7. *If* A *is self-adjoint then* $\sigma(A)$ *is real and* $\sigma_r(A)$ *is empty.*

Proof. First, let us observe that

(3.7) (Au, u) is real $\forall u \in D(A)$,

if A is self-adjoint. This follows easily from (3.5). Next, we claim the following inequality:

(3.8) $\|A_\lambda u\| \geq |\mathfrak{Im}\,\lambda|\,\|u\|$ $\forall u \in D(A), \forall \lambda \in \mathbb{C}$.

To this end, let us put $f = A_\lambda u = Au - \lambda u$. Taking inner product of f with u and taking the imaginary part, we obtain the relation

$$\mathfrak{Im}\,(f, u) = -(\mathfrak{Im}\,\lambda)\|u\|^2.$$

It now suffices to apply Cauchy-Schwarz's Inequality to the left side of the above relation. This yields (3.8).

It follows from (3.8) that $\mathcal{N}(A_{\bar\lambda}) = \{0\}$ if $\mathfrak{Im}\,\lambda \neq 0$. Applying Theorem 3.4 to A_λ, we conclude that $\overline{\mathcal{R}(A_\lambda)} = H$ if $\mathfrak{Im}\,\lambda \neq 0$. An easy computation with (3.8) shows that furthermore $\mathcal{R}(A_\lambda)$ is closed. Hence $\mathcal{R}(A_\lambda) = H$. Consequently $\lambda \in \rho(A)$ if $\mathfrak{Im}\,\lambda \neq 0$ (i.e., $\sigma(A)$ is real) and (3.8) provides an estimate for the inverse A_λ^{-1}, namely

(3.9) $\|A_\lambda^{-1}\| \leq \dfrac{1}{|\mathfrak{Im}\,\lambda|}.$

To prove the second part of the corollary, observe that since λ is real and A_λ is injective then $R(A_\lambda)$ is dense in H by virtue of Theorem 3.6. ∎

◇ **Exercise 3.2.** Show that if A is unitary then A is bounded and $\|A\| = 1$. Show further that $\sigma(A)$ is contained in the unit circle and that $\sigma_r(A)$ is empty. ∎

◇ **Exercise 3.3.** Prove that eigenvectors corresponding to different eigenvalues of a self-adjoint operator are orthogonal. ∎

We have seen, in the proof of Proposition 3.2, that the spectrum of an arbitrary bounded operator A is contained in the closed ball $\overline{B(0, \|A\|)}$. If A is in addition self-adjoint, we have the following more precise result:

Proposition 3.8. *Let $A \in \mathcal{L}(H)$ be self-adjoint. We define*

(3.10) $m_-(A) = \inf\limits_{\|u\|=1} (Au, u)$ *and* $m_+(A) = \sup\limits_{\|u\|=1} (Au, u)$.

Then $\sigma(A) \subset [m_-, m_+]$, $m_- \in \sigma(A)$ and $m_+ \in \sigma(A)$.
Moreover, $\|A\| = \max\{|m_-|, |m_+|\}$.

Proof. Let $\lambda \in \mathbb{R}$ be such that $\lambda > m_+$. From the definition of m_+, it follows that

$$(Au, u) \leq m_+ \|u\|^2 \quad \forall u \in H.$$

Therefore, the operator $(-A_\lambda)$ satisfies the following H-ellipticity inequality with the constant $(\lambda - m_+) > 0$:

$$(-A_\lambda u, u) \geq (\lambda - m_+)\|u\|^2 \quad \forall u \in H.$$

By Lax-Milgram Lemma, A_λ is an isomorphism and $\lambda \in \rho(A)$. Since A is self-adjoint, this proves that $\sigma(A) \subset (-\infty, m_+]$. The other inclusion, namely that $\sigma(A) \subset [m_-, \infty)$ is proved in a similar fashion.

Now, let us prove that $m_- \in \sigma(A)$. The sesquilinear form

$$a(u, v) \stackrel{\text{def}}{=} (A_{m_-} u, v)$$

is Hermitian and non-negative. Applying Cauchy-Schwarz's Inequality with respect to $a(\cdot, \cdot)$, we have

$$|(A_{m_-} u, v)| \leq (A_{m_-} u, u)^{1/2} (A_{m_-} v, v)^{1/2} \quad \forall u, v \in H.$$

Taking supremum over v such that $\|v\| = 1$, we get

$$(3.11) \qquad \|A_{m_-} u\| \leq c(A_{m_-} u, u)^{1/2} \quad \forall u \in H$$

where $c = \sup\{(A_{m_-} v, v)^{1/2} \mid \|v\| = 1\}$.

Let $\{u_n\}$ be a minimizing sequence for m_- (i.e., $\|u_n\| = 1$ and $(Au_n, u_n) \longrightarrow m_-$). We see, from (3.11), that $A_{m_-} u_n \longrightarrow 0$ in H and so A_{m_-} cannot be invertible. Hence $m_- \in \sigma(A)$.

Replacing A by $(-A)$ and applying the above argument to $(-A)$, we conclude that $m_+ \in \sigma(A)$.

Finally, since $(A^2 u, u) = (Au, Au)$, we have for arbitrary $t > 0$

$$\|Au\|^2 = \frac{1}{4}\left[\left(A\left(tu + \frac{1}{t}Au\right), tu + \frac{1}{t}Au\right) - \left(A\left(tu - \frac{1}{t}Au\right), tu - \frac{1}{t}Au\right)\right]$$

$$\leq \frac{1}{4}\left[m\left\|tu + \frac{1}{t}Au\right\|^2 + m\left\|tu - \frac{1}{t}Au\right\|^2\right]$$

$$= \frac{1}{2}m\left[t^2\|u\|^2 + \frac{1}{t^2}\|Au\|^2\right],$$

where $m = \max\{|m_-|, |m_+|\}$. The minimum of the right side is attained when

$$t^2 = \frac{\|Au\|}{\|u\|},$$

which gives

$$\|Au\|^2 \leq m\|Au\|\|u\|.$$

This proves that $\|A\| \leq m$ since u in arbitrary. The reverse inequality, namely that $m \leq \|A\|$ being obvious, this concludes the proof. ∎

Corollary 3.9. *Let $A \in \mathcal{L}(H)$ be a self-adjoint operator with $\sigma(A) = \{0\}$. Then $A = 0$.*

Proof. It follows immediately from Proposition 3.8. ∎

Let us end this paragraph with some examples of self-adjoint operators borrowed from the earlier section.

◇ **Example 3.4.** (Green's operator for the drum problem) Let us recall that $G: L^2(\Omega) \longrightarrow L^2(\Omega)$ is defined by $Gh = z$, where z is the unique solution of (2.8). It is easily seen that G is bounded via Poincaré Inequality and so G^* is also bounded and $D(G^*) = L^2(\Omega)$. The relation defining G^* is

$$\int_\Omega G^* f \bar{h} dx = \int_\Omega f \overline{Gh} dx \quad \forall f, h \in L^2(\Omega).$$

To prove that G is self-adjoint, it suffices to show that

$$(3.12) \qquad \int_\Omega G f \bar{h} dx = \int_\Omega f \overline{Gh} dx \quad \forall f, h \in L^2(\Omega).$$

Now, taking $\phi = Gh$ in the relation defining Gf and taking $\phi = Gf$ in the relation defining Gh (see (2.8)), we see that (3.12) follows easily. ∎

◇ **Example 3.5.** The Green's operators associated with Neumann, Robin and mixed boundary conditions defined in Paragraphs 2.2 and 2.4 are all easily verified to be bounded and self-adjoint. We leave the details of proof to the reader. ∎

◇ **Example 3.6.** (Green's operator for the Steklov problem) Another interesting exercise for the reader is to verify that the Green's operator $G: L^2(\Gamma) \longrightarrow L^2(\Gamma)$, $Gg = u_{|\Gamma}$, where u is the unique solution of (2.21) is bounded and self-adjoint. ∎

◇ **Example 3.7.** (Laplace operator in $\mathbb{R}^N$) This is the case of an unbounded operator whose domain $D(-\Delta)$ is the space $H^2(\mathbb{R}^N)$:

$$-\Delta: H^2(\mathbb{R}^N) \longrightarrow L^2(\mathbb{R}^N).$$

To identify $(-\Delta)^*$, we appeal to the very definition of adjoint (see (3.3)). Now, $v \in D((-\Delta)^*)$ iff $v \in L^2(\mathbb{R}^N)$ and there exists $f \in L^2(\mathbb{R}^N)$ such that

$$-\int_{\mathbb{R}^N} v \Delta \bar{u} dx = \int_{\mathbb{R}^N} f \bar{u} dx \quad \forall u \in H^2(\mathbb{R}^N).$$

Taking $u \in \mathcal{D}(\mathbb{R}^N)$, it follows that $-\Delta v = f$ in $\mathbb{R}^N$ and so $v \in H^2(\mathbb{R}^N; \Delta) = H^2(\mathbb{R}^N)$. This shows that $D((-\Delta)^*) \subset H^2(\mathbb{R}^N)$. The other inclusion follows easily from Cauchy-Schwarz's Inequality and Green's formula:

$$(3.13) \qquad \int_{\mathbb{R}^N} v\Delta \bar{u} dx = \int_{\mathbb{R}^N} \bar{u}\Delta v dx \quad \forall u, v \in H^2(\mathbb{R}^N).$$

The above formula also shows that $(-\Delta)^* v = -\Delta v \ \forall v \in H^2(\mathbb{R}^N)$. Thus $-\Delta$ is a self-adjoint operator in $L^2(\mathbb{R}^N)$. With the terminology introduced in this paragraph, we can now make Proposition 2.1 more accurate and say that $\sigma(-\Delta) = \sigma_c(-\Delta) = [0, \infty)$. ∎

◇ **Example 3.8.** The unbounded operator $A = -\partial^2/\partial x_1^2$ defined in $L^2(\Omega)$ with domain $D(A)$ given by (2.40) is also self-adjoint. The proof follows the lines of the previous example and is left to the reader. Our discussion in Paragraph 2.5 amounts to say that if $\ell(x_2) \equiv$ constant, then $\sigma(A) = \sigma_p(A) = \{\lambda_n \mid n \geq 0\}$, where λ_n is defined by (2.41) and if $\ell(x_2)$ is a smooth, strictly monotone positive function, then $\sigma(A) = \sigma_p(A) \cup \sigma_c(A)$ with $\sigma_p(A) = \{0\}$ and $\sigma_c(A) = \{\lambda_n \mid n \geq 1 \text{ and } 0 < \gamma < 1\}$, where λ_n is now given by (2.42). ∎

◇ **Example 3.9.** Here we take $H = L^2(0,1)$,

$$D(A) = \left\{ u \in H^1(0,1) \mid u(1) = e^{i\theta} u(0) \right\},$$

where θ is a real parameter and $Au = iu'(x) \ \forall u \in D(A)$. We ask the reader to compute $D(A^*)$ and A^*. For which values of θ, A is self-adjoint? ∎

◇ **Example 3.10.** Again we take $H = L^2(0,1)$ but $D(A) = H^1(0,1)$ or $D(A) = H_0^1(0,1)$ and $Au(x) = iu'(x) \ \forall u \in D(A)$. Prove that in neither case A is self-adjoint. On the other hand, with the following choices, namely $H = L^2(\mathbb{R})$, $D(A) = H^1(\mathbb{R})$ and $Au(x) = iu'(x)$, we can easily verify that A is self-adjoint. The latter case gives what is called *momentum operator* in Quantum Mechanics. ∎

3.3. Compact Operators

In this paragraph, we are interested in a subclass of bounded operators called *compact operators*. Such operators arise very naturally in elliptic boundary-value problems. For instance, if we are in a situation where Rellich's Lemma is applicable, then the corresponding Green's operator is compact. In fact, history shows that Green's operators were the first examples of compact operators studied by Ivar Fredholm at the end of the last century. Nowadays, there are a multitude of applications of this theory and the topic is extremely well covered in the literature. See for instance, BREZIS [1983],

HUTSON and PYM [1980], KATO [1966] and RIESZ and NAGY [1955]. Our aim in this paragraph is to analyze the structure of the spectrum and the corresponding eigenspaces for general compact operators and to prove the *Spectral Decomposition Theorem* for compact operators which are self-adjoint. As a corollary of this analysis, we deduce a variational characterization of the eigenvalues called the *mini-max principle*. Let us start with the following definition:

Definition. A linear operator $A: H \longrightarrow H$ is called *compact* if the image of the unit ball $B(0,1)$ in H under A is relatively compact in H. The space of all compact operators is denoted as $\mathcal{K}(H)$. ∎

Easy examples of compact operators are linear operators in finite dimensional spaces and more generally, operators with *finite rank*, i.e., operators whose range is finite dimensional. Moreover, $\mathcal{K}(H)$ can be obtained as the closure of operators of finite rank in the operator norm (see BREZIS [1983] pp. 89–90). Thus we see that compact operators are natural candidates of study when one passes from finite to infinite dimensions.

The following results are very useful in applications in checking compactness of operators:

Proposition 3.10. *Let $A \in \mathcal{K}(H)$ and $B \in \mathcal{L}(H)$. Then AB and BA are in $\mathcal{K}(H)$.*

Proof. It is enough to apply the definition. ∎

Proposition 3.11. *An operator $A \in \mathcal{L}(H)$ is compact iff for all sequences $\{u_n\}$ such that $u_n \rightharpoonup u$ weakly in H, then $Au_n \to Au$ strongly in H.*

Proof. It is enough to apply the definition once more. ∎

Theorem 3.12. (Schauder) *An operator $A \in \mathcal{L}(H)$ is compact iff its adjoint A^* is compact.*

Proof. Let $A \in \mathcal{K}(H)$ be given and pose $M = \overline{A(B(0,1))}$. Since A is compact, M is a compact metric space. Let us now regard the elements of $\overline{B(0,1)}$ as continuous functions on M via the inner product. Let us denote this set as $\bar{B}$. Evidently $\bar{B}$ is bounded and for all $f \in \bar{B}$ we have

$$|(f, u_1) - (f, u_2)| \leq \|f\| \|u_1 - u_2\| \leq \|u_1 - u_2\| \quad \forall u_1, u_2 \in M,$$

and so $\bar{B}$ is equi-continuous. Therefore, from the classical *Arzelà-Ascoli Theorem*, a given sequence in $\bar{B}$ has a subsequence, $\{f_n\}$ say, convergent in $\mathcal{C}^0(M)$ and hence Cauchy, so that

$$\sup_{u \in B(0,1)} |(f_n, Au) - (f_m, Au)| \longrightarrow 0 \quad \text{as } m, n \to \infty.$$

Using the relation (3.5), this amounts to say that $\{A^* f_n\}$ is a Cauchy sequence and therefore convergent, i.e., $A^*(\bar{B})$ is relatively compact.

The converse statement follows from the direct part of the theorem and the fact that $A = A^{**}$. ∎

Let us now consider the resolvent equation

$$(3.14) \qquad\qquad A_\lambda u = Au - \lambda u = f$$

where $A: H \longrightarrow H$ is linear. If H is finite dimensional, then there are two possibilities which are mutually exclusive: Either λ is not an eigenvalue of A and in this case A_λ^{-1} exists and (3.14) has a unique solution for every f in H, or λ is an eigenvalue of A and in this case (3.14) has a solution iff f belongs to $\mathcal{N}(A_\lambda^*)^\perp = \mathcal{N}(A^* - \bar{\lambda}I)^\perp$ (see Theorem 3.6.) which amounts to a finite number of *orthogonality conditions* for f. The *Fredholm's Alternative* gives a remarkable generalization of this statement to infinite dimensions with a compact operators A and $\lambda \neq 0$. This is the object of our next theorem.

Theorem 3.13. (Fredholm's Alternative) *Let $A \in \mathcal{K}(H)$ and $\lambda \neq 0$. Then*
(i) $\mathcal{N}(A_\lambda)$ *is finite dimensional.*
(ii) $\mathcal{R}(A_\lambda)$ *is closed and consequently $\mathcal{R}(A_\lambda) = \mathcal{N}(A^* - \bar{\lambda}I)^\perp$.*
(iii) $\mathcal{N}(A_\lambda) = \{0\}$ *iff* $\mathcal{R}(A_\lambda) = H$.

Proof. (i) If possible, let $\mathcal{N}(A_\lambda)$ be infinite dimensional. Choose an orthonormal set $\{e_n\}$ in $\mathcal{N}(A_\lambda)$. We have $Ae_n = \lambda e_n$ $\forall n \geq 1$. Since $\{e_n\}$ is bounded and A is compact, up to a subsequence, $\{Ae_n\}$ is convergent. Since $\lambda \neq 0$, this implies that $\{e_n\}$ is convergent too. But this is not possible because $\|e_n - e_m\| = \sqrt{2}$ for all $m \neq n$.

(ii) We must prove that if $\{f_n\}$ is any convergent sequence in $\mathcal{R}(A_\lambda)$, then its limit, f say, also lies in $\mathcal{R}(A_\lambda)$. Certainly, there are elements u_n such that $f_n = Au_n - \lambda u_n$. We would to have the sequence $\{u_n\}$ bounded. If it is not, we will construct from $\{u_n\}$ another sequence $\{w_n\}$ which is bounded and satisfying

$$(3.15) \qquad\qquad Aw_n - \lambda w_n = f_n \quad \forall n \geq 1.$$

Indeed let us consider the orthogonal decomposition

$$H = \mathcal{N}(A_\lambda) \oplus \mathcal{N}(A_\lambda)^\perp.$$

Let w_n be the orthogonal projection of u_n on $\mathcal{N}(A_\lambda)^\perp$. Obviously, w_n satisfies (3.15). Suppose that $\{w_n\}$ is not bounded. Then for a subsequence (again denoted by n) we will have $\|w_n\| \to \infty$. We then pose $w'_n = (w_n/\|w_n\|)$. As $\|w'_n\| = 1$, the sequence $\{Aw'_n\}$ has a subsequence which is convergent since A is compact. Still denoting this subsequence by n and dividing (3.15) by

$\|w_n\|$ we conclude that $\{w'_n\}$ is convergent since $\lambda \neq 0$. If w' is its limit, we see that w' satisfies $Aw' - \lambda w' = 0$. Thus $w' \in \mathcal{N}(A_\lambda)$ and hence $w' = 0$ since it also belongs to $\mathcal{N}(A_\lambda)^\perp$. This contradicts the fact that $\|w'_n\| = 1$ for all n. We conclude therefore that $\{w_n\}$ is bounded.

The rest of the proof is easy. Since A is compact, using (3.15) and the fact that $\lambda \neq 0$, we deduce again that up to a subsequence $\{w_n\}$ is convergent. If w is its limit, then

$$f = \lim_{n\to\infty} f_n = \lim_{n\to\infty} (Aw_n - \lambda w_n) = Aw - \lambda w.$$

This proves that $f \in \mathcal{R}(A_\lambda)$ which is what we set out to prove.

(iii) First, let us prove that if $\mathcal{N}(A_\lambda) = \{0\}$ then $\mathcal{R}(A_\lambda) = H$. If not, let $H_1 = A_\lambda(H) \neq H$. We know that H_1 is closed from (ii). Then $A_{|H_1} \in \mathcal{K}(H_1)$ and $H_2 = A_\lambda(H_1)$ is a closed subspace of H_1. We claim that $H_2 \neq H_1$. If $H_2 = H_1$, i.e., if $\mathcal{R}(A_\lambda^2) = \mathcal{R}(A_\lambda)$, then for all $u \in H$, there exists $v \in H$ such that $A_\lambda^2 v = A_\lambda u$. Since A_λ is injective, this implies that for all $u \in H$, there exists $v \in H$ such that $u = A_\lambda v$, i.e., $\mathcal{R}(A_\lambda) = H$ which was supposed to be not true. Hence our claim.

Therefore, if we put $H_n = A_\lambda^n(H)$, we get a strictly decreasing sequence of closed subspaces. For all n we can choose $e_n \in H_n$ such that $e_n \perp H_{n+1}$ and $\|e_n\| = 1$. Clearly, $\text{dist}(e_n, H_{n+1}) \geq 1$. Thus we have

$$Ae_n - Ae_m = A_\lambda e_n - A_\lambda e_m + \lambda(e_n - e_m).$$

Then if $n > m$, $H_{n+1} \subset H_n \subset H_{m+1} \subset H_m$ and consequently

$$(A_\lambda e_n - A_\lambda e_m + \lambda e_n) \in H_{m+1}.$$

Therefore, $\|Ae_n - Ae_m\| \geq |\lambda| > 0$. This is absurd since A is compact and $\|e_n\| = 1$. Thus $\mathcal{R}(A_\lambda) = H$.

Conversely, if $\mathcal{R}(A_\lambda) = H$ then $\mathcal{N}(A^* - \bar{\lambda}I) = \{0\}$ by Theorem 3.6. Since A^* is also compact, applying the implication just shown, we get

$$\mathcal{R}(A^* - \bar{\lambda}I) = H.$$

Using again Theorem 3.6, we have $\mathcal{N}(A_\lambda)^\perp = \mathcal{R}(A^* - \bar{\lambda}I) = H$. Therefore $\mathcal{N}(A_\lambda) = \{0\}$. ∎

Our subsequent results reveal the structure of the spectrum of an arbitrary compact operator in infinite dimensions.

Theorem 3.14. *Let H be an infinite dimensional Hilbert space and A in $\mathcal{K}(H)$. Then $0 \in \sigma(A)$ and non-zero elements $\lambda \in \sigma(A)$ are eigenvalues of A, and the corresponding eigenspace is finite dimensional.*

Proof. If $0 \notin \sigma(A)$ then A^{-1} exists and is bounded. By Proposition 3.10, $I = AA^{-1}$ is compact since A is compact. This is not possible in an infinite dimensional space (argue with an orthonormal set $\{e_n\}$ as in the proof of Theorem 3.13, part (i)). The other two assertions follow from Fredholm's Alternative (Theorem 3.13). ∎

Theorem 3.15. *Let $A \in \mathcal{K}(H)$. Then $\sigma(A)$ is either finite or countably infinite and its only possible limit point is zero.*

Proof. We argue by contradiction. Suppose therefore that for some $\varepsilon > 0$, there exist infinitely many λ_n of distinct eigenvalues of A with $|\lambda_n| \geq \varepsilon$. Let u_n be an eigenvector corresponding to λ_n and E_n be the space spanned by $\{u_1, ..., u_n\}$. It is easily checked, using the fact that *Vandermonde's determinant* is not zero, that u_n's are linearly independent and so E_n is a proper subspace of E_{n+1}. As in the proof of Theorem 3.11, we pick up a sequence $\{e_n\}$ such that $\|e_n\| = 1$, $e_n \in E_n$ and dist $(e_n, E_{n-1}) \geq 1$. We have

$$Ae_n - Ae_m = \lambda_n e_n + (Ae_n - \lambda_n e_n - Ae_m) = \lambda_n \left(e_n - \frac{1}{\lambda_n} f \right)$$

where $f = Ae_m - A\lambda_n e_n$. If $m < n$ then $f \in E_{n-1}$ for $Ae_m \in E_m$ and $A\lambda_n e_n \in E_{n-1}$. Therefore, $\|Ae_n - Ae_m\| \geq |\lambda_n| \geq \varepsilon_n$ and so $\{Ae_n\}$ cannot have any convergent subsequence which contradicts our hypothesis that A is compact. ∎

The above result might tempt the reader to think that there are always eigenvalues for compact operators. As the following example shows, this is not the case.

◇ **Example 3.11.** Let $H = L^2(0,1)$ and A be the bounded operator defined by

$$Av(x) = \int_0^x v(\tau)d\tau.$$

First, observe that the image of A is precisely the space

$$V = \{f \in H^1(0,1) \mid f(0) = 0\}.$$

It follows that A is compact by Rellich's Lemma. Clearly, zero is not an eigenvalue of A. If $\lambda \neq 0$ is an eigenvalue of A, then any eigenvector must satisfy the equation $v(x) = \lambda v'(x)$. Then $v(x) = v(0) \exp(x/\lambda)$. Since v must belong to V, $v(0) = 0$ and hence $v \equiv 0$. This proves that $\sigma_p(A)$ is empty and so $\sigma(A) = \{0\}$. Moreover, since V is dense in H, we conclude that $\sigma(A) = \sigma_c(A) = \{0\}$. ∎

This terminates our analysis of the spectrum of general compact operators.

3.4. Compact Self-Adjoint Operators

In order to say more on the properties of eigenvalues and eigenspaces, we will require the compact operator A to be self-adjoint. The central result we shall prove below is the so-called *Hilbert-Schmidt Theorem* which states that the eigenvectors of A can be chosen to form an orthonormal basis for the Hilbert space H.

For simplicity, we will assume H to be separable and let $A\colon H \longrightarrow H$ be a compact self-adjoint operator. From Corollary 3.7, Proposition 3.2 and Theorem 3.15, we know that the spectrum of A is real, bounded and at most countable. Let $\{\lambda_n\}$ be the set of eigenvalues of A (it can be finite or infinite and it can include zero or not) which we shall arrange in the decreasing order in modulus, i.e.,

$$(3.16) \qquad |\lambda_1| \geq \cdots \geq |\lambda_n| \geq |\lambda_{n+1}| \geq \cdots$$

This is possible since zero is the only possible limit point of the spectrum. Moreover, the eigenspaces corresponding to non-zero eigenvalues are finite dimensional and so we repeat each eigenvalue according to its multiplicity. Let $\{v_n\}$ be the corresponding set of eigenvectors which we can choose to be an orthonormal set; this is always possible as eigenvectors corresponding to different eigenvalues are orthogonal and the eigenvectors corresponding to the same eigenvalue can be orthonormalized via *Gram-Schmidt procedure*.

Theorem 3.16. (Hilbert-Schmidt) *Let $A\colon H \longrightarrow H$ be a compact self-adjoint operator. Then its eigenvectors form a basis for H. More precisely, every $u \in H$ can be written as*

$$(3.17) \qquad u = \sum_n (u, v_n) v_n,$$

where the convergence of the series takes place in H.

Proof. Let F be the closure of the linear span of $\{v_n\}$. Consider the orthogonal complement $F^\perp$ of F in H. Since A is self-adjoint and F is invariant under A (i.e., $A(F) \subset F$), we see that $F^\perp$ is also invariant under A and the restriction of A to $F^\perp$, denoted as A_0, is self-adjoint and compact. Obviously we have $\sigma(A_0) = \{0\}$. By virtue of Corollary 3.9, it follows that $A_0 \equiv 0$, i.e., $A(F^\perp) = \{0\}$. This implies that $F^\perp \subset \mathcal{N}(A) \subset F$ which is not possible unless $F^\perp = \{0\}$. This concludes the proof. ∎

We remark that the resolution of the identity (3.17) is classically known as the *Fourier series* of u in the basis $\{v_n\}$ and the *Fourier coefficients* (u, v_n) satisfy the so-called *Parseval's Identity*:

$$(3.18) \qquad \|u\|^2 = \sum_n |(u, v_n)|^2.$$

Our next result asserts that a compact self-adjoint operator can be diagonalized in a suitable basis. This is a generalization of a well-known result for Hermitian matrices and itself admits a far-reaching generalization for arbitrary self-adjoint operators which will be considered in the subsequent paragraph.

Theorem 3.17. *Let $A: H \longrightarrow H$ be a compact self-adjoint operator. Then, for all $u \in H$,*

(3.19)
$$Au = \sum_n \lambda_n(u, v_n)v_n.$$

Proof. To obtain (3.19), it suffices to apply A to the relation (3.17). ∎

◇ **Example 3.12.** (Green's operator for the drum problem revisited)
Let us recall that, in Example 3.4, we saw that the Green's operator G for the drum problem is bounded and self-adjoint. Since the range of G is contained in $H_0^1(\Omega)$, we see that $G: L^2(\Omega) \longrightarrow L^2(\Omega)$ is compact via Rellich's Lemma. Thus, we can apply Hilbert-Schmidt Theorem to G. Since we have $(Gh, h) > 0 \ \forall h \in L^2(\Omega)$, $h \neq 0$ (to see this, it suffices to take $\phi = z$ in (2.8)), zero is not an eigenvalue of G (even though it is an element of the spectrum of G) and the eigenvalues of G are positive. They form then a sequence which converges to zero. The corresponding eigenvectors $\{v_n\}$ form a basis in $L^2(\Omega)$. Using (2.8) we can also conclude that the eigenvectors form a basis in $H_0^1(\Omega)$ also. ∎

◇ **Exercise 3.13.** Applying Hilbert-Schmidt Theorem to the operators defined in Examples 3.5 and 3.6, reach a conclusion similar to that of the preceding example. ∎

◇ **Exercise 3.14.** The purpose of this exercise is to generalize the examples given just above and put them in an abstract general framework. Let $A: D(A) \subset H \longrightarrow H$ be an unbounded, densely defined self-adjoint operator on an infinite dimensional separable Hilbert space H. Assume that A has a compact inverse $A^{-1} = G$, say. Denote by $\{\mu_n\}$ and $\{v_n\}$ the sets of eigenvalues and corresponding eigenvectors of G. Prove that (a) G is self-adjoint, (b) zero is not an eigenvalue of G, (c) $\sigma(A) = \{\lambda_n\}$, where $\lambda_n = \mu_n^{-1}$ for all n and the eigenvectors of A is the set $\{v_n\}$, (d) $|\lambda_n| \to \infty$ as $n \to \infty$, (e) and finally, prove that $u \in D(A)$ iff

$$\sum_n \lambda_n^2 |(u, v_n)|^2 < \infty$$

and that in this case one has the following expression for Au:

$$Au = \sum_n \lambda_n(u, v_n)v_n \quad \forall u \in D(A). \ ∎$$

Having analyzed the structure of the spectrum and the eigenspaces, let us now pass on to another characterization of eigenvalues which involves only the operator in question and not its eigenvectors. This characterization is formulated as a variational principle known as *mini-max principle* where the functional involved is the so-called *Rayleigh quotient*.

Definition. Let $A \in \mathcal{L}(H)$ be self-adjoint. For all $u \in H$, $u \neq 0$, we introduce the *Rayleigh quotient* by

$$(3.20) \qquad R(u) = \frac{(Au, u)}{\|u\|^2}. \quad \blacksquare$$

It is worth to point out that $R(u)$ is a real number since A is supposed to be self-adjoint.

We place ourselves in the framework of Hilbert-Schmidt Theorem. In particular, we assume that A is a self-adjoint compact operator and without loss of generality, that $m_- = m_-(A) \geq 0$ (see (3.10) for the definition of $m_-(A)$); otherwise, one can decompose the space H and A into two parts, $H = H^+ \oplus H^-$ and $A = A^+ \oplus A^-$ respectively, where $A^\pm$ are defined by

$$(3.21) \qquad \begin{cases} A^+ u = \displaystyle\sum_{n;\lambda_n \geq 0} \lambda_n (u, v_n) v_n, & H^+ = \text{span}\{v_n \mid \lambda_n \geq 0\}, \\ A^- u = \displaystyle\sum_{n;\lambda_n < 0} \lambda_n (u, v_n) v_n, & H^- = \text{span}\{v_n \mid \lambda_n < 0\} \end{cases}$$

and repeat the analysis that follows for A^+ and $(-A^-)$ separately†.

With this choice, we remark that $\sigma(A) \subset [0, m_+]$ and let us arrange the eigenvalues of A in the decreasing order as indicated below:

$$(3.22) \qquad \lambda_1 \geq \cdots \geq \lambda_n \geq \cdots \longrightarrow 0.$$

Denote by $\{v_n\}$ an orthonormal basis in H formed by eigenvectors of A. We now set out to prove the following variational characterization for the first eigenvalue λ_1:

$$(3.23) \qquad \lambda_1 = \max_{u \in H, u \neq 0} R(u).$$

Towards this end, observe first that we have

$$(3.24) \qquad R(v_n) = \lambda_n \quad \forall n \geq 1.$$

---

† Observe that $A^\pm \in \mathcal{L}(H^\pm)$ and they are self-adjoint and compact.

Consequently, we deduce that λ_1 is less or equal than the right hand side of (3.23). To prove the reverse inequality, let us take $u \in H$ arbitrary. By virtue of (3.17), (3.18) and (3.19), we get

$$R(u) = \frac{\sum\limits_n \lambda_n |(u, v_n)|^2}{\sum\limits_n |(u, v_n)|^2} \leq \lambda_1,$$

which completes the proof of (3.23).

In order to obtain a variational characterization for λ_m analogous to (3.23), we introduce E_m to be the linear span of $\{v_1, ..., v_m\}$. Let $E_m^{\perp}$ denote the orthogonal complement of E_m in H. If $u \in E_{m-1}^{\perp}$ then

$$R(u) = \frac{\sum\limits_{n \geq m} \lambda_n |(u, v_n)|^2}{\sum\limits_{n \geq m} |(u, v_n)|^2} \leq \lambda_m,$$

which implies, together with (3.24), that

$$(3.25) \qquad \lambda_m = \max_{u \in E_{m-1}^{\perp}, u \neq 0} R(u).$$

This is the source of the other characterizations formulated in the following result:

Theorem 3.18. (Courant-Fischer) *Let A be a compact self-adjoint operator in a Hilbert space H. Then, for all $m \geq 1$, we have*

$$(3.26) \qquad \lambda_m = \max_{F_m \in \mathcal{H}_m} \min_{u \in F_m, u \neq 0} R(u)$$

$$(3.27) \qquad \lambda_m = \min_{F_{m-1} \in \mathcal{H}_{m-1}} \max_{u \in F_{m-1}^{\perp}, u \neq 0} R(u),$$

where $\mathcal{H}_m$ is the set of subspaces of H of dimension m. Moreover, in (3.26) the maximum is attained at E_m and in (3.27) the minimum is attained at E_{m-1}.

Proof. Take first $F_m = E_m$. If $u \in F_m$, we have

$$R(u) = \frac{\sum\limits_{n=1}^{m} \lambda_n |(u, v_n)|^2}{\sum\limits_{n=1}^{m} |(u, v_n)|^2} \geq \lambda_m,$$

which implies that λ_m is lower or equal than the right side of (3.26). To prove the reverse inequality, it suffices to show that for all $F_m \in \mathcal{H}_m$, we have

$$\lambda_m \geq \min_{u \in F_m, u \neq 0} R(u).$$

We can always choose $u \in F_m$, $u \neq 0$ such that $u \in E_{m-1}^{\perp}$ since the dimension of F_m is m. By virtue of (3.25), we see that $\lambda_m \geq R(u)$ and so the proof of (3.26) is completed.

Towards the proof of (3.27), we first note that (3.25) immediately implies that λ_m is greater or equal than the right side of (3.27). The proof of the reverse inclusion follows the arguments analogous to the earlier case except that now, given $F_{m-1} \in \mathcal{H}_{m-1}$, we look for an element $u \in F_{m-1}^{\perp} \cap E_m$, $u \neq 0$ which is obviously possible. ∎

◇ **Exercise 3.15.** (Exercise 3.14 revisited) Let us go back to the situation considered in Exercise 3.14 which corresponds to the case where A is an unbounded self-adjoint operator in H with a compact inverse G. To simplify matters, assume that A is non-negative in the sense that

$$(3.28) \qquad\qquad (Au, u) > 0 \quad \forall u \in D(A), \ u \neq 0.$$

According to Exercise 3.14, the spectrum of A consists then of a sequence of positive numbers $\{\lambda_m\}$ which converges to ∞. Arrange them in the increasing order, i.e.,

$$(3.29) \qquad\qquad 0 < \lambda_1 \leq \cdots \leq \lambda_m \leq \cdots \longrightarrow \infty.$$

Following the proof of Theorem 3.18 step by step deduce the following characterizations for the λ_m's:

$$(3.30) \qquad\qquad \lambda_1 = \min_{u \in D(A), u \neq 0} R(u),$$

and if E_m is the linear span of $\{v_1, ..., v_m\}$, then

$$(3.31) \qquad\qquad \lambda_m = \min_{\substack{u \in D(A) \cap E_{m-1}^{\perp} \\ u \neq 0}} R(u).$$

Deduce further that

$$(3.32) \qquad\qquad \lambda_m = \min_{F_m \in \mathcal{H}_m} \ \max_{\substack{u \in D(A) \cap F_m \\ u \neq 0}} R(u)$$

and

$$(3.33) \qquad\qquad \lambda_m = \max_{F_m \in \mathcal{H}_{m-1}} \ \min_{\substack{u \in F_{m-1}^{\perp} \cap D(A) \\ u \neq 0}} R(u). \ ∎$$

Comparing with (3.26) and (3.27), we observe that max-min procedure is reversed. This is due to the fact that the eigenvalues in (3.29) are arranged in the increasing order.

◇ **Exercise 3.16.** Here we take $H = L^2(\Omega)$, where Ω is a bounded open set in $\mathbb{R}^N$. Define the unbounded operator A in H by

(3.34)
$$\begin{cases} D(A) = H(\Omega, \Delta) \cap H_0^1(\Omega), \\ \quad Au = -\Delta u. \end{cases}$$

Prove that A is invertible and its inverse G is none other than the Green's operator for the drum problem (see Examples 3.4 and 3.12). Verify that A satisfies all the hypotheses of the previous exercise. Using the variational characterization (3.30), deduce that $(1/\sqrt{\lambda_1})$ is the best constant with which *Poincaré Inequality* holds (see Theorem 1.3). ■

3.5. Spectral Decomposition of Self-Adjoint Operators

This paragraph is devoted to a generalization of the *Hilbert-Schmidt Theorem* to general unbounded operators which are self-adjoint. We concentrate our attention to bounded operators and give guidelines to cover the unbounded ones. These results are by now classical and different proofs can be found in various books cited in the introduction of this section; see for instance, DUNFORD and SCHWARTZ [1964], KATO [1966] and RIESZ and NAGY [1955]. The *Spectral Decomposition Theorem* for bounded self-adjoint operators is due to D. Hilbert. Several other approaches to the same problem were undertaken by different authors, notably by F. Riesz, J. von Neumann and M. Stone. In the sequel, we follow the method introduced by F. Riesz and presented in RIESZ and NAGY [1955]. We recommend the same reference for the treatment of general unbounded self-adjoint operators also. It can be pointed out that it was J. von Neumann who succeeded first in giving a proof of the spectral decomposition result for general self-adjoint operators.

To motivate the things to come, suppose first that $A \in \mathcal{L}(H)$ is compact and self-adjoint. The power of the *Hilbert-Schmidt Theorem* rests on the fact that it provides a decomposition of the identity operator in terms of eigenvectors of A. If the sequence $\{\lambda_k\}$ denotes all distinct eigenvalues of A and $\{M_k\}$ the corresponding eigenspaces, we can write

(3.35)
$$H = \bigoplus_k M_k,$$

i.e., every $u \in H$ can be decomposed as

(3.36)
$$u = \sum_k P_k u \quad (I = \sum_k P_k),$$

where P_k is the orthogonal projection operator on M_k. Moreover, we also have as a consequence that

(3.37)
$$Au = \sum_k \lambda_k P_k u \quad (A = \sum_k \lambda_k P_k),$$

There is no hopeful sign of generalizing this relation as such to an operator A which is self-adjoint but not compact. This is because A need not have any eigenvector at all and can exhibit continuous spectrum (see Examples 3.7 and 3.8). However, there is another way of decomposing I which will allow us to overcome these difficulties. To do this, we reconsider the compact operator A and introduce for all $\lambda \in \mathbb{R}$

(3.38)
$$\begin{cases} G_\lambda = \bigoplus_{k;\lambda_k \leq \lambda} M_k \quad \text{and} \\ E_\lambda, \text{ the orthogonal projection on } G_\lambda. \end{cases}$$

As $\sigma(A) \subset [m_-, m_+]$, we see that $E_\lambda = 0$ for $\lambda < m_-$ and $E_\lambda = I$ for $\lambda \geq m_+$. Therefore, for any interval $[a, m_+]$ with $a < m_-$ and for any partition $\Delta = \{a = \mu_0 < \mu_1 < \cdots < \mu_n = m_+\}$ of the interval $[a, m_+]$, we have trivially the decomposition

(3.39)
$$I = \sum_{j=1}^{n} (E_{\mu_j} - E_{\mu_{j-1}}).$$

We can pass to the limit in the above relation as the partition Δ is refined and conclude the following integral representation for I:

(3.40)
$$I = \int_a^{m_+} dE_\lambda.$$

The above integral is understood in the sense of Riemann-Stieltjes. For the sake of completeness, let us recall its definition. Consider a finite interval $[a, b]$ of $\mathbb{R}$ and a Banach space $\mathcal{B}$. Suppose f and E are respectively a bounded real-valued and a $\mathcal{B}$-valued functions defined on $[a, b]$. Taking a finite partition $\Delta = \{a = \mu_0 < \mu_1 < \cdots < \mu_n = b\}$ of the interval $[a, b]$ we define, as usual, the lower and the upper sums as follows:

$$L(f, \Delta) = \sum_{j=1}^{n} \left[\inf_{\mu \in [\mu_{j-1}, \mu_j]} f(\mu) \right] (E_{\mu_j} - E_{\mu_{j-1}}),$$

$$U(f, \Delta) = \sum_{j=1}^{n} \left[\sup_{\mu \in [\mu_{j-1}, \mu_j]} f(\mu) \right] (E_{\mu_j} - E_{\mu_{j-1}}).$$

Definitions. A function f is said to be *Riemann-Stieltjes integrable* with respect to E if there exists $g \in \mathcal{B}$ such that

$$\lim_{|\Delta| \to 0} L(f, \Delta) = \lim_{|\Delta| \to 0} U(f, \Delta) = g,$$

where $|\Delta|$ is the maximum length of the subintervals $[\mu_{j-1}, \mu_j]$. The element g is called the *Riemann-Stieltjes integral* of f and we write

$$(3.41) \qquad\qquad g = \int_a^b f(\lambda) dE_\lambda.$$

Finally, if the function f is complex-valued, then its *Riemann-Stieltjes integral* can be defined by splitting f into its real and imaginary parts. ∎

The goal of this paragraph is to generalize the integral representation (3.40) to the case of a bounded self-adjoint operator and then indicate the main changes necessary to cover the case of an unbounded operator which is self-adjoint. Let then A be a bounded self-adjoint operator. The first stage is to construct the family $\{E_\lambda\}_{\lambda \in \mathbb{R}}$ of projections associated with A. E_λ will be defined as $e_\lambda(A)$, where e_λ is the step function

$$(3.42) \qquad\qquad e_\lambda(t) = \begin{cases} 1 & \text{if } t \le \lambda, \\ 0 & \text{if } t > \lambda. \end{cases}$$

This requires to define nonlinear functions of A, a task which we will achieve by the so-called *monotonicity method*. It is easy to see that polynomial functions $p(A)$ of A are well-defined. By a passage to the limit, using *Weierstrass Theorem*, we will then be able to define $f(A)$ for all continuous functions $f(\lambda)$. Another approximation argument will then lead us to the definition of $f(A)$ where $f(\lambda)$ is just an upper semi-continuous function. In order to carry out this program, we begin by defining a partial ordering in the set of bounded self-adjoint operators.

Definition. Let $A, B \in \mathcal{L}(H)$ be self-adjoint. We shall write $A \le B$ or $B \ge A$ if $(Au, u) \le (Bu, u) \; \forall u \in H$. A is said to be *non-negative definite* if $A \ge 0$ and *positive definite* if $(Au, u) > 0 \; \forall u \ne 0$. Finally, A is called *uniformly positive definite* if $A \ge \alpha I$ for some $\alpha > 0$. ∎

The method of monotonicity is based on the following result:

Lemma 3.19. *Let $\{A_n\}$ be a sequence of bounded self-adjoint operators and assume that the sequence is monotonically decreasing (i.e., $A_{n+1} \le A_n, n = 1, 2, ...$) and bounded below by a self-adjoint operator $B \in \mathcal{L}(H)$. Then $\{A_n\}$ converges strongly to a bounded self-adjoint operator A. Moreover, if C in $\mathcal{L}(H)$ commutes with each A_n then C commutes with A also.*

Proof. Without loss of generality, we assume that

$$0 \leq A_n \leq I \quad \forall n \geq 1.$$

Take $n \leq m$ and note that $0 \leq A_n - A_m \leq I$. Applying Proposition 3.8, we get $\|A_n - A_m\| = m_+(A_n - A_m)$. But trivially $0 \leq m_+(A_n - A_m) \leq 1$ and so $\|A_n - A_m\| \leq 1$. On the other hand, we have

$$\|[A_n - A_m]u\|^4 = ([A_n - A_m]u, [A_n - A_m]u)^2$$
$$= ([A_n - A_m]^2 u, u)^2$$

and applying Cauchy-Schwarz's Inequality for the inner product induced by the self-adjoint non-negative operator $(A_n - A_m)$, we obtain

$$\|(A_n - A_m)u\|^4 \leq ([A_n - A_m]^2 u, [A_n - A_m]u) \ ([A_n - A_m]u, u) \leq$$
$$\leq \|u\|^2 [(A_n u, u) - (A_m u, u)],$$

since $\|[A_n - A_m]^2\| \leq \|A_n - A_m\| \leq 1$.

As the sequence $\{(A_n u, u)\}$ is non-increasing and bounded, it is convergent. It follows that the sequence $\{A_n u\}$ is Cauchy and hence convergent. Let us call the limit Au. A is clearly linear and it is bounded because,

$$\|Au\| = \lim_{n \to \infty} \|A_n u\| \leq \limsup_{n \to \infty} \|A_n\| \|u\| \leq \|u\|.$$

Also, using (3.5), we conclude easily that A is self-adjoint. The commutativity follows from the relations

$$CAu = C(\lim_{n \to \infty} A_n u) = \lim_{n \to \infty} CA_n u = \lim_{n \to \infty} A_n Cu = ACu. \quad \blacksquare$$

The next step is to define the operator $f(A)$ for any real-valued function f, continuous on $[m_-, m_+]$. If f is a polynomial $p(\lambda)$ with real coefficients, say

$$p(\lambda) = \alpha_0 + \alpha_1 \lambda + \cdots + \alpha_n \lambda^n,$$

then the operator $p(A)$ has an obvious definition

$$p(A) = \alpha_0 I + \alpha_1 A + \cdots + \alpha_n A^n.$$

Moreover, $p(A)$ thus defined is self-adjoint and this association $p(\lambda) \longmapsto p(A)$ is of *positive type* in the sense that if $p(\lambda) \geq 0$ on $[m_-, m_+]$ then $p(A) \geq 0$. To see this, let us rewrite $p(\lambda)$ as follows:

$$p(\lambda) = a \prod_j (\lambda - a_j) \prod_k (b_k - \lambda) \prod_\ell \left[(\lambda - c_\ell)^2 + d_\ell^2 \right]$$

with $a > 0$, the first two products containing the simple real roots so that $a_j \leq m_-$ and $b_j \geq m_+$ and the last containing the complex and repeated roots. Replacing λ by A and using the fact that $m_- I \leq A \leq m_+ I$, we see that each factor will be a non-negative operator. Since the factors commute, it is easily seen that $p(A) \geq 0$. Arguing the same way as before, we can show that if $p(\lambda) \geq q(\lambda)$ on $[m_-, m_+]$, then $p(A) \geq q(A)$. Thus the association $p(\lambda) \longmapsto p(A)$ is monotone.

Now, let $f(\lambda)$ be a real-valued function which is continuous on $[m_-, m_+]$. By classical *Weierstrass Theorem*, there is a sequence $\{p_n(\lambda)\}$ of polynomials with real coefficients such $p_n(\lambda) \longrightarrow f(\lambda)$ uniformly on $[m_-, m_+]$. Therefore, given $\varepsilon > 0$, there exists n_0 such that

$$-\varepsilon \leq p_n(\lambda) - p_m(\lambda) \leq \varepsilon \quad \forall n, m \geq n_0, \forall \lambda \in [m_-, m_+],$$

and by monotonicity

$$-\varepsilon I \leq p_n(A) - p_m(A) \leq \varepsilon I.$$

Applying Proposition 3.8, we get

$$\|p_n(A) - p_m(A)\| \leq \varepsilon.$$

The sequence $\{p_n(A)\}$ is therefore Cauchy and hence convergent and the limit of which we denote by $f(A)$. It is straightforward to verify that $f(A)$ does not depend on the particular choice of the sequence $\{p_n(\lambda)\}$ and therefore $f(A)$ is well-defined. The above defined operator $f(A)$ inherits the nice properties of $p(A)$. More precisely, we have

Lemma 3.20. *Let $A \in \mathcal{L}(H)$ be self-adjoint and $f \in \mathcal{C}^0([m_-, m_+])$ be real-valued. Then the operator $f(A)$ defined above enjoys the following properties:*
(i) *$f(A)$ is self-adjoint.*
(ii) *$f(A)$ is bounded and in fact we have $\|f(A)\| \leq \|f\|_{\infty, [m_-, m_+]}$.*
(iii) *The association $f \longmapsto f(A)$ preserves order, i.e., if $f \geq g$ on $[m_-, m_+]$, then $f(A) \geq g(A)$.*
(iv) *If $g \in \mathcal{C}^0([m_-, m_+])$ also, then*

$$(fg)(A) = f(A)g(A) \quad and \quad (f+g)(A) = f(A) + g(A).$$

(v) *$f(A)$ commutes with every bounded operator that commutes with A.*

Proof. (i), (iv) and (v) easily follow from the definition of $f(A)$. To prove (ii), let us set $\varepsilon_n = \|f - p_n\|_{\infty,[m_-,m_+]}$ where $\{p_n\}$ is a sequence of polynomials converging to f uniformly on $[m_-, m_+]$. Therefore, $\varepsilon_n \to 0$ and

$$p_n(\lambda) \geq \inf_{\lambda \in [m_-,m_+]} f(\lambda) - \varepsilon_n.$$

Hence $p_n(A) \geq [\inf f(\lambda) - \varepsilon_n]I$ and so for all $u \in H$

$$(p_n(A)u, u) \geq [\inf f(\lambda) - \varepsilon_n]\|u\|^2,$$

whence

$$(f(A)u, u) = (p_n(A)u, u) + ([f(A) - p_n(A)]u, u) \geq$$
$$\geq [\inf f(\lambda) - \varepsilon_n]\|u\|^2 - \|f(A) - p_n(A)\|\|u\|^2.$$

Note that the right hand side of the above inequality converges to $[\inf f(\lambda)]\|u\|^2$ because $\varepsilon_n \to 0$ and $p_n(A)$ converges to $f(A)$ in the operator norm. Therefore we obtain

$$(3.43) \qquad (f(A)u, u) \geq \left[\inf_{\lambda \in [m_-,m_+]} f(\lambda)\right]\|u\|^2 \quad \forall u \in H.$$

In a similar manner, we can establish

$$(3.44) \qquad (f(A)u, u) \leq \left[\sup_{\lambda \in [m_-,m_+]} f(\lambda)\right]\|u\|^2 \quad \forall u \in H.$$

Now we observe that (ii) follows from these estimates via Proposition 3.8.

Finally, we note that (iii) is an easy consequence of the fact that if $f \geq 0$ on $[m_-, m_+]$, then $f(A) \geq 0$ which follows immediately from (3.43). ∎

The third step in the construction is to define $f(A)$ where f is a non-negative real-valued and upper semi-continuous function on $[m_-, m_+]$. Such a function can be point-wise approximated by a monotonically decreasing sequence $\{f_n\}$ of real-valued continuous functions on $[m_-, m_+]$. From the result just established above, we see that the sequence of operators $\{f_n(A)\}$ is monotonically decreasing and it clearly fulfills all the hypotheses of Lemma 3.19. Thus $f_n(A)$ converges strongly to a bounded self-adjoint operator $f(A)$. The independence of this definition of $f(A)$ on the choice of the sequence $\{f_n\}$, i.e., the well definedness of $f(A)$ is a consequence of the classical *Dini's Theorem*: If a sequence of continuous functions on a compact Hausdorff space converges monotonically and point-wise to a continuous function, then the convergence is uniform. With this definition, it is not hard to see that $f(A)$ enjoys the properties stated in Lemma 3.20. It is also worth to note that in the definition of $f(A)$ only the values of f on the interval $[m_-, m_+]$ are relevant.

Since the step function e_λ defined in (3.42) is clearly upper semi-continuous and non-negative, $e_\lambda(A)$ is well-defined as a bounded self-adjoint operator. Moreover, since $e_\lambda^2 = e_\lambda$ we have $e_\lambda(A)^2 = e_\lambda(A)$ and so we conclude that $e_\lambda(A)$ is an orthogonal projection onto a closed subspace of H. The map

$$\lambda \longmapsto E_\lambda \overset{\text{def}}{=} e_\lambda(A)$$

which is constructed above is referred to as the *spectral family* associated to the self-adjoint operator $A \in \mathcal{L}(H)$. When no confusion can arise, we shall use the notation $\{E_\lambda\}_\lambda$ or simply E_λ for the *spectral family* associated to A. When there is need to explicit its dependence on A, we use the notation $\{E_\lambda(A)\}_\lambda$ or simply $E_\lambda(A)$. The characterizing properties of this family are listed in the result below.

Theorem 3.21. (Spectral Decomposition for Bounded Operators) *Let H be a complex Hilbert space and $A \in \mathcal{L}(H)$ be a self-adjoint operator. Then there is a unique family $\{E_\lambda\}_\lambda$ of self-adjoint projections with the following properties:*
 (i) *The map $\lambda \longmapsto E_\lambda$ is strongly right continuous.*
 (ii) *$E_\lambda = 0$ if $\lambda < m_-$ and $E_\lambda = I$ if $\lambda \geq m_+$.*
 (iii) *The map $\lambda \longmapsto E_\lambda$ is monotonically non-decreasing, i.e., $E_\lambda \geq E_\mu$ if $\lambda \geq \mu$.*
 (iv) *With the integral being understood in the Riemann-Stieltjes sense, we have*

(3.45)
$$A = \int_a^{m_+} \lambda \, dE_\lambda,$$

where $a < m_-$ is arbitrary. Moreover, each projection E_λ in the spectral family of A commutes with any bounded operator B that commutes with A and for any function f which is continuous on an open set containing $\sigma(A)$, we have

(3.46)
$$f(A) = \int_a^{m_+} f(\lambda) \, dE_\lambda.$$

Proof. *First Step*: Here we show that the family $E_\lambda = e_\lambda(A)$ constructed above satisfies indeed the conditions (i)–(iii). Since $e_\mu(t)e_\lambda(t) = e_\mu(t)$ for $\mu < \lambda$, we have $E_\mu E_\lambda = E_\lambda E_\mu = E_\mu$ and therefore (iii) is true. (ii) is an easy consequence of the observation that $e_\lambda(t) = 0$ on $[m_-, m_+]$ if $\lambda < m_-$ and $e_\lambda(t) = 1$ on $[m_-, m_+]$ if $\lambda \geq m_+$ and the fact that the definition of $f(A)$ depends only on the values of f on $[m_-, m_+]$. To prove the strong right continuity of $\lambda \longmapsto E_\lambda$, let us fix λ and construct a sequence $\{f_n\}$ in $C^0(\mathbb{R})$

which decreases monotonically to e_λ on the interval $[m_-, m_+]$ and which is such that $f_n(t) \geq e_{\lambda+\frac{1}{n}}(t)$ for all t. Then we have

$$f_n(A) \geq E_{\lambda+\frac{1}{n}} \geq E_\lambda.$$

Since $f_n(A) \longrightarrow E_\lambda$ strongly as $n \to \infty$, it follows that $E_{\lambda+\frac{1}{n}} \longrightarrow E_\lambda$ for $n \to \infty$ and hence, thanks to the monotonicity of E_λ with respect to λ, we have also $E_{\lambda+\varepsilon} \longrightarrow E_\lambda$ strongly as $\varepsilon \to 0^+$.

Second Step: We are going to establish the decomposition (3.46). Since f can be decomposed into its real and imaginary parts, we can assume that f is real-valued. Let us take a partition $\Delta = \{a = \mu_0 < \mu_1 < \cdots < \mu_n\}$ with $\mu_0 < m_-$ and $m_+ \leq \mu_n$ and consider the functions defined by

$$L(f, \Delta, t) = \sum_{k=1}^{n} \left[\inf_{s \in [\mu_{k-1}, \mu_k]} f(s) \right] (e_{\mu_k}(t) - e_{\mu_{k-1}}(t)),$$

$$U(f, \Delta, t) = \sum_{k=1}^{n} \left[\sup_{s \in [\mu_{k-1}, \mu_k]} f(s) \right] (e_{\mu_k}(t) - e_{\mu_{k-1}}(t)).$$

We have evidently

$$-w \leq U(f, \Delta, t) - f(t) \leq w$$
$$-w \leq f(t) - L(f, \Delta, t) \leq w,$$

where $w = w(|\Delta|)$ is the modulus of continuity of f. It follows that

$$\|U(f, \Delta, A) - f(A)\| \leq w,$$
$$\|L(f, \Delta, A) - f(A)\| \leq w.$$

Since $w(|\Delta|) \longrightarrow 0$ as $|\Delta| \to 0$, we see that $U(f, \Delta, A)$ and $L(f, \Delta, A)$ both converge to $f(A)$ and therefore (3.46) follows because the common limit of $U(f, \Delta, A)$ and $L(f, \Delta, A)$ is, by definition, the integral on the right side of (3.46).

Third Step: The proof of the uniqueness of the spectral family is based on *Riesz Representation Theorem*. Let $\{E_\lambda\}_\lambda$ be any family of projections of self-adjoint operators satisfying the conditions (i)–(iv) of Theorem 3.21. Then for any integer $r \geq 0$, we have

$$\sum_{k=1}^{n} \left[\inf_{\lambda \in [\mu_{k-1}, \mu_k]} \lambda^r \right] (E_{\mu_k} - E_{\mu_{k-1}}) = \left[\sum_{k=1}^{n} \left[\inf_{\lambda \in [\mu_{k-1}, \mu_k]} \lambda \right] (E_{\mu_k} - E_{\mu_{k-1}}) \right]^r$$

and a similar relation holds for the upper sum associated with the function λ^r. These relations hold because the projections $\{(E_{\mu_k} - E_{\mu_{k-1}})\}_{k=1,n}$ are pair-wise orthogonal, which is a consequence of (iii). The above relations show that the lower and upper sums associated with λ^r have a common limit as $|\Delta| \to 0$ which coincides with A^r. Thus (3.45) extends to all polynomial functions p and so we have

$$(p(A)u, v) = \int_a^{m+} p(\lambda)d(E_\lambda u, v) \quad \forall u, v \in H.$$

We can now pass from polynomials to any continuous function f by Weierstrass Theorem and conclude that the integrals

$$\int_a^{m+} f(\lambda)d(E_\lambda u, v)$$

are uniquely determined by the operator A.

Since the function $\lambda \longmapsto (E_\lambda u, v)$ is right continuous by virtue of Riesz Representation Theorem, we deduce that this function is completely determined up to an additive constant. This constant is also fixed by property (ii) of Theorem 3.21. Thus the spectral family E_λ is completely determined by A.

Fourth Step: Finally, we remark that the commutativity property of E_λ announced in this theorem is a trivial consequence of Lemma 3.20 (v). ∎

◇ **Exercise 3.17.** (Spectral family of the inverse) Let A be a bounded self-adjoint operator on H which is *uniformly positive definite*, i.e., $A \geq \alpha I$ for some $\alpha > 0$. Show that A^{-1} is a bounded self-adjoint operator and its spectral family F_λ is given by

$$F_\lambda = \begin{cases} I - \tilde{E}_{\lambda^{-1}} & \text{if } \lambda > 0, \\ 0 & \text{if } \lambda \leq 0, \end{cases}$$

where the tilde ˜ indicates that the spectral family is taken to be strongly continuous from the left instead of the usual convention established in Theorem 3.21 (i). ∎

   With the above theorem and its subsequent exercise, we have accomplished what we set out to prove in the case of bounded self-adjoint operators. However, the restriction that the operator is bounded is too severe in most applications and in particular, it is not satisfied by differential operators (see Examples 3.7 and 3.8). Therefore, we take efforts to give some guidelines on the natural modifications to be done in order to extend the above theorem to the case of an unbounded self-adjoint operator. The most significant change arises from the fact that the spectrum may be not bounded and the spectral family E_λ will then vary over an unbounded interval and much more care is needed in the convergence of the integral. In particular, we need the notion of improper Riemann-Stieltjes integral with values in H.

Definition. Suppose f and h are respectively a real-valued and a H-valued functions defined on $\mathbb{R}$. If f is Riemann-Stieltjes integrable on each compact subinterval on $\mathbb{R}$ with respect to h, we define the *improper Riemann-Stieltjes integral* by

$$(3.47) \qquad \int_{-\infty}^{\infty} f(\lambda)dh_\lambda = \lim_{\substack{a \to -\infty \\ b \to +\infty}} \int_a^b f(\lambda)dh_\lambda,$$

if the limit on the right hand side exists in the norm of H. ∎

Another change involves the notion of commutativity.

Definition. Let $A: D(A) \subset H \longrightarrow H$ be an unbounded operator and B be a bounded operator. We say that A *commutes* with B if $BA \subset AB$, i.e., $D(BA) \subset D(AB)$ and AB is an extension of BA. ∎

One of the methods to obtain the spectral decomposition for a general self-adjoint operator A is to "approximate" it by a sequence of bounded operators $\{A_n\}$; for the details of this method and also for the proof of the following theorem, we refer to RIESZ and NAGY [1955] pp. 310–316.

Theorem 3.22. (Spectral Decomposition for Unbounded Operators) *Let H be a complex Hilbert space and $A: D(A) \subset H \longrightarrow H$ be a self-adjoint operator. Then there exists a unique spectral family $\{E_\lambda\}_\lambda$ of self-adjoint projections with the following properties:*
  (i) *The map $\lambda \longmapsto E_\lambda$ is strongly right continuous.*
  (ii) *$E_\lambda \longrightarrow 0$ strongly as $\lambda \to -\infty$, $E_\lambda \longrightarrow I$ strongly as $\lambda \to +\infty$.*
  (iii) *The map $\lambda \longmapsto E_\lambda$ is monotonically non-decreasing.*
  (iv) *For all $u \in H$, we have*

$$(3.48) \qquad u = \int_{-\infty}^{\infty} dE_\lambda u \quad (in \ the \ sense \ of \ (3.47)),$$

and if $u \in D(A)$, then we have

$$(3.49) \qquad Au = \int_{-\infty}^{\infty} \lambda dE_\lambda u \quad (in \ the \ sense \ of \ (3.47)).$$

Moreover, each projection E_λ in the spectral family of A commutes with any bounded operator that commutes with A (in particular, $\mathcal{R}(E_\lambda)$ is contained in $D(A)$ for all $\lambda \in \mathbb{R}$) and $D(A)$ can be characterized as

$$D(A) = \left\{ u \in H \ \middle| \ \int_{-\infty}^{\infty} \lambda^2 d(E_\lambda u, u) < \infty \right\},$$

where the last integral is understood in the sense of the usual Riemann-Stieltjes improper integral. ∎

The main difference between the bounded and the unbounded cases is the following: while the integrals in (3.45), (3.46) converge in the operator norm those in (3.48), (3.49) converge strongly. One of the useful applications of the Spectral Decomposition Theorem is the so-called *spectral calculus* which allows us to define the operator $f(A)$ where f is a continuous complex-valued function.

Definition. Let A be unbounded and self-adjoint and f be a continuous function defined on an open set containing $\sigma(A)$. We introduce

$$(3.50) \qquad D(f(A)) = \left\{ u \in H \ \Big| \ \int_{-\infty}^{\infty} |f(\lambda)|^2 d(E_\lambda u, u) < \infty \right\},$$

and we define $f(A)$ by requiring that

$$(3.51) \qquad f(A)u = \int_{-\infty}^{\infty} f(\lambda) dE_\lambda u \quad \forall u \in D(f(A)). \ \blacksquare$$

With these definitions, we have

Theorem 3.23. (Spectral Calculus) *Let A be an unbounded self-adjoint operator and f, g be complex-valued functions continuous on an open set containing $\sigma(A)$. Then*
(i) $(f + g)(A) = f(A) + g(A)$ *and if g is bounded on $\sigma(A)$ then*

$$(fg)(A) = f(A)g(A).$$

(ii) *If f is real-valued then $f(A)$ is self-adjoint.*
(iii) *The map $f \longmapsto f(A)$ preserves order.*
(iv) *If f is bounded on $\sigma(A)$ then $f(A)$ is bounded and $\|f(A)\| = \|f\|_{\infty,\sigma(A)}$.*
(v) *For all $u \in D(A)$, we have (Generalized Parseval's Identity):*

$$\|f(A)u\|^2 = \int_{-\infty}^{\infty} |f(\lambda)|^2 d(E_\lambda u, u). \ \blacksquare$$

The definition (3.51) of $f(A)$ which holds for functions f which are continuous in a neighbourhood of $\sigma(A)$ can be extended for functions f which are just upper semi-continuous. For the details, we refer the reader once more to RIESZ and NAGY [1955], pp. 337–347. In particular, let us briefly show how the projections $\{E_\lambda\}_\lambda$ which, we know, already exist can be obtained via spectral calculus. Indeed E_λ is nothing but the operator $e_\lambda(A)$ which can be defined using spectral calculus as follows. We approximate, as usual, the function $e_\lambda(\cdot)$ by a monotonically decreasing sequence $\{f_n\}$ of real-valued bounded continuous functions on $\mathbb{R}$. From Theorem 3.23, we see that $\{f_n(A)\}$ is a monotonically decreasing sequence of bounded operators which clearly fulfills all the hypotheses of Lemma 3.19. Arguing as in the case of bounded operators, we define $e_\lambda(A)$ as the strong limit in $\mathcal{L}(H)$ of the sequence $\{f_n(A)\}$. One can also verify that $e_\lambda(A)$ is a projection and the following integral representation holds:

$$(3.52) \qquad e_\lambda(A)u = \int_{-\infty}^{\lambda} dE_\mu u \quad \forall u \in H.$$

$\Diamond$ **Example 3.18.** Let us consider the situation of Paragraph 2.5 and Example 3.7, namely the case of $(-\Delta)$ in $\mathbb{R}^N$. It was shown in Example 3.7 that this operator is self-adjoint and so we can apply Spectral Decomposition Theorem and deduce that there is a spectral family $\{E_\lambda\}_\lambda$ associated to $(-\Delta)$. The purpose here is to give an explicit expression for E_λ. We shall show, in fact, that $E_\lambda u$ is defined in terms of the Fourier transform of u as follows:

$$(3.53) \qquad (E_\lambda u)(x) = \begin{cases} \dfrac{1}{(2\pi)^{N/2}} \displaystyle\int_{\{|\xi|^2 \leq \lambda\}} \hat{u}(\xi)e^{i\xi \cdot x} d\xi & \text{if } \lambda > 0, \\[4mm] 0 & \text{if } \lambda \leq 0 \end{cases}$$

for all $u \in L^2(\mathbb{R}^N)$. First, let us verify that $E_\lambda u \in L^2(\mathbb{R}^N)\ \forall u \in L^2(\mathbb{R}^N)$ and $\lambda > 0$. By Parseval's Identity, it suffices to prove that $\widehat{E_\lambda u} \in L^2(\mathbb{R}^N)$. But from the definition of $E_\lambda u$, it follows that

$$(3.54) \qquad \widehat{E_\lambda u}(\xi) = \begin{cases} \hat{u}(\xi) & \text{if } |\xi|^2 \leq \lambda, \\ 0 & \text{if } |\xi|^2 > \lambda, \end{cases}$$

and hence $\widehat{E_\lambda u} \in L^2(\mathbb{R}^N)$ and

$$(3.55) \qquad \|E_\lambda u\|_{0,\mathbb{R}^N}^2 = \|\widehat{E_\lambda u}\|_{0,\mathbb{R}^N}^2 = \int_{\{|\xi|^2 \leq \lambda\}} |\hat{u}(\xi)|^2 d\xi.$$

Similar computations using Plancherel's Identity establish also that

$$(3.56) \qquad (E_\lambda u, v) = \int_{\{|\xi|^2 \leq \lambda\}} \hat{u}(\xi)\hat{\bar{v}}(\xi)d\xi \quad \forall u, v \in L^2(\mathbb{R}^N)$$

which implies that E_λ is self-adjoint. Moreover, using (3.54), we have for all $u, v \in L^2(\mathbb{R}^N)$,

$$(E_\lambda^2 u, v) = (E_\lambda u, E_\lambda v) = (\widehat{E_\lambda u}, \widehat{E_\lambda v}) = \int\limits_{\{|\xi|^2 \leq \lambda\}} \hat{u}(\xi)\hat{\bar{v}}(\xi)d\xi.$$

This shows that $E_\lambda^2 = E_\lambda$ thanks to (3.56), i.e., the family $\{E_\lambda\}_\lambda$ consists of projections.

Let us now check that E_λ has all the properties listed in Theorem 3.22 which characterize the spectral family associated with a given unbounded self-adjoint operator. First, properties (ii) and (iii) are easily verified from the definition of E_λ. Regarding property (i), we shall prove something stronger, namely that the function $F: \lambda \longmapsto (E_\lambda u, v)$ is absolutely continuous for all $u, v \in L^2(\mathbb{R}^N)$. More precisely, we claim that F is locally integrable and its distributional derivative is given by a function in $L^1(\mathbb{R})$. The first claim follows from the fact that F is locally bounded, which is obvious. Regarding the second one, we have the following expression for F' which is an easy consequence of *Fubini's Theorem*:

$$(3.57) \qquad \langle F', \phi \rangle = \int\limits_{\mathbb{R}^N} \hat{u}(\xi)\hat{\bar{v}}(\xi)\phi(|\xi|^2)d\xi \quad \forall \phi \in \mathcal{D}(\mathbb{R}).$$

Here we make a change of variables in the integral found on the right side of (3.57): $\xi \longmapsto [\lambda, \eta]$, where $\lambda = |\xi|^2$ and $\eta = \lambda^{-1/2}\xi$. (3.57) is then transformed to

$$\langle F', \phi \rangle = \int\limits_0^\infty \lambda^{1/2(N-1)}\frac{1}{2}\lambda^{-1/2} \int\limits_{S^{N-1}} \hat{u}(\lambda^{1/2}\eta)\hat{\bar{v}}(\lambda^{1/2}\eta)d\eta\, d\lambda$$

and so, for almost all λ in $\mathbb{R}$, we obtain

$$(3.58) \qquad F'(\lambda) = \begin{cases} \dfrac{1}{2}\lambda^{1/2(N-2)} \displaystyle\int\limits_{S^{N-1}} \hat{u}(\lambda^{1/2}\eta)\hat{\bar{v}}(\lambda^{1/2}\eta)d\eta & \text{if } \lambda > 0, \\[4mm] 0 & \text{if } \lambda \leq 0. \end{cases}$$

Since the right hand side defines an integrable function of λ, we see that the second claim is proved. This concludes the proof of property (i) in Theorem 3.22.

Now, let us prove the formula

$$(3.59) \qquad (u, v) = \int\limits_{-\infty}^\infty d(E_\lambda u, v) \quad \forall u, v \in L^2(\mathbb{R}^N),$$

i.e., (3.48). Indeed, integrating (3.58) with respect to λ in $\mathbb{R}$, we see that the right side of (3.59) is equal to

$$\frac{1}{2}\int_0^\infty \lambda^{1/2(N-2)} \int_{S^{N-1}} \hat{u}(\lambda^{1/2}\eta)\hat{\bar{v}}(\lambda^{1/2}\eta)d\eta d\lambda$$

which re-expressed in the original variable ξ gives

$$\int_{\mathbb{R}^N} \hat{u}(\xi)\hat{\bar{v}}(\xi)d\xi.$$

Thus by *Plancherel's Identity*, this integral is nothing but (u,v) which establishes (3.59). Similar computations show that

$$(3.60) \qquad (-\Delta u, v) = \int_{-\infty}^\infty \lambda d(E_\lambda u, v) \quad \forall u \in H^2(\mathbb{R}^N), \forall v \in L^2(\mathbb{R}^N).$$

This is just (3.49).

By appealing to the uniqueness part of Theorem 3.22, we conclude that the family of projections $\{E_\lambda\}_\lambda$ defined by (3.53) is the spectral family associated to $(-\Delta)$. ∎

Remark. A natural perturbation of the problem studied in the previous example is the propagation of waves in an exterior domain Ω of the form $\Omega = \mathbb{R}^N \backslash \bar{\mathcal{O}}$, where $\mathcal{O}$ is a bounded open set having the regularity specified by (1.5). This consideration leads us to consider $A = -\Delta$ as an unbounded operator with dense domain in $L^2(\Omega)$. More precisely, in case of a Dirichlet condition on $\partial\Omega$, we have

$$D(A) = \{v \in L^2(\Omega) \mid \Delta v \in L^2(\Omega), \; v = 0 \quad \text{on} \quad \partial\Omega\}.$$

Following a technique similar to the one in Example 3.7, we can show that A is self-adjoint. So, we are in a position to apply Spectral Decomposition Theorem to A and deduce the existence of a spectral family $\{E_\lambda\}_\lambda$ associated to A. It can be proved that A admits no eigenvalues and in fact $\sigma(A)$ has a structure similar to the one studied in Example 3.18, namely that $\sigma(A) = \sigma_c(A) = [0, \infty)$. The proofs of these facts are not trivial and so is the construction of the associated spectral projections $\{E_\lambda\}_\lambda$. Recall that the plane waves $e^{i\xi \cdot x}$ were used to achieve this in Example 3.18. To construct the waves which serve the same purpose in this case, one has to introduce the so-called *Sommerfeld radiation condition* and the associated *scattering problem*. Since these techniques lie outside the scope of this book, we refer the reader for the details to SÁNCHEZ-HUBERT and SÁNCHEZ-PALENCIA [1989] Chapter VIII. ∎

Remark. We will see another interesting example of perturbation of Example 3.18 in Paragraph 2.1 of Chapter III. The spectral decomposition of *Schrödinger operator* with periodic potential will be obtained by using the so-called *Bloch waves* which replace plane waves of Example 3.18. ∎

◇ **Exercise 3.19.** Consider the situation of Paragraph 2.6 and Example 3.8, i.e., the case of the unbounded operator $A = -\partial^2/\partial x_1^2$ defined in $L^2(\Omega)$ with domain defined in (2.40). We adopt the notations introduced in the paragraph just quoted.

(i) Show that if $\ell(x_2) \equiv$ constant, then the spectral family associated with A is defined, for $u \in L^2(\Omega)$, by

$$
(3.61) \qquad E_\lambda u = \begin{cases} 0 & \text{if } \lambda < 0, \\ \displaystyle\sum_{n;\lambda_n \leq \lambda} P_n u & \text{if } \lambda > 0, \end{cases}
$$

where P_n is the orthogonal projection onto the eigenspace (which is infinite dimensional) corresponding to the eigenvalue λ_n described by (2.41).

(ii) If $\ell(x_2)$ is a smooth strictly monotone positive function, then the definition of the spectral family requires the decomposition (2.47) of $u \in L^2(\Omega)$, i.e., $u(x_1, x_2) = v(x_1, x_2) + \mathcal{M}(u)(x_2)$, where $\mathcal{M}(u)(x_2)$ is the average of u with respect to x_1 (see (2.48)) and the average of v with respect to x_1 is zero. In this situation, show that $E_\lambda u$ is defined, for $u \in L^2(\Omega)$, by

$$
(3.62) \qquad E_\lambda u = \begin{cases} 0 & \text{if } \lambda < 0, \\ Q_0 \mathcal{M}(u) & \text{if } \lambda = 0, \\ \displaystyle Q_0 \mathcal{M}(u) + \int_0^1 \Big(\sum_{n;\lambda_n(\gamma) \leq \lambda} Q_n(\gamma)v \Big) d\gamma & \text{if } \lambda > 0, \end{cases}
$$

where Q_0 is the orthogonal projection in $L^2(0,1)$ onto the eigenspace (which is infinite dimensional) corresponding to the eigenvalue $\lambda_0 = 0$ and $Q_n(\gamma)$ $(n \geq 1)$ is the orthogonal projection in $L_0^2(0, \ell(\gamma))$. ∎

◇ **Example 3.20.** Let us consider a vibrating system whose state $u(t)$ at the instant t is governed by the differential equation

$$
(3.63) \qquad u''(t) + Au(t) = f(t),
$$

where A is an unbounded self-adjoint operator in a Hilbert space H. It is understood that $u(t) \in D(A)$ $\forall t$. If $\{E_\lambda\}_\lambda$ is the spectral family of A, the above equation can be put in the equivalent form

$$
\frac{d^2}{dt^2} \int_{-\infty}^{\infty} d(E_\lambda u, v) + \int_{-\infty}^{\infty} \lambda d(E_\lambda u, v) = \int_{-\infty}^{\infty} d(E_\lambda f, v) \quad \forall v \in H,
$$

i.e., for all t and λ in $\mathbb{R}$, we have

(3.64) $$\frac{d^2}{dt^2}dE_\lambda u + \lambda dE_\lambda u = dE_\lambda f.$$

To simplify matters, we assume that A is uniformly positive definite with constant $\alpha > 0$ so that its spectrum is contained in $[\alpha, \infty[$. Supposing zero initial conditions, we obtain the following expression for the solution via variation of parameters,

$$dE_\lambda u(t) = \frac{1}{\sqrt{\lambda}} \int_0^t \cos\left[\sqrt{\lambda}(t-s)\right] dE_\lambda f(s) ds.$$

Applying Spectral Decomposition Theorem, we get

(3.65) $$u(t) = \int_0^\infty \frac{1}{\sqrt{\lambda}} \int_0^t \cos\left[\sqrt{\lambda}(t-s)\right] dE_\lambda f(s) ds.$$

Now, to make things more explicit, we assume that the function $\lambda \longmapsto E_\lambda f(t)$ is absolutely continuous and denote by $g_\lambda(t)$ its derivative. With this assumption, (3.65) can be rewritten as follows:

(3.66) $$u(t) = \int_0^\infty \frac{1}{\sqrt{\lambda}} \int_0^t \cos\left[\sqrt{\lambda}(t-s)\right] g_\lambda(s) ds d\lambda.$$

Therefore, we are tempted to argue as in Paragraphs 2.1 and 2.4 and conclude that if the function $t \longmapsto g_\lambda(t)$ is sinusoidal with frequency $\sqrt{\lambda}$ then there will be resonance. However, the presence of an integral in λ variable has consequences. In fact, if we have $g_\lambda(s) = \cos\sqrt{\lambda}s$, then elementary computations show that $u(t)$ is bounded as $t \to \infty$. Thus apparently the kind of phenomenon described in Paragraphs 2.1 and 2.4 seems to be absent. We will come back to this example in later paragraphs and explain more clearly the phenomenon involved (see Remark in Paragraph 3.5).

Let us recall that the expression (3.65) for the solution $u(t)$ of (3.63) corresponds to the case when the initial conditions $u(0)$ and $u'(0)$ vanish. On the other hand, if the right hand side f is identically zero, but the initial conditions are given by

(3.67) $$u(0) = a, \quad u'(0) = b,$$

then we can express formally the solution of (3.63), by using spectral calculus, as follows:

(3.68) $$u(t) = \left[\cos(A^{1/2}t)\right] a + \left[\sin A^{1/2}t\right] A^{-1/2}b.$$

To make this rigorous, we have to assume the following conditions on a and b so that the above expression makes sense:

(3.69) $a \in D(\cos A^{1/2}), \quad b \in D(A^{-1/2}) \quad \text{and} \quad A^{-1/2}b \in D(\sin A^{1/2}).$ ∎

3.6. Elementary Properties of Spectral Families

In this paragraph, we study some properties of the spectral family associated with a self-adjoint operator A in a Hilbert space. These properties indeed characterize, in a different but equivalent way, the various spectra associated with A already introduced. This will lead us in a natural fashion to introduce another classification of the spectrum of A.

Let us begin by giving a result which characterizes the resolvent $\rho(A)$, the complement of the spectrum $\sigma(A)$, in terms of the spectral family.

Theorem 3.24. *Let A be a self-adjoint unbounded operator with domain $D(A)$ in a complex Hilbert space H and $\{E_\lambda\}_\lambda$ be its spectral family. Then a real number λ_0 belongs to $\rho(A)$ iff there is a neighbourhood of λ_0 on which the map $\lambda \longmapsto E_\lambda$ is constant.*

Proof. Let $\lambda_0 \in \rho(A)$ be given. For all $v \in D(A)$ we then have $\|Av - \lambda_0 v\| \geq c\|v\|$ with $c^{-1} = \|(A - \lambda_0)^{-1}\|$. Choose $\varepsilon < c$ and suppose that the map $\lambda \longmapsto E_\lambda$ is not constant on $[\lambda_0 - \varepsilon, \lambda_0 + \varepsilon]$. Then $E = E_{\lambda_0 + \varepsilon} - E_{\lambda_0 - \varepsilon}$ is a non-zero projection since $\{E_\lambda\}_\lambda$ is a non-decreasing family of projections. Therefore, there exists $u \in H$ such that $\|u\| = 1$ and $Eu = u$. On the other hand, by spectral calculus we have (see (3.52))

$$(3.70) \qquad E = \int_{\lambda_0 - \varepsilon}^{\lambda_0 + \varepsilon} dE_\lambda \quad (\text{i.e., } E = \chi_{]\lambda_0 - \varepsilon, \lambda_0 + \varepsilon]}(A))$$

and consequently

$$(3.71) \qquad (A - \lambda_0 I)Ev = \int_{\lambda_0 - \varepsilon}^{\lambda_0 + \varepsilon} (\lambda - \lambda_0) dE_\lambda v \quad \forall v \in D(A).$$

Therefore, we see that $(A - \lambda_0 I)E$ is bounded and

$$(3.72) \qquad \|(A - \lambda_0 I)E\| \leq \varepsilon.$$

In particular, we obtain $\|(A - \lambda_0 I)u\| \leq \varepsilon\|u\|$ since $Eu = u$. This contradicts our choice of ε.

Suppose next that $\lambda_0 \in \mathbb{R}$ is such that the map $\lambda \longmapsto E_\lambda$ is constant on a neighbourhood $U = [\lambda_0 - \varepsilon, \lambda_0 + \varepsilon]$ of λ_0 with $\varepsilon > 0$. Let f be any bounded real-valued continuous function such that $f(\lambda) = (\lambda - \lambda_0)^{-1}$ for all $\lambda \notin U$. By the spectral calculus, the operator $f(A)$ is well-defined. Since the support of the map $\lambda \longmapsto dE_\lambda$ is disjoint from $\overset{\circ}{U}$ (the interior of U), we see that

$$f(A)(A - \lambda_0 I) = (A - \lambda_0 I)f(A) = \int_{\mathbb{R} \setminus \overset{\circ}{U}} f(\lambda)(\lambda - \lambda_0) dE_\lambda = I$$

which proves that $f(A)$ is the inverse of $(A - \lambda_0 I)$ and so $\lambda_0 \in \rho(A)$. ∎

Let us define $E_{\lambda-}$ to be the strong limit of E_μ as $\mu \to \lambda-$; the existence of this limit is guaranteed by an obvious variant of Lemma 3.19 and the non-decreasing property of the family E_λ. We are now ready for one of the principal results of this paragraph.

Theorem 3.25. *Let* $A\colon D(A) \subset H \longrightarrow H$ *be an unbounded self-adjoint operator. Then* $\lambda_0 \in \sigma_p(A)$ *iff* $E_{\lambda_0} \neq E_{\lambda_0-}$. *The eigenspace* $\mathcal{N}(A - \lambda_0 I)$ *corresponding to* λ_0 *is the range of* $(E_{\lambda_0} - E_{\lambda_0-})$. *Moreover, the continuous spectrum* $\sigma_c(A)$ *of* A *consists of those points at which the map* $\lambda \longmapsto E_\lambda$ *is continuous, but which do not have any neighbourhood on which the above map is constant.*

Proof. Suppose that $\varepsilon, \eta \geq 0$ and set $E_{\varepsilon,\eta} = E_{\lambda_0+\eta} - E_{\lambda_0-\varepsilon}$. The proof of (3.72) shows that we have

$$(3.73) \qquad \|(A - \lambda_0 I)E_{\varepsilon,\eta}\| \leq \max(\varepsilon, \eta).$$

Assume that $E_{\lambda_0} \neq E_{\lambda_0-}$. Then the map $\lambda \longmapsto E_\lambda$ is not constant on any interval $[\lambda_0 - \varepsilon, \lambda_0]$. Choose any $u \in \mathcal{R}(E_{\lambda_0} - E_{\lambda_0-})$ such that $\|u\| = 1$. Evidently, $E_{\varepsilon,0}u = u$ because $\lambda \longmapsto E_\lambda$ is non-decreasing. From (3.73), it follows that $\|(A - \lambda_0)u\| = \|(A - \lambda_0)E_{\varepsilon,0}u\| \leq \varepsilon$. Letting $\varepsilon \to 0$ we see that $u \in \mathcal{N}(A - \lambda_0 I)$. This proves the inclusion $\mathcal{R}(E_{\lambda_0} - E_{\lambda_0-}) \subset \mathcal{N}(A - \lambda_0 I)$.

To prove the other inclusion, let $u \in \mathcal{N}(A - \lambda_0 I)$ be such that $\|u\| = 1$. Take ν large enough and $\eta > 0$ such that $\lambda_0 + \eta < \nu$. As in Theorem 3.24, using spectral calculus, we have that $A(E_\nu - E_{\lambda_0+\eta})$ is bounded and

$$A(E_\nu - E_{\lambda_0+\eta}) = \int_{\lambda_0+\eta}^{\nu} \lambda dE_\lambda \geq (\lambda_0 + \eta)(E_\nu - E_{\lambda_0+\eta}).$$

Since E_λ commute with A, it follows that

$$(\lambda_0 + \eta)([E_\nu - E_{\lambda_0+\eta}]u, u) \leq ([E_\nu - E_{\lambda_0+\eta}]Au, u) = \lambda_0([E_\nu - E_{\lambda_0+\eta}]u, u).$$

But $[E_\nu - E_{\lambda_0+\eta}] \geq 0$, therefore $([E_\nu - E_{\lambda_0+\eta}]u, u) = 0$. Letting $\nu \to \infty$, we conclude that $E_{\lambda_0+\eta}u = u$ by Cauchy-Schwarz's Inequality. By a similar argument $E_{\lambda_0-\varepsilon}u = 0$. Letting $\varepsilon, \eta \to 0$ and using the right continuity of the map $\lambda \longmapsto E_\lambda$, we obtain $(E_{\lambda_0} - E_{\lambda_0-})u = u$. This proves the reverse inclusion.

To conclude, it suffices to note that the characterization of $\sigma_c(A)$ is a consequence of Theorem 3.24 and the earlier part of this theorem. ∎

We now present another classical classification of the spectrum $\sigma(A)$ and characterize it in terms of the spectral family of A.

Definition. Let $A\colon D(A) \subset H \longrightarrow H$ be an unbounded self-adjoint operator in an infinite dimensional Hilbert space H. The simplest subset of $\sigma(A)$ is that of isolated eigenvalues of finite multiplicity. We will call it the *discrete spectrum* of A and it will be denoted by $\sigma_d(A)$. The *essential spectrum* $\sigma_e(A)$ is defined to be the complement of $\sigma_d(A)$ in $\sigma(A)$, i.e., $\sigma_e(A)$ consists of the set of eigenvalues of infinite multiplicity, accumulation points of eigenvalues and the continuous spectrum. ∎

It follows immediately from the definition that $\sigma_e(A)$ is a closed subset of $\mathbb{R}$. As a consequence of Theorems 3.24 and 3.25, we obtain the following characterization of $\sigma_e(A)$ in terms of the spectral family $\{E_\lambda\}_\lambda$ of A:

Corollary 3.26. *We have*

$$\sigma_e(A) = \left\{ \lambda \in \sigma(A) \mid \dim \mathcal{R}(E_{\lambda+\varepsilon} - E_{\lambda-\varepsilon}) = +\infty \quad \forall \varepsilon > 0 \right\}. \blacksquare$$

There is another characterization of the essential spectrum which does not use the spectral family but based directly on the operator in question. This is classically known as *Weyl's criterion.*

Theorem 3.27. (Weyl) *A necessary and sufficient condition for λ to belong to $\sigma_e(A)$ is that there exists a sequence $\{u_n\}$ in $D(A)$ such that*

$$(3.74) \qquad \begin{cases} \|u_n\| = 1, \ u_n \longrightarrow 0 \quad in \quad H\text{-}weakly, \\ f_n \overset{\text{def}}{=} (A - \lambda I)u_n \longrightarrow 0 \quad in \quad H\text{-}strongly. \end{cases}$$

Proof. Let $\lambda_0 \in \sigma_e(A)$. Making use of Corollary 3.26, we pick up a sequence $\{\varepsilon_n\}$ of positive numbers tending to zero and a sequence $\{u_n\}$ of orthonormal elements of H such that $u_n \in \mathcal{R}(E_{\lambda_0+\varepsilon_n} - E_{\lambda_0-\varepsilon_n})$. Setting $E_n = E_{\lambda_0+\varepsilon_n^-} E_{\lambda_0-\varepsilon_n}$ and using our usual representation formula (3.71), we get

$$\|(A - \lambda_0 I)u_n\| = \|(A - \lambda_0)E_n u_n\| \leq \int_{\lambda_0-\varepsilon_n}^{\lambda_0+\varepsilon_n} |\lambda - \lambda_0| dE_\lambda u_n$$

which implies that $\|(A - \lambda_0 I)u_n\| \leq \varepsilon_n$. Since $\varepsilon_n \to 0$, we get (3.74).

Conversely, let u_n in $D(A)$ satisfy (3.74). First, observe that the existence of such a sequence implies that $(A - \lambda I)$ is not invertible and hence $\lambda \in \sigma(A)$. It suffices to show that $\lambda \notin \sigma_d(A)$. Supposing the contrary, let $\mathcal{N}(A_\lambda)$ be the eigenspace corresponding to λ and P be the orthogonal projection onto this space. Since this space is finite dimensional, we see that

$Pu_n \longrightarrow 0$ in H-strongly and so the sequence defined by $v_n = u_n - Pu_n$ satisfies

(3.75)
$$\begin{cases} \|v_n\| \longrightarrow 1, \ v_n \longrightarrow 0 \quad \text{in} \quad H\text{-weakly}, \\ (A - \lambda I)v_n \longrightarrow 0 \quad \text{in} \quad H\text{-strongly}, \\ v_n \in D(A) \cap \mathcal{N}(A_\lambda)^\perp. \end{cases}$$

On the other hand, since λ is an isolated point in $\sigma(A)$, it follows that $(A - \lambda I)$ is invertible on the space $\mathcal{N}(A_\lambda)^\perp$. Under this situation, a sequence satisfying (3.75) cannot exist. This contradiction completes the proof of the theorem. ∎

Remark. Weyl's criterion throws light on the *phenomenon of resonance* in the presence of the essential spectrum. Let us recall that we have already encountered this phenomenon in the case of discrete spectrum (see Paragraph 2.1). On the other hand, (3.74) shows that if $\lambda \in \sigma_e(A)$ then the *stationary problem* $u \in D(A)$, $(A - \lambda I)u = f$ has "large" solutions u with "small" data f. Putting $v = e^{i\omega t}u$ and $g = e^{i\omega t}f$ with $\omega^2 = \lambda$, we see that the following evolution equation $v''(t) + Av(t) = g$ can have "small" data g with "large" solution v. This means that if we excite a vibrating system with an external force with a frequency which is "natural" to the system, then the amplitude of its vibrations can be as "large" as we please when compared to the external force. ∎

Comments

There is yet another way of classifying the elements of the spectrum $\sigma(A)$ of a self-adjoint operator. This is different from the ones we have introduced earlier. For simplicity, we assume that H is separable. Let us first define H_p to be the closed subspace of H spanned by all eigenvectors of A. Obviously, H_p is invariant under A and the spectrum of $A_{|H_p}$ is exactly $\bar\sigma_p(A)$. Let us now consider $H_p^\perp$, the orthogonal complement of H_p in H. Since A is self-adjoint, $H_p^\perp$ is also left invariant by A. The spectrum of $A_{|H_p^\perp}$ will be written as a union of two pieces denoted as

$$\sigma_{ac}(A) \quad \text{and} \quad \sigma_{sc}(A)$$

which are not necessarily disjoint. To obtain this, we decompose the space $H_p^\perp$ into two closed orthogonal subspaces:

$$H_p^\perp = H_{ac} \oplus H_{sc}$$

where

$$H_{ac} \overset{\text{def}}{=} \{u \in H_p^\perp \mid \lambda \longmapsto (E_\lambda u, u) \text{ is absolutely continuous on } \mathbb{R}\},$$

$$H_{sc} \overset{\text{def}}{=} \{u \in H_p^{\perp} \mid \lambda \longmapsto (E_\lambda u, u) \text{ is singularly continuous on } \mathbb{R}\}.$$

One can also characterize these spaces in terms of the associated measure dE_λ. Indeed, $H_{ac} = \{u \in H_p^{\perp} \mid$ the measure $d(E_\lambda u, u)$ is absolutely continuous with respect to Lebesgue measure on $\mathbb{R}\}$ and $H_{sc} = \{u \in H_p^{\perp} \mid$ the measure $d(E_\lambda u, u)$ is singular with respect to Lebesgue measure on $\mathbb{R}\}$. These spaces are invariant under A and the corresponding spectra $\sigma(A_{|H_{ac}})$ and $\sigma(A_{|H_{sc}})$ are respectively called the *absolutely continuous spectrum* of A and the *singularly continuous spectrum* of A. They are denoted by $\sigma_{ac}(A)$ and $\sigma_{sc}(A)$ respectively. It is important to note that $\sigma_{ac}(A) \cup \sigma_{sc}(A)$ does not coincide with $\sigma_c(A)$ introduced in Paragraph 3.1. We have the following relations among the various subsets of $\sigma(A)$ defined so far:

$$\begin{cases} \sigma(A) = \bar{\sigma}_p(A) \cup \sigma_{ac}(A) \cup \sigma_{sc}(A), \\ \sigma_e(A) \supset \sigma_{ac}(A) \cup \sigma_{sc}(A), \\ \sigma_d(A) \subset \sigma_p(A). \end{cases}$$

The consideration of the space $H_p^{\perp}$ and the spectrum of A restricted to it is an original idea due to D. Hilbert whereas the notion of continuous spectrum introduced in Paragraph 3.1 follows the ideas of M. Stone and F. Nagy. ∎

§4. Effects of Perturbations

This section is devoted to some special topics of the theory of perturbations with emphasis on eigenvalues and eigenvectors associated with a *compact self-adjoint operator* A. The general situation in perturbation theory could be described as follows: An operator A (bounded or unbounded) is given in a Hilbert space H. Suppose A is perturbed, i.e., we consider $(A+B_\varepsilon)$ where $\{B_\varepsilon\}_\varepsilon$ is a family of operators in H depending on a complex parameter ε. One is then interested in the relations between the spectrum $\sigma(A+B_\varepsilon)$ of $(A+B_\varepsilon)$ and that of A. What are the properties of $\sigma(A+B_\varepsilon)$ as ε varies? Such a situation is very familiar, for example, in (a) Quantum Mechanics where A represents the Hamiltonian of a *free particle* and $(A+B_\varepsilon)$ represents the Hamiltonian of a particle moving in a *potential field* (b) Numerical Analysis too where one tries to compute the spectrum of an operator A by approaching it by a sequence $\{A_h\}$ of finite dimensional operators.

In Chapter II we will encounter several other situations in fluid-solid structures which can be considered as perturbations from classical ones. Our purpose in this section is therefore to develop a suitable framework and present some results which will be needed in the sequel.

The classical perturbation theory which deals with perturbations small in operator norm is a vast subject which is essentially due to F. Rellich and T. Kato (see KATO [1966]). We confine ourselves to the case where the operator A and the perturbation B_ε are compact self-adjoint operators. More exactly, in Paragraph 4.1, we will be concerned with perturbations of the form $B_\varepsilon = \varepsilon B$, where B is a fixed compact self-adjoint operator. From the analysis of such perturbations, we deduce the continuity properties of the eigenvalues of $(A + \varepsilon B)$ as functions of ε. Later on in Paragraph 4.2, we will treat the case $B_\varepsilon = B$, B being now a self-adjoint operator of finite rank. We are then interested in deducing the so-called *interlacing inequalities* between the eigenvalues of A and that of $(A + B)$. Paragraph 4.3 is devoted to the application of the material developed in the earlier paragraphs to *quadratic eigenvalue problems*. Finally, Paragraph 4.4 examines the convergence of the spectral families associated with a sequence of operators which converges strongly or weakly.

We end this introduction by setting up the necessary functional framework and by recalling some notations. H will denote an infinite dimensional separable Hilbert space. The inner product in H and the corresponding norm are denoted as $(\cdot, \cdot)$ and $\|\cdot\|$ respectively. By $\mathcal{L}(H)$, we understand the space of bounded operators on H. For $A \in \mathcal{L}(H)$, let us recall the definition of $\|A\|$:

$$\|A\| = \sup_{\substack{u \in H \\ u \neq 0}} \frac{\|Au\|}{\|u\|}.$$

4.1. Continuity Properties of Eigenvalues

Let $A \in \mathcal{L}(H)$ be a compact self-adjoint operator on H. For the moment, we assume that A is non-negative definite, i.e.,

(4.1) $$(Au, u) \geq 0 \quad \forall u \in H.$$

Let us arrange the eigenvalues of A in decreasing order as indicated below:

(4.2) $$\lambda_1 \geq \cdots \geq \lambda_j \geq \cdots \longrightarrow 0.$$

It is understood that each eigenvalue is repeated as many times as its multiplicity. Denote by $\{v_j\}$ an orthonormal basis formed by the corresponding eigenvectors of A.

Theorem 4.1. *Let $B \in \mathcal{L}(H)$ be another compact self-adjoint operator on H. Assume that $(A+B)$ is non-negative definite and arrange its eigenvalues as*

(4.3)
$$\mu_1 \geq \cdots \geq \mu_j \geq \cdots \longrightarrow 0.$$

Then for all $j \geq 1$, we have

(4.4)
$$\max\{0, \lambda_j + m_-\} \leq \mu_j \leq \lambda_j + m_+,$$

where $m_- = m_-(B)$, $m_+ = m_+(B)$ are the quantities introduced in (3.10). ∎

In this theorem, it can be observed that μ_j and λ_j play a symmetric role. Considering A as a perturbation of $(A+B)$, i.e., $A = (A+B)+(-B)$, it is easily seen that (4.4) amounts to say that for all $j \geq 1$,

(4.5)
$$\max\{0, \mu_j - m_+\} \leq \lambda_j \leq \mu_j - m_-.$$

Proof of Theorem 4.1. Let E_j be the linear span of $\{v_1, ..., v_j\}$. According to Courant-Fischer characterization (see Theorem 3.18) of the eigenvalues μ_j of $(A+B)$, we have

$$\mu_j = \max_{F_j \in \mathcal{H}_j} \min_{u \in F_j, u \neq 0} R_{A+B}(u),$$

where $\mathcal{H}_j$ is the set of subspaces of H of dimension j and $R_{A+B}(\cdot)$ is the Rayleigh quotient for the operator $(A+B)$ defined in (3.20). Choosing $F_j = E_j$, we have

$$\mu_j \geq \min_{u \in E_j, u \neq 0} R_{A+B}(u) \geq \min_{u \in E_j, u \neq 0} R_A(u) + \min_{u \in E_j, u \neq 0} R_B(u).$$

The second term on the right side of the above inequality is obviously bounded below by m_- while the first one equals λ_j by Theorem 3.18. Therefore, we get $\mu_j \geq \lambda_j + m_-$ and this proves the first inequality in (4.4) because $\mu_j \geq 0$.

To prove the other inequality in (4.4), we interchange the roles of A and $(A+B)$ in the above arguments. We arrive at the reverse inequality, i.e.,

$$\lambda_j \geq \mu_j - m_+.$$

This concludes the proof of (4.4) and hence the theorem. ∎

If A does not satisfy (4.1), we decompose $A = A^+ + A^-$ where A^+ and A^- are the operators defined in (3.21). One can repeat the arguments of the above theorem for each of the perturbations (A^++B) and $(B-A^-)$ separately with the hypothesis that $(A^+ + B) \geq 0$ and $(B - A^-) \geq 0$. Combining these two results, we once again get the estimate (4.4).

An immediate consequence of the foregoing result and Proposition 3.8 is the following estimate:

$$(4.6) \qquad\qquad |\lambda_j - \mu_j| \leq \|B\| \qquad \forall j \geq 1.$$

Quite often in applications, the perturbation B_ε is of the form $B_\varepsilon = \varepsilon B$ where B is a fixed compact self-adjoint operator and ε is a real parameter. Theorem 4.1 and estimate (4.6) then allow us to conclude that the eigenvalues of $(A + \varepsilon B)$ depend continuously on ε. More precisely, we have

Corollary 4.2. *Let $A, B \in \mathcal{L}(H)$ be compact self-adjoint operators. Assume A is non-negative definite and denote by E the subset of reals defined by*

$$E = \{\varepsilon \in \mathbb{R} \mid (A + \varepsilon B) \geq 0\}.$$

For $\varepsilon \in E$, arrange the eigenvalues of $(A + \varepsilon B)$ in decreasing order as follows:

$$(4.7) \qquad\qquad \mu_1(\varepsilon) \geq \cdots \geq \mu_j(\varepsilon) \geq \cdots \longrightarrow 0.$$

Then for all $\varepsilon \in E$ and for all $n \geq 1$, we have

$$|\mu_j(\varepsilon) - \lambda_j| \leq |\varepsilon| \|B\|,$$

i.e., the eigenvalues are Lipschitz functions of the perturbations. ∎

Though the perturbations depend analytically (in fact linearly) on ε, then we can only say that the *eigenvalue-functions* are Lipschitz continuous. This is because of the ordering (4.7) imposed on the eigenvalues. However, if we dispense with the ordering then the eigenvalues define locally analytic branches as shown by the classical theorem of Rellich (see KATO [1966] pp. 392–393).

◇ **Example 4.1.** Very simple examples show that the result of Corollary 4.2 is optimal. Indeed, consider the family of matrices $(I + \varepsilon B)$ where

$$B = \begin{pmatrix} 0 & 0 & 0 \\ 0 & 0 & 1 \\ 0 & 1 & 0 \end{pmatrix}.$$

The eigenvalues of $(I + \varepsilon B)$ are $\mu_1(\varepsilon) = 1 + \varepsilon$, $\mu_2(\varepsilon) = 1$ and $\mu_3(\varepsilon) = 1 - \varepsilon$. They are analytic functions which coincide at $\varepsilon = 0$. We note that smoothness is lost when we ask for the customary ordering $\mu_1 \leq \mu_2 \leq \mu_3$, for this requires that

$$\mu_1(\varepsilon) = 1 - |\varepsilon|, \quad \mu_2(\varepsilon) = 1, \quad \mu_3(\varepsilon) = 1 + |\varepsilon|. \quad\blacksquare$$

◇ **Exercise 4.2.** Under the hypotheses of Corollary 4.2, show that if $B \geq 0$, then $E = \mathbb{R}$ and $\mu_j(\varepsilon)$ is a non-decreasing function of ε, for all $j \geq 1$. ∎

◇ **Exercise 4.3.** Consider the eigenvalue problem (2.16) of Paragraph 2.2 with $b \geq 0$ constant. To simplify matters, take $c^2 = 1$ and rewrite the problem as follows: Find $\lambda \in \mathbb{C}$ for which there exists a non-zero $v \in H^1(\Omega)$ such that

$$\begin{cases} -\Delta v = \lambda v & \text{in} \quad \Omega, \\ \dfrac{\partial v}{\partial n} + bv = 0 & \text{on} \quad \Gamma. \end{cases}$$

We know from Example 3.5 and Exercise 3.13 that for all $b \geq 0$, the above problem possesses a sequence of eigenvalues $\{\lambda_n(b)\}_n$ such that $0 \leq \lambda_1 \leq \cdots \leq \lambda_n \leq \cdots \longrightarrow \infty$. Prove that for any fixed n, $\lambda_n(b)$ is a non-decreasing continuous function of b. In particular

$$\lambda_n(0) \leq \lambda_n(b) \leq \lambda_n(\infty),$$

where $\{\lambda_n(\infty)\}_n$ are the eigenvalues corresponding to a Dirichlet boundary condition on Γ. ∎

4.2. Perturbations by Operators of Finite Rank

As in the previous paragraph, A denotes a compact self-adjoint operator on H satisfying (4.1) whose eigenvalues $\{\lambda_j\}$ are arranged in the decreasing order according to (4.2) and the corresponding eigenvectors are denoted $\{v_j\}$. The perturbation operator B is now assumed to be self-adjoint and of finite rank equal to r. Observe that r equals also the number of non-zero eigenvalues of B including their multiplicities. On the perturbed operator $(A+B)$, we put the same restriction as on A. We assume that $(A + B) \geq 0$ and we arrange its eigenvalues $\{\mu_j\}$ as in (4.3). The corresponding eigenvectors $\{w_j\}$ form an orthonormal basis of H.

In such conditions, we are interested in studying how the spectrum changes by the presence of B. We shall see below that B causes the eigenvalues of A to rise or to fall not more than r levels. More precisely, each positive eigenvalue of B causes the eigenvalues of A to rise at most one level and each negative eigenvalue, to fall at most one level. This fact is proved in the form of *interlacing inequalities* between $\{\mu_j\}$ and $\{\lambda_j\}$ in the following

Theorem 4.3. *Let $r_- \geq 0$ (resp. $r_+ \geq 0$) be the number of negative (resp. positive) eigenvalues of B (of course $r = r_+ + r_-$). Then*

$$(4.8) \qquad \begin{cases} \lambda_{j+r_-} \leq \mu_j \leq \lambda_{j-r_+} & \forall j \geq r_+ + 1, \\ \lambda_{j+r_-} \leq \mu_j & \forall j = 1, ..., r_+, \end{cases}$$

under the convention that the second set of inequalities in (4.8) is ignored if $r_+ = 0$. ∎

Note that λ_j and μ_j enter into (4.8) in a symmetric way, i.e., (4.8) can also be written as

$$(4.9) \qquad \begin{cases} \mu_{j+r_+} \leq \lambda_j \leq \mu_{j-r_-}, & \forall j \geq r_- + 1, \\ \mu_{j+r_+} \leq \lambda_j & \forall j = 1, ..., r_-, \end{cases}$$

where the second set of inequalities in (4.9) is ignored if $r_- = 0$.

Proof of Theorem 4.3. The proof is based on the Courant-Fischer characterization (Theorem 3.18) of the eigenvalues of compact self-adjoint operators. First, we prove the required result in case $r = r_+ = 1$ or $r = r_- = 1$. Next, we suppose $r > 1$ and prove that B can be decomposed into a sum of r operators of rank one. We terminate the proof by applying the result of the first step.

First Step: Assume that $r = 1$. Without loss of generality, we will suppose then that the unique non-zero eigenvalue is positive, i.e., $r = r_+ = 1$. Then $(Bu, u) \geq 0 \ \forall u \in H$ and the inequalities in (4.8) become

$$(4.10) \qquad \begin{cases} \lambda_j \leq \mu_j \leq \lambda_{j-1} & \forall j \geq 2, \\ \lambda_1 \leq \mu_1. \end{cases}$$

Fix $j \geq 1$. From Theorem 3.18, we have

$$\lambda_j = \min_{F_{j-1} \in \mathcal{H}_{j-1}} \max_{u \in F_{j-1}^\perp, u \neq 0} R_A(u).$$

Since $B \geq 0$, it follows that $R_A(u) \leq R_{A+B}(u)$ and so $\lambda_j \leq \mu_j$. To finish the proof of (4.10), it remains to prove that $\mu_j \leq \lambda_{j-1}$, for all $j \geq 2$.

To this end, let $z \in H, \|z\| = 1$ be an element spanning the image of B. Since B is self-adjoint, we have the following orthogonal decomposition:

$$(4.11) \qquad \begin{cases} H = \langle z \rangle \oplus \mathcal{N}(B), \\ \langle z \rangle^\perp = \mathcal{N}(B). \end{cases}$$

Fix now $j \geq 2$. By Theorem 3.18, μ_j is characterized as

$$\mu_j = \min_{F_{j-1} \in \mathcal{H}_{j-1}} \max_{u \in F_{j-1}^\perp, u \neq 0} R_{A+B}(u).$$

Choosing $F_{j-1} = V_{j-1}$ which we define as the linear span of $\{z, v_1, ..., v_{j-2}\}$ with the v_j's being the first $(j-2)$ eigenvectors of A. If $j = 2$, we simply put $V_1 = \langle z \rangle$. First, let us consider the case where the dimension of V_{j-1} is equal to $(j-1)$. Then we obtain

$$\mu_j \leq \max_{u \in V_{j-1}^\perp, u \neq 0} R_{A+B}(u).$$

From (4.11), it follows that $V_{j-1}^\perp \subset \mathcal{N}(B)$ and hence,

$$\mu_j \leq \max_{u \in \mathcal{N}(B) \cap \langle v_1, ..., v_{j-2} \rangle^\perp} R_A(u) \leq \max_{u \in \langle v_1, ..., v_{j-2} \rangle^\perp} R_A(u).$$

But Theorem 3.18 shows that the right side of the above inequality is nothing but λ_{j-1}. This yields the desired inequality provided the dimension of V_{j-1} is $(j-1)$. On the other hand, if the dimension of V_{j-1} is $(j-2)$ then z is a linear combination of $\{v_1, ..., v_{j-2}\}$. Therefore, the orthogonal complement of the linear span of $\{v_1, ..., v_{j-2}\}$ and hence that of $\{v_1, ..., v_{j-1}\}$ are contained in $\mathcal{N}(B)$. By Theorem 3.18, we then obtain

$$\mu_j \leq \max_{u \in \langle v_1, ..., v_{j-1} \rangle^\perp} R_A(u) = \lambda_j.$$

Since $\lambda_j \leq \lambda_{j-1}$, we see that the required inequality holds also in this case.

Second Step: Here, the rank of B is assumed to be greater than one. It is enough to prove that every self-adjoint operator B of rank r can be decomposed into a sum of r self-adjoint operators and to apply the first step r times.

Let us denote by $\xi_1 \leq \cdots \leq \xi_{r_-} < 0 < \xi_{r_-+1} \leq \cdots \leq \xi_r$ be the non-zero eigenvalues of B. The corresponding eigenvectors are denoted by $z_1, ..., z_r$ which can be chosen to be an orthonormal set. Let V be the linear span of $\{z_1, ..., z_r\}$. From the Hilbert-Schmidt Theorem, we have

(4.12)
$$\begin{cases} H = V \oplus \mathcal{N}(B), \\ V^\perp = \mathcal{N}(B). \end{cases}$$

Moreover, each $u \in H$ can be uniquely written as

$$u = u_0 + \sum_{k=1}^r (u, z_k) z_k, \quad u_0 \in \mathcal{N}(B)$$

and so $Bu = \sum_{k=1}^r (u, z_k) \xi_k z_k$. Therefore, if we define $B_k \in \mathcal{L}(H)$ by

$$B_k u = (u, z_k) \xi_k z_k \quad \forall u \in H, \; k = 1, ..., r,$$

we see that B_k is obviously self-adjoint and of rank 1. Furthermore, we have also the decomposition

$$(4.13) \qquad\qquad B = \sum_{k=1}^{r} B_k.$$

The proof of Theorem 4.3 will be completed if we apply the first step successively to the operators $(A + B_1), (A + B_1) + B_2, \ldots$ and so on. ∎

If H is a finite dimensional Hilbert space, the result of Theorem 4.3 and several other results in this direction have been known for a long time. For a detailed study of this subject we refer, for example, to the work of THOMPSON [1976] and the references therein.

4.3. Quadratic Eigenvalue Problems

In this paragraph, we study a class of non-standard eigenvalue problems in Hilbert spaces. We call this class as *quadratic eigenvalue problems* because the eigenvalue equation is quadratic in the parameter λ. In this sense, this type of problems is different from the ones considered so far. The motivation to study such problems comes from the applications in fluid-solid structures to be seen in the next chapter (see Paragraph 2.3 of Chapter II concerning the Stokes model).

Here we shall deal with existence questions and bounds on the eigenvalues. Our approach is to regard such quadratic problems as perturbations from the standard ones and therefore we will, naturally, be led to apply perturbation techniques introduced in the previous paragraphs. We shall begin by looking at the finite dimensional case and then generalize the main results to infinite dimensions.

Let H be a finite dimensional Hilbert space over $\mathbb{C}$ with dimension $n \geq 1$. Let $R, S \in \mathcal{L}(H)$ be two self-adjoint operators and suppose that R is positive definite in the sense that $(Ru, u) > 0 \ \forall u \in H, u \neq 0$. As usual, we arrange its eigenvalues as

$$(4.14) \qquad\qquad \alpha_1 \geq \cdots \geq \alpha_n > 0$$

and the corresponding eigenvectors are denoted $\{v_1, \ldots, v_n\}$. They can be chosen to form an orthonormal basis for H and we shall do so.

On the other hand, let r be the rank of S ($1 \leq r \leq n$) and assume that S is non-negative definite. Our goal in this paragraph is to study the following quadratic eigenvalue problem associated with R and S: Find $\lambda \in \mathbb{C}$ for which there exists $u \in H, u \neq 0$ such that

$$(4.15) \qquad\qquad Ru - \lambda u + \lambda^2 Su = 0.$$

The idea is to regard (4.15) as a perturbation of the equation $(Ru - \lambda u) = 0$ by the quadratic term $\lambda^2 Su$. As will be seen, problem (4.15) can admit non-real eigenvalues even if R and S are self-adjoint operators. The main results of this paragraph show that the above problem has at least $(n - r)$ real solutions λ and exactly $(n + r)$ solutions in total; therefore the number of imaginary (i.e., non-real) eigenvalues cannot exceed twice the rank of S. Further, there are some *interlacing inequalities* between the real eigenvalues and those of R. More exactly, we have

Theorem 4.4. *There exist p real numbers $\lambda_1, ..., \lambda_p$, $(n + r - p)$ non-real numbers $\lambda_{p+1}, ..., \lambda_{n+r}$ (the $\{\lambda_j\}_{j=1}^{n+r}$ are not necessarily distinct) and $(n+r)$ vectors $u_1, ..., u_{n+r}$ in H with the following properties:*
(i) *$(n - r) \leq p \leq (n + r)$.*
(ii) *For each $j = 1, ..., n + r$, the couple (λ_j, u_j) is a solution of problem (4.15).*
(iii) *The elements $\{u_j\}_{j=1}^{n+r}$ satisfy the orthonormality conditions*

$$(4.16) \qquad \begin{cases} \|u_j\| = 1, \\ (u_j, u_k) = 0 \quad \forall j \neq k \quad such\ that \quad \lambda_j = \lambda_k. \end{cases}$$

(iv) *The pairs $\{(\lambda_j, u_j)\}_{j=1}^{n+r}$ are all the solutions of problem (4.15) in the following sense: if $(\tilde{\lambda}, u) \in \mathbb{C} \times H$ is any solution of (4.15), then there exists at least one $j, 1 \leq j \leq q$ such that $\lambda = \lambda_j$ and u can be written as a linear combination of u_k's for which $\lambda_k = \lambda$.*
(v) *The real numbers $\{\lambda_j\}_{j=1}^{n-r}$ and the eigenvalues $\{\alpha_j\}_{j=1}^{n}$ of R satisfy the following interlacing inequalities:*

$$(4.17) \qquad \alpha_{j+r} \leq \lambda_j \leq \alpha_j \quad \forall j = 1, ..., (n - r). \quad \blacksquare$$

The foregoing theorem describes the spectrum of problem (4.15). It can be noted that eigenvectors corresponding to different eigenvalues are not necessarily orthogonal and that they need not form a basis of H. On the other hand, properties (4.16) imply that the multiplicity of each eigenvalue is at most n. The hypothesis that $S \geq 0$ in the above result is completely irrelevant (see CONCA and PUSCHMANN [1993]).

Proof of Theorem 4.4. This is based on the fact that problem (4.15) can be reduced to a *generalized eigenvalue problem* which, in turn, can be transformed to a standard eigenvalue problem for a linear operator acting on $H \times H$.
  Indeed, let (λ, u) be a solution of (4.15). First, it is clear that λ is not zero because R is non-singular. Thus we can define v by

$$(4.18) \qquad v = u - \frac{1}{\lambda} Ru$$

and we see that (4.15) becomes

(4.19) $$v = \lambda S u.$$

Writing these two equations as a simultaneous system of equations in $H \times H$, we have

(4.20) $$\begin{bmatrix} R & 0 \\ 0 & I \end{bmatrix} \begin{bmatrix} u \\ v \end{bmatrix} = \lambda \begin{bmatrix} I & -I \\ S & 0 \end{bmatrix} \begin{bmatrix} u \\ v \end{bmatrix},$$

where I is the identity operator on H. Conversely, it is straightforward to see that if $(\lambda, u, v) \in \mathbb{C} \times H \times H, [u, v] \neq [0, 0]$ is a solution of (4.20), then $\lambda \neq 0$, (λ, u) is a solution of (4.15) and v satisfies (4.18). Thus (4.15) is equivalent to the generalized eigenvalue problem (4.20).

The left and right hand sides of (4.20) define linear continuous operators in $H \times H$ and we denote them by E and F respectively. Since R is non-singular, so is E. On the other hand, it is clear that the rank of F is $(n+r)$. Thus problem (4.20) can be transformed to the following standard eigenvalue problem for the operator $G = E^{-1}F$ acting on $H \times H$:

(4.21) $$G[u, v] = \frac{1}{\lambda}[u, v].$$

From the properties of E and F, it follows that G has rank $(n+r)$. Further, using Proposition 2.2 in CONCA and PUSCHMANN [1993] p.10, we deduce that the characteristic polynomial of G is of degree exactly $(n + r)$. Let $\lambda_1^{-1}, ..., \lambda_{n+r}^{-1}$ be the non-zero eigenvalues (including multiplicities) of G and denote by $[u_1, v_1], ..., [u_{n+r}, v_{n+r}]$ the corresponding eigenvectors in $H \times H$. It is clear that the set of pairs $\{(\lambda_j, u_j)\}_{j=1}^{n+r}$ satisfy statement (ii).

Let us next prove that they also verify (iii). Obviously, the vectors $\{u_j\}_{j=1}^{n+r}$ can be chosen to be of unit norm since $u_j \neq 0$ for all j. Assume that $\lambda^{-1} = \lambda_j^{-1}$ is an eigenvalue of G of multiplicity $s \geq 1$ with corresponding eigenvectors $\{[u_j^\ell, v_j^\ell]\}_{\ell=1}^{s}$. Since they are linearly independent and

$$v_j^\ell = u_j^\ell - \frac{1}{\lambda} A u_j^\ell \quad \ell = 1, ..., s,$$

it is clear that the vectors $\{u_j^\ell\}_{\ell=1}^{s}$ are also linearly independent. Next, by the standard Gram-Schmidt's procedure, they can be orthogonalized and this yields (iii).

To prove (iv), let $(\lambda, u) \in \mathbb{C} \times H$ be any solution of (4.15). If we define v by (4.18), then (λ, u, v) is a solution of (4.20). Thus λ^{-1} is an eigenvalue of G with corresponding eigenvector $[u, v]$. This implies that there exists $j, 1 \leq j \leq n + r$ such that $\lambda = \lambda_j$ and $[u, v]$ can be written as a linear combination of $[u_k, v_k]$'s for which $\lambda_j = \lambda_k$. This obviously implies (iv).

To conclude, let us prove the existence of real eigenvalues. To this end, we introduce, for $\varepsilon \geq 0$, the operator T_ε defined by $T_\varepsilon = R + \varepsilon S$. It is obvious that T_ε is positive definite and denote its eigenvalues by

$$(4.22) \qquad \Lambda_1(\varepsilon) \geq \cdots \geq \Lambda_n(\varepsilon) > 0.$$

From Corollary 4.2, we know that $\Lambda_j(\varepsilon)$ are Lipschitz continuous functions of ε. Moreover, they are non-decreasing (see Exercise 4.2) and by virtue of Theorem 4.3, we have

$$(4.23) \qquad \begin{cases} \alpha_j \leq \Lambda_j(\varepsilon) \leq \alpha_{j-r} & \forall j = r+1, ..., n, \\ \alpha_j \leq \Lambda_j(\varepsilon) & \forall j = 1, ..., r. \end{cases}$$

Now, let us fix j such that $r + 1 \leq j \leq n$. From (4.23), it follows that the graph of functions $\varepsilon \mapsto \Lambda_j(\varepsilon)$ and $\varepsilon \mapsto \varepsilon^{1/2}$ intersect at least at one point, the first of which is denoted by ε_j, i.e., (see Figure 4.1)

$$(4.24) \qquad \varepsilon_j = \min\{\varepsilon \geq 0 \mid \Lambda_j(\varepsilon) = \varepsilon^{1/2}\}.$$

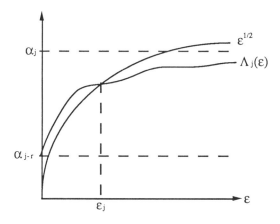

Figure 4.1: *Intersection between Λ_j and $\varepsilon^{1/2}$.*

Let w_j be an eigenvector of T_{ε_j} associated with $\Lambda_j(\varepsilon_j)$. It is clear that $(\Lambda_j(\varepsilon_j), w_j)$ is a solution of (4.15) where $\Lambda_j(\varepsilon_j)$ is real and satisfies

$$\alpha_j \leq \Lambda_j(\varepsilon_j) \leq \alpha_{j-r} \quad \forall j = r+1, ..., n.$$

The above process thus provides $(n - r)$ real solutions of (4.15) which will be numbered $\lambda_1, ..., \lambda_{n-r}$, i.e., we put $\lambda_k = \Lambda_{k+r}(\varepsilon_{k+r})$, $k = 1, ..., n - r$. It is a consequence of the ordering (4.22) that $\{\varepsilon_j\}_{j=r+1}^{n}$ is decreasing and hence so is $\{\lambda_k\}_{k=1}^{n-r}$.

To show that the eigenvectors obtained above can be modified to satisfy (4.16), assume that there exist k and ℓ such that $1 \leq k < \ell \leq n - r$ and $\lambda_k = \lambda_\ell = \lambda^*$ (say). Then λ^* is clearly an eigenvalue of T_{ε^*} with $\varepsilon^* = (\lambda^*)^2$ and its multiplicity is at least $(\ell - k + 1)$. Therefore one can associate with λ^* at least two eigenvectors u_k and u_ℓ of T_{ε^*} which satisfy (4.16). This completes the proof of (iv) concerning the real eigenvalues $\{\lambda_j\}_{j=1}^{n-r}$. Of course, problem (4.15) can admit other real solutions which will be numbered $\lambda_{n-r+1}, ..., \lambda_p$ and hence p satisfies (i) (see Exercise 4.5). ∎

Let us now pass from the case of finite dimensional space to that of infinite dimensional space H which, in addition, is supposed to be separable. We present first the hypotheses on the operators R and S: R is a compact self-adjoint operator which is positive definite. Arrange the eigenvalues of R as follows:

$$\alpha_1 \geq \cdots \geq \alpha_j \geq \cdots \longrightarrow 0.$$

On the other hand, S is assumed to be a self-adjoint operator with finite rank $(= r)$ and which is non-negative definite. Denote its non-zero eigenvalues by

$$\beta_1 \geq \cdots \geq \beta_r > 0$$

and the corresponding eigenvectors by $s_1, ..., s_r$. They form a basis for the image of S which we denote by L.

Theorem 4.5. *There exists an infinite sequence of complex numbers (not necessarily distinct) $\{\lambda_j\}$ which converges to zero and a sequence $\{u_j\}$ of elements of H with the following properties:*

(i) *For each $j \geq 1$, the couple (λ_j, u_j) is a solution of (4.15).*

(ii) *Each eigenvalue λ_j has finite multiplicity and the corresponding eigenvector u_j can be chosen to satisfy (4.16).*

(iii) *The pairs $\{(\lambda_j, u_j)\}$ are all the solutions of problem (4.15) in the sense of Theorem 4.4 (iv).*

(iv) *There exists a real subsequence $\{\lambda_{j'}\}$ of $\{\lambda_j\}$ satisfying the interlacing inequalities*

(4.25) $\alpha_{j'+r} \leq \lambda_{j'} \leq \alpha_{j'} \quad \forall j' \geq 1.$

(v) *The number of non-real elements in $\{\lambda_j\}$ including multiplicities is at most $2r$.*

Proof. We shall not go into the details because it is quite similar to Theorem 4.4. We will only highlight the new issues involved in the infinite dimensional case. Following the proof of Theorem 4.4., it is easily checked that (4.15) is equivalent to the following generalized eigenvalue problem in $H \times L$: Find $\lambda \in \mathbb{C}$ for which there exists $[u, v] \in H \times L, [u, v] \neq [0, 0]$ such that

$$(4.26) \qquad \begin{bmatrix} R & 0 \\ 0 & I_L \end{bmatrix} \begin{bmatrix} u \\ v \end{bmatrix} = \lambda \begin{bmatrix} I_H & -I_L \\ S & 0 \end{bmatrix} \begin{bmatrix} u \\ v \end{bmatrix}.$$

In order to transform this into a standard eigenvalue problem, denote by E the operator on the left side of (4.26) and by F the right side operator. Let us admit for the moment the following result:

Proposition 4.6. *The operator F has a linear continuous inverse F^{-1}.* ∎

Therefore (4.15) is in fact equivalent to the standard eigenvalue problem for the operator $G \overset{\text{def}}{=} F^{-1}E$, acting on $H \times L$:

$$(4.27) \qquad G[u, v] = \lambda[u, v].$$

Since R is compact and the rank of I_L is finite, E is compact and so G is compact. Applying the general results on the spectrum of compact operators (see Theorems 3.14 and 3.15), we see that the spectrum of G consists of a sequence $\{\lambda_j\}$ of complex numbers whose only possible point of accumulation is zero. Each eigenvalue has a finite multiplicity and the corresponding eigenvectors of G are denoted by $\{[u_j, v_j]\}$. Since for all j the triplet (λ_j, u_j, v_j) is a solution of (4.26), it follows that $\lambda_j \neq 0$ and (λ_j, u_j) is a solution of (4.15). Thus the pairs $\{(\lambda_j, u_j)\}$ satisfy statements (i) and (ii) of Theorem 4.5. The fact that they satisfy (iii) also follows step by step the finite dimensional case.

Repeating the perturbation technique followed in the proof of the previous theorem, one can construct a subsequence $\{\lambda_{j'}\}$ of real solutions of (4.15) satisfying the interlacing inequalities (4.25). Since there are infinitely many $\alpha_{j'}$ and they converge to zero, it follows that $\{\lambda_{j'}\}$ is also infinite and it converges to zero. Thus the whole sequence $\{\lambda_j\}$ converges to zero.

It only remains to prove (v). To this end, let us consider for all $m \geq 1$ the subspace of H spanned by $\{u_1, ..., u_m\}$. Denote by d_m the dimension of H_m; clearly $d_m \leq m$. On the other hand, let $P_m : H \longrightarrow H_m$ be the orthogonal projection onto H_m and consider the following approximation of problem (4.15): Find $\lambda \in \mathbb{C}, u \in H_m, u \neq 0$ such that

$$(4.28) \qquad R_m u - \lambda u + \lambda^2 S_m u = 0,$$

where $R_m, S_m : H_m \longrightarrow H_m$ are defined by

$$(4.29) \qquad \begin{cases} R_m = P_m R = P_m R P_m, \\ S_m = P_m S = P_m S P_m. \end{cases}$$

Since the rank of S is r, we see that S_m has a finite rank r_m which is estimated by

$$(4.30) \qquad\qquad r_m \leq \min\{d_m, r\} \leq r.$$

Further, it is easily verified that R_m and S_m fulfill all the requirements of Theorem 4.4 and hence the number of non-real solutions of problem (4.28) is at most $2r_m$. Let now (λ_j, u_j) be any solution of (4.15). For all $m \geq j$, u_j belongs to H_m and so (λ_j, u_j) is solution of (4.28) also. From this fact and (4.30), the conclusion stated in (v) follows.

To conclude the proof of Theorem 4.5, let us prove Proposition 4.6.

Proof of Proposition 4.6. Note that F can be written as

$$F = I_{H \times L} - D,$$

where $D \in \mathcal{L}(H \times L)$ is given by

$$D = \begin{bmatrix} 0 & I_L \\ -S & I_L \end{bmatrix}.$$

It is clear that D has finite rank and so it is compact. By virtue of Fredholm's Alternative (see Theorem 3.13), to prove that F is bijective, it is enough to show that F is injective. Let us therefore take $[u, v] \in \mathcal{N}(F)$. Then $u = v$ and $Su = 0$. Hence $u \in L$ and u can be expressed as a linear combination of eigenvectors $\{s_j\}_{j=1}^r$ of S:

$$u = \sum_{j=1}^{r} a_j s_j, \quad a_j \in \mathbb{C}.$$

Applying S to the above linear combination, we obtain

$$0 = \sum_{j=1}^{r} a_j \beta_j s_j$$

which implies $a_j \beta_j = 0 \ \forall j$. However, $\beta_j \neq 0 \ \forall j$. Thus $a_j = 0 \ \forall j$ and so $u = 0$. Consequently $v = 0$. This finishes the proof of the injectivity as well as the surjectivity of F. The fact that F^{-1} is continuous comes by a simply application of the Open Mapping Theorem. ∎

◇ **Exercise 4.4.** Show that the requirement in Theorem 4.4 that S is self-adjoint is essential for the existence of real eigenvalues of (4.15). (*Hint:* Chose n even. Let

$$
\tilde{R} = \begin{pmatrix} a_1 & -1 & & & \\ -1 & a_2 & -1 & \ddots & \\ & -1 & a_3 & \ddots & \\ & & \ddots & \ddots \end{pmatrix}, \quad R = \tilde{R} + aI
$$

and

$$
S = \begin{pmatrix} 0 \cdots & 0 & b \\ 0 \cdots & 0 & 0 \\ \vdots & \vdots & \vdots \\ 0 & 0 & 0 \end{pmatrix}.
$$

Prove that if a is not an eigenvalue of $\tilde{R}$ and b is large enough, the quadratic eigenvalue problem (4.15) has no real eigenvalues). ■

◇ **Exercise 4.5.** Show, by means of an example, that there can be other real eigenvalues of problem (4.15) than those constructed explicitly in the proof of Theorem 4.4. (*Hint:* Chose $n = 4$ and let

$$
R = \begin{pmatrix} 1 & 0 & 0 & 0 \\ 0 & 101 & 99 & 0 \\ 0 & 99 & 101 & 0 \\ 0 & 0 & 0 & 201 \end{pmatrix}, \quad S = \text{diag}(0, b, 0, 0).
$$

Prove that the eigenvalues of R are $\{201, 200, 2, 1\}$ and that the quadratic eigenvalue problem (4.15) has 5 eigenvalues all of them real. Among them, identify those constructed by the method of Theorem 4.4). ■

4.4. Convergence of Spectral Families

In our analysis so far, we have studied the effects of perturbation of an operator on its spectrum. In practice, one is also motivated to study the behaviour of eigenvectors, or more generally, of the projections on the eigenspaces if they exist. If the perturbation is small in the operator norm then there are well-known convergence results and even asymptotic expansions for eigenvalues and eigenvectors (see KATO [1966]). Remember that there are many interesting physical situations where the point spectrum is completely absent. As we have seen in §3, the appropriate object then to be studied is the spectral family associated with the operator which exists at least in the case of self-adjoint operators thanks to the general Spectral

Decomposition Theorems (see Theorems 3.21 and 3.22). It is therefore proposed in this paragraph to study the influence of perturbations on the spectral family. More exactly, we will see how a particular type of convergence of a sequence of operators can be translated in terms of a suitable convergence of the corresponding spectral family. In this context, we will see a classical result of F. Rellich and a recent result due to E. Sánchez-Palencia useful especially in the presence of continuous spectrum at the limit. In the examples of interest for us, we will not have convergence in the operator norm and there exists only strong convergence. In such situations, the criterion of E. Sánchez-Palencia comes very handy. For both of these, we borrow the proofs from SÁNCHEZ-PALENCIA [1980].

As one can easily imagine, convergence of spectral families is not very useful from a practical point of view because it does not provides relevant information about point-wise convergence of the corresponding spectra. An alternative approach to analyze which kind of convergence for the eigenvalues one can derive from the strong convergence of a sequence of operators has been recently proposed in ALLAIRE and CONCA [1995]. We will conclude this paragraph presenting the main result on which their method is based.

Consider a sequence $\{A_n\}$ of bounded self-adjoint operators on a Hilbert space H such that they *converge strongly* to another operator A, i.e.,

$$(4.31) \qquad\qquad A_n u \longrightarrow Au \quad \text{in} \quad H \quad \forall u \in H.$$

From *Banach-Steinhaus Theorem*, it follows immediately that A is bounded and that there exists a constant c, independent of n, such that

$$(4.32) \qquad\qquad \|A_n\| < c \quad \text{and} \quad \|A\| < c.$$

It is easily checked that A is self-adjoint too. Let $\{E_\lambda^n\}_\lambda$ and $\{E_\lambda\}_\lambda$ be the spectral families associated to the operators A_n and A respectively. Recall that they are projection operators on H enjoying the properties stated in Theorem 3.21. The following result, due to F. Rellich, describes the behaviour of the spectral families under the circumstances described above:

Theorem 4.7. *If $A_n \longrightarrow A$ strongly and if $\mu \in \mathbb{R}$ is not an eigenvalue of A, then $E_\mu^n \longrightarrow E_\mu$ strongly, i.e., as $n \to \infty$*

$$(4.33) \qquad\qquad E_\mu^n u \longrightarrow E_\mu u \quad \text{in} \quad H \quad \forall u \in H.$$

Proof. Without loss of generality, we can assume that $\mu = 0$; otherwise it suffices to consider the sequence of operators $\{(A_n - \mu I)\}$ and its limit $(A - \mu I)$. From our hypothesis (4.31), it follows easily that

$$(4.34) \qquad\qquad p(A_n) \longrightarrow p(A)$$

strongly, for all polynomial p. Next, we observe that (4.32) implies that the spectra of A_n and A are all contained in the interval $[-c, c]$. Therefore, we have the following representations in terms of the spectral families:

$$f(A_n) = \int_{-c}^{c} f(\lambda)dE_\lambda^n, \quad f(A) = \int_{-c}^{c} f(\lambda)dE_\lambda,$$

where f is any function, continuous on $[-c, c]$. Writing

$$f(A_n)u - f(A)u = [f(A_n)u - p(A_n)u] + [p(A_n)u - p(A)u] + [p(A)u - f(A)u]$$

and using the classical *Weierstrass Approximation Theorem*, we deduce that

$$(4.35) \qquad\qquad f(A_n) \longrightarrow f(A)$$

strongly, for all function f, continuous on $[-c, c]$. In particular, considering the continuous function f defined by

$$f(\lambda) = \begin{cases} \lambda & \text{if} \quad \lambda \le 0, \\ 0 & \text{if} \quad \lambda > 0, \end{cases}$$

we get

$$(4.36) \qquad f(A_n) = \int_{-c}^{c} f(\lambda)dE_\lambda^n = \int_{-c}^{0} \lambda dE_\lambda^n = E_0^n A_n;$$

the last equality being a consequence of the definition of E_0^n and the spectral calculus. Needless to mention that an analogous formula holds in the case of A. Combining this with (4.35), we obtain

$$(4.37) \qquad\qquad E_0^n A_n u \longrightarrow E_0 A u \quad \text{in} \quad H \quad \forall u \in H.$$

On the other hand, it is easily seen from (4.31) that

$$(4.38) \qquad\qquad \| E_0^n A_n u - E_0^n A u \| \longrightarrow 0 \quad \forall u \in H.$$

From these two convergence results namely (4.37) and (4.38), we conclude that

$$(4.39) \qquad\qquad E_0^n v \longrightarrow E_0 v$$

for all v in the image of A. However, as A is injective and self-adjoint, it follows that the image of A is dense in H (see Theorem 3.6). Therefore, the convergence (4.39) takes place for all $v \in H$ since the operators E_0^n are all uniformly bounded. This concludes the proof of Theorem 4.7. ∎

Remark. There is a more general version of Theorem 4.7 which covers the case of unbounded self-adjoint operators. The interested reader can consult RIESZ and NAGY [1955] §135. ∎

◊ **Example 4.6.** In general, one cannot say that the spectral projections E_λ^n *converges uniformly* to E_λ (i.e., with respect to the norm in $\mathcal{L}(H)$) even if $A_n \longrightarrow A$ uniformly. As an example, let us consider the sequence of operators A_n defined on $L^2(0,1)$ by

$$A_n f(x) = \left(x - \frac{1}{n}\right) f(x), \quad x \in]0,1[.$$

This sequence converges uniformly to A where $Af(x) = xf(x)$, $x \in]0,1[$. It is easily checked that the associated spectral projections are given by

$$\begin{cases} E_\lambda^n f(x) = e_{\lambda + \frac{1}{n}}(x)f(x), \\ E_\lambda f(x) = e_\lambda(x)f(x), \quad x \in]0,1[, \end{cases}$$

where $e_\lambda(x)$ is the characteristic function of $]-\infty,\lambda]$. Note that $(E_\lambda^n - E_\lambda)$ is the orthogonal projection onto the subspace of functions vanishing outside $]\lambda, \lambda + \frac{1}{n}[$ and so $\|E_\lambda^n - E_\lambda\| = 1$ for $0 \le \lambda < 1$ and for all n. ∎

One can also reach the conclusion of Theorem 4.7 from the strong convergence of the inverses of the operators. More precisely, we have

Theorem 4.8. *Let A_n, A be bounded self-adjoint operators on H which are uniformly positive definite, i.e., $A_n, A \ge \alpha I$, for some $\alpha > 0$. Assume, moreover, that*

(4.40) $A_n^{-1} \longrightarrow A^{-1}$ *strongly.*

If $\mu > 0$ is not an eigenvalue of A, then

(4.41) $E_\mu^n \longrightarrow E_\mu$ *strongly.*

Proof. From our hypothesis, it follows that A_n^{-1} and A^{-1} exist as bounded operators and their norms are bounded by α^{-1}. Since μ^{-1} is not an eigenvalue of A^{-1}, we get by Theorem 4.7,

(4.42) $F_{\mu^{-1}}^n u \longrightarrow F_{\mu^{-1}} u$ in H, $\forall u \in H$,

where F_λ^n and F_λ denote the spectral families of A_n^{-1} and A^{-1} respectively. Using the spectral calculus and the uniqueness of the spectral family, we can check that (see Exercise 3.17)

(4.43) $F_\lambda^n = I - E_{\lambda^{-1}-}^n$ and $F_\lambda = I - E_{\lambda^{-1}-}$ for all $\lambda > 0$, where we recall that $E_{\lambda-}$ is defined to be the strong limit of E_λ as $\lambda \longrightarrow \lambda-$. Combining (4.42) and (4.43), we reach the desired conclusion. ∎

The above two results are useful if one has a priori information on the spectrum of the limit operator. Quite often, in applications, this is not the case and both theorems can be unless in a large subset of the real axis. Therefore, we present another result which is more general but weaker in its conclusion. More precisely, we will show that $E_\lambda^n \longrightarrow E_\lambda$ in the sense of distributions on $\mathbb{R}$ in the variable λ under a hypothesis weaker than in Theorem 4.7. The idea of the proof has two steps. The first one is to consider the general vibrations described by the wave equations associated with the operators and prove the convergence of the corresponding solutions of the initial-value problems. The next step is to translate this convergence in terms of the associated spectral families using Fourier transform and spectral calculus.

Definition. Let $\{A_n\}$ be a sequence of bounded self-adjoint operators. $\{A_n\}$ is said to *converge weakly* if $\{A_n u\}$ converges weakly for all $u \in H$. If we denote the limit by Au, then A defines a self-adjoint operator on H which is called the *weak limit* of $\{A_n\}$. We use the notation $A_n \rightharpoonup A$ weakly. ∎

If $\{A_n\}$ is weakly convergent, it is uniformly bounded thanks to Banach-Steinhauss Theorem. Also its weak limit A is bounded and

$$\|A\| \le \liminf_{n \to \infty} \|A_n\|.$$

Moreover, if $u_n \to u$ in H then

(4.44) $$A_n u_n \rightharpoonup Au \quad \text{weakly in } H.$$

Theorem 4.9. *Let $A_n, A \in \mathcal{L}(H)$ be bounded self-adjoint operators on H which are uniformly positive definite, i.e., $A_n, A \ge \alpha I$, for some $\alpha > 0$. Assume further that*

(4.45) $$A_n \rightharpoonup A \quad \text{weakly.}$$

Then the spectral family of A_n converges to that of A in the following sense: For every $v, w \in H$,
$$(E_\lambda^n v, w) \longrightarrow (E_\lambda v, w)$$

in the space of tempered distributions on $\mathbb{R}$.

Proof. *First Step*: Given $v \in H$, we consider the following initial-value problem associated to A_n: Find $u_n \in L^2_{loc}(\mathbb{R}; H)$ such that $u'_n \in L^2_{loc}(\mathbb{R}; H)$ and

$$(4.46) \qquad \begin{cases} u''_n(t) + A_n u_n(t) = 0 & \forall t \in \mathbb{R}, \\ u_n(0) = 0, \\ u'_n(0) = v. \end{cases}$$

A weak formulation of the above problem is the following: Find u_n in $L^2_{loc}(\mathbb{R}; H)$ such that $u'_n \in L^2_{loc}(\mathbb{R}; H)$ and

$$(4.47) \qquad \begin{cases} -\int_0^\infty (u'_n(t), \psi'(t))dt + \int_0^\infty (A_n u_n(t), \psi(t))dt - (v, \psi(0)) = 0, \\ u_n(0) = 0 \quad \text{and} \quad u_n(t) = -u_n(-t) \quad \forall t \in \mathbb{R}, \end{cases}$$

where the test function $\psi(t)$ is of the form $\psi(t) = \phi(t)z$ with $\phi \in \mathcal{D}([0, \infty[)$ and $z \in H$ arbitrary. It is well-known that there exists a unique solution to (4.46). Moreover, the solution u_n belongs to $C^\infty(\mathbb{R}; H)$ (for details see LIONS [1968] or SÁNCHEZ-PALENCIA [1980]). The proof is, in fact, based on the conservation of energy associated with problem (4.46). To get this, we multiply the equation in (4.46) by $u'_n(t)$ and integrate by parts. We obtain

$$\frac{1}{2}\frac{d}{dt}\|u'_n(t)\|^2 + (A_n u_n(t), u'_n(t)) = 0.$$

Since A_n is self-adjoint, we can rewrite the above identity as follows:

$$\frac{1}{2}\frac{d}{dt}[\|u'_n(t)\|^2 + (A_n u_n(t), u_n(t))] = 0.$$

We can thus conclude that

$$(4.48) \qquad \|u'_n(t)\|^2 + (A_n u_n(t), u_n(t)) = \|v\|^2 \quad \forall t \in \mathbb{R}$$

Using now our hypothesis that the operators A_n are uniformly positive definite, we see that there exists a constant c independent of n such that

$$(4.49) \qquad \|u_n\|_{L^\infty(\mathbb{R};H)} \le c \quad \text{and} \quad \|u'_n\|_{L^\infty(\mathbb{R};H)} \le c.$$

We can therefore extract a subsequence of n (again denoted by n) such that

$$(4.50) \qquad \begin{cases} u_n \rightharpoonup u & \text{in} \quad L^\infty(\mathbb{R}; H) \text{ weak*}, \\ u'_n \rightharpoonup u' & \text{in} \quad L^\infty(\mathbb{R}; H) \text{ weak*}, \end{cases}$$

for some element u of $L^\infty(\mathbb{R}; H)$. Using Corollary 1.6, we see that we also have

$$u_n(t) \longrightarrow u(t) \quad \text{in} \quad H \quad \forall t \in \mathbb{R}.$$

It is now a simple matter to check, by means of (4.44) and Lebesgue's Dominated Convergence Theorem, that u is the solution of the initial-value problem associated with the wave equation for A:

(4.51)
$$\begin{cases} u''(t) + Au(t) = 0 & \forall t \in \mathbb{R}, \\ u(0) = 0, \\ u'(0) = v. \end{cases}$$

Because of the uniqueness of u, the convergence (4.50) holds good for the full sequence. This completes the first step.

Second Step: We begin by remarking that the solutions of (4.46) and (4.51) can be represented, using spectral calculus, as follows:

(4.52)
$$\begin{cases} u'_n(t) = \cos(A_n^{1/2}t)v, \\ u'(t) = \cos(A^{1/2}t)v. \end{cases}$$

Thus (4.50) yields, for all $w \in H$

(4.53) $\quad \left(\cos(A_n^{1/2}t)v, w\right) \to \left(\cos(A^{1/2}t)v, w\right) \quad \text{in} \quad L^\infty(\mathbb{R}) \text{ weak*}.$

We remark that the convergence (4.53) is stronger than the convergence in the space $\mathcal{S}'$ of tempered distributions. Therefore, by Fourier transformation, we get

(4.54) $\quad \left(\cos(A_n^{1/2}t)v, w\right)^\wedge(\tau) \to \left(\cos(A^{1/2}t)v, w\right)^\wedge(\tau)$

in the sense of tempered distributions (τ denotes the variable dual to t in Fourier transformation).

On the other hand, let us denote by F_λ^n, F_λ the spectral families associated to $A_n^{1/2}$ and $A^{1/2}$ respectively. Then by spectral calculus, we have

$$\cos(A_n^{1/2}t) = \frac{1}{2}\int_0^\infty [e^{i\tau t} + e^{-i\tau t}]dF_\tau^n = \frac{1}{2}\int_{-\infty}^{+\infty} e^{i\tau t}d[F_\tau^n - F_{-\tau}^n].$$

Therefore, we also have

(4.55) $\quad \left(\cos(A_n^{1/2}t)v, w\right) = \frac{1}{2}\int_{-\infty}^\infty e^{i\tau t}d([F_\tau^n - F_{-\tau}^n]v, w).$

Certainly, an analogous relation holds for the operator A. We note that the right hand side of the above equality is nothing but the inverse of the Fourier transform of (modulo a multiplicative constant)

$$\frac{d}{d\tau}\left([F_\tau^n - F_{-\tau}^n]v, w\right).$$

Thus (4.54) is equivalent to the following convergence:

$$\frac{d}{d\tau}\left([F_\tau^n - F_{-\tau}^n]v, w\right) \to \frac{d}{d\tau}\left([F_\tau - F_{-\tau}]v, w\right).$$

in the sense of tempered distributions. Since the distributions $\frac{d}{d\tau}(F_\tau^n v, w)$ have supports in $[\alpha^{1/2}, \infty[$, it follows that

$$(4.56) \qquad\qquad \frac{d}{d\tau}(F_\tau^n v, w) \to \frac{d}{d\tau}(F_\tau v, w)$$

in the sense of tempered distributions. Next, using again the information on the supports, we can pass from the convergence of the derivatives to that of the distributions themselves. Hence

$$(4.57) \qquad\qquad (F_\tau^n v, w) \to (F_\tau v, w)$$

in the sense of tempered distributions. Finally, the above convergence of F_τ^n can be easily translated into that of E_τ^n if we take into account the following relation:

$$(4.58) \qquad\qquad E_\tau^n = \begin{cases} 0 & \text{if } \tau < 0, \\ F_{\tau^{1/2}}^n & \text{if } \tau \geq 0. \end{cases}$$

This completes the proof of Theorem 4.9. ∎

Remark. The hypothesis of Theorem 4.9 that the operators A_n, A are uniformly positive definite is not indispensable. What we need is that they are bounded below uniformly, i.e., there exists $\alpha \in \mathbb{R}$ such that $A_n, A \geq \alpha I$. In such a case, the translated operators

$$B_n \overset{\text{def}}{=} A_n + MI, \qquad B \overset{\text{def}}{=} A + MI$$

for some constant M large enough will be uniformly positive definite for which the conclusions of Theorem 4.9 apply. From the convergence of the spectral family of B_n, one can deduce that of A_n since they are related by

$$(4.59) \qquad\qquad E_\lambda(B_n) = E_{\lambda - M}(A_n) \quad \forall \lambda \in \mathbb{R}. \ \blacksquare$$

In the Introduction of Paragraph 4.4 we mentioned that the convergence of the spectral families associated with a sequence of operators which converges strongly (or weakly) gives very little information about point-wise convergence of the spectra. As it is described in the following result, which is due to G. Allaire and C. Conca (see the reference quoted in the Introduction), from the strong convergence of a sequence of operators one can only conclude that the elements of the limit spectrum $\sigma(A)$ can be approximated by elements in $\sigma(A_n)$ but, unfortunately, there may exist sequences of elements in $\sigma(A_n)$ which converge to elements outside $\sigma(A)$. We have

Theorem 4.10. *Let $A_n, A \in \mathcal{L}(H)$ be bounded self-adjoint operators. If $A_n \longrightarrow A$ then*

(i) *For all $\mu \in \sigma(A)$ there exists a sequence $\mu_n \in \sigma(A_n)$ such that $\mu_n \longrightarrow \mu$.*

(ii) *If μ_n is a sequence in the discrete spectrum $\sigma_d(A_n)$ of A_n such that $\mu_n \longrightarrow \mu$, where μ does not belong to $\sigma(A)$, then any sequence of normalized eigenvectors associated with the eigenvalues $\{\mu_n\}$ converges weakly to zero in H.*

Proof. Let $\mu \in \sigma(A)$ be given. Assume that μ is not the limit of any sequence of elements in $\sigma(A_n)$. Then there exists $\delta > 0$ such that, for sufficiently large n, and for any element $\mu_n \in \sigma(A_n)$, one has

$$|\mu_n - \mu| \geq \delta.$$

This implies that $\|A_n u - \mu u\| \geq \delta$ for all $u \in H$. Since the convergence of A_n to A is strong, one can pass to the limit in this inequality and obtain

$$(4.60) \qquad \|Au - \mu u\| \geq \delta \quad \forall u \in H.$$

This is clearly a contradiction with the fact that μ belongs to $\sigma(A)$ and completes the proof of (i).

To prove (ii), let us consider a sequence of eigenvalues $\mu_n \in \sigma_d(A_n)$ converging to a limit μ outside $\sigma(A)$ and let u_n be an associated sequence of eigenvectors in H. We have

$$(4.61) \qquad A_n u_n = \mu_n u_n.$$

Thanks to the strong convergence of A_n, we can pass to the limit (up to a subsequence) in (4.61) and denoting by u the weak limit of the sequence u_n we deduce that $Au = \mu u$. This is not possible except if $u = 0$. This finishes the proof. ∎

Chapter II

Spectral Problems in Fluid-Solid Structures

Four sections constitute this chapter. The first one deals with modelling aspects of the phenomenon of vibrations in fluid-solid structures. Various mathematical models are presented depending on the nature of the fluid and other physical conditions. As we shall see, these models constitute eigenvalue problems for certain partial differential operators with non-standard boundary conditions. Throughout the text, we shall essentially be concerned with properties of these models. More precisely, in Section 2, we start our investigations on the question of existence of solutions which is necessary to validate the models. This is done using the general framework developed in Chapter I. We continue our study of these models in Section 3 by presenting certain qualitative properties of their solutions and we conclude this chapter in Section 4, where some of the numerical methods, popular among engineers, are presented.

§1. Mathematical Models of Vibrations of Fluid-Solid Structures

The problem of vibrations of a solid immersed in a fluid has been of interest to many researchers in the last several years. This problem has considerable importance in engineering as this occurs naturally in the design and simulation of heat exchangers, condensers, fuel assemblies and cores of nuclear reactors. Since the pioneering work of CONNORS [1970], in the last few years much effort has been devoted to experimental and theoretical researches of these coupled structures. From the engineering point of view, let us point out the works by BLEVINS [1977], CHEN [1975], [1987], CONCA, PLANCHARD, THOMAS and VANNINATHAN [1994], PAIDOUSSIS [1966] and PLANCHARD [1980], [1982], [1985], [1987], [1988]. Among the many ways of modelling the physical phenomena involved, we have chosen the one introduced by one of the authors in the early 80's. This is because of its virtues in the simplicity of the construction of models and their amenability to a nice mathematical analysis as we shall see.

Throughout this book, we will concentrate our attention to the vibrations of a tube bundle immersed in a fluid inside a cavity. In this section, we shall deal with modelling aspects of these vibrations taking into account various physical effects and conditions involved. When an elastically mounted vibrating tube bundle is immersed in a fluid, both the tubes and the fluid vibrate. The result is a non-stationary partial differential equation (pde) in the fluid region coupled with a system of ordinary differential equations (ode) which represents the oscillations of the tubes. The solution in the fluid region measures the amplitude of the vibrations of the fluid. The fluid-solid interactions are taken into account by the coupling between the pde and the system of ode's.

Depending on the kind of fluid in which the solid part is immersed, these models involve different well-known pde's. For example, we will be concerned with the Laplace equation in the case of a perfect incompressible fluid, with wave equation if the fluid is assumed to be perfect and compressible and with an evolutionary Stokes system of equations in the case of a viscous incompressible fluid (or in the case of a slightly compressible fluid). On the other hand, the equations which model the vibrations of the tubes depend on the kind of assumptions we make on them. Usually they will be considered as solid masses which vibrate and sometimes they are allowed to make significant displacements apart from vibrations. One can also consider them to be elastic bodies undergoing deformations.

To arrive at an eigenvalue problem, we seek, as usual, solutions which are sinusoidal in time and this leads us, because of the coupling, to a class of non-standard eigenvalue problems which incorporates a *non-local boundary condition*. Apart from this, there are also other qualitative differences with the classical models introduced in Chapter I. This will be seen below.

Let us begin with a reference to the geometry of the domain on which the problem is posed and with some preliminary physical considerations behind the models. To this end, let us imagine a homogeneous fluid contained in a three-dimensional cylindrical cavity with a constant planar section $\Omega_0 \subset \mathbb{R}^2$. Within the cavity there is a solid structure immersed in the fluid whose projection on Ω_0 is also a constant region which we shall assume to be multiconnected with components $\{\mathcal{O}_j\}_{j=1}^K$. We can consider it to be a bundle of K cylindrical tubes of same length whose generating lines are parallel to each other and perpendicular to Ω_0. In this chapter, the tubes need not have a circular cross section and neither be identical.

The interaction between the fluid and this solid structure will be studied on the basis of the following standing assumptions: (i) The ends of each tube are fixed to two opposite surfaces of the cavity in such a way that each tube can be likened to a bar (of section $\mathcal{O}_j$) which can move transversally but which does not allow movement perpendicular to its section and (ii) The tubes are long enough. These permit us to ignore three-dimensional effects and the problem can be studied in any cross section perpendicular to the tubes. Under such conditions, let us fix some notations. We shall denote the part of Ω_0 occupied by the fluid as Ω, i.e.,

$$\Omega = \Omega_0 \setminus \bigcup_{j=1}^K \bar{\mathcal{O}}_j.$$

We assume that Ω_0, Ω and $\mathcal{O}_j$ are all bounded connected open subsets of $\mathbb{R}^2$ with locally Lipschitz boundaries. We denote the boundaries of Ω_0 and $\mathcal{O}_j$ by Γ_0 and Γ_j respectively. For obvious physical reasons, the sets $\{\bar{\mathcal{O}}_j\}_{j=1}^K$ are assumed to be mutually disjoint and disjoint from Γ_0 (see Figure 1.1).

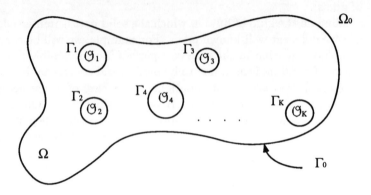

Figure 1.1: *The regions Ω_0 and Ω.*

These are some of the general features of the physical assumptions and models to be treated in this chapter. The special ones that are found in each individual case will be explained in the corresponding section.

1.1. The Case of a Perfect Incompressible Fluid

The simplest model we shall study in this chapter has been proposed by several authors (see for example PLANCHARD [1982], [1987] or IBNOU-ZAHIR and PLANCHARD [1983a]). In this model, the tubes are assumed to be rigid and only small oscillations of the fluid around the state of rest are allowed. Moreover, the velocity $\mathbf{u}_0$ of the fluid is assumed to derive from a *potential function* $\phi_0 = \phi_0(x,t)$. Using the fact that the fluid is incompressible, we have the following equation

(1.1) $$\Delta\phi_0 = 0 \quad \text{in} \quad \Omega \times \mathbb{R}$$

which governs the motion of the fluid. The fluid is not allowed to escape the cavity and so ϕ_0 satisfies

(1.2) $$\frac{\partial\phi_0}{\partial n} = 0 \quad \text{on} \quad \Gamma_0 \times \mathbb{R}.$$

On Γ_j, the normal velocity of the fluid should coincide with the normal component of the tube's velocity. Thus we have

$$(1.3) \qquad \frac{\partial \phi_0}{\partial n} = \frac{d\mathbf{r}_{0j}}{dt} \cdot \mathbf{n} \quad \text{on} \quad \Gamma_j \times \mathbb{R}, \ j = 1, ..., K,$$

where $\mathbf{n} = (n_1, n_2)$ is the outward unit normal on the boundary of Ω, $\partial/\partial n$ denotes the derivation along that normal and $\mathbf{r}_{0j}(t)$ is the *transverse displacement vector* at the instant t of the jth tube. Throughout the text, we use the notation $\mathbf{a} \cdot \mathbf{b}$ to understand the sum $\sum_j a_j b_j$ for complex vectors $\mathbf{a} = (a_j)$ and $\mathbf{b} = (b_j)$. It is furthermore assumed that the motion of tube j obeys a simple harmonic oscillator with a forcing term modelled by its interaction with the fluid†. Since the fluid is assumed to be perfect, this term depends only on the pressure p_0 of the fluid. More exactly, $\mathbf{r}_{0j}$ satisfies the following second order ode:

$$(1.4) \qquad m_j \frac{d^2 \mathbf{r}_{0j}}{dt^2} + k_j \mathbf{r}_{0j} = \int_{\Gamma_j} p_0(x, t) \mathbf{n} d\gamma \quad \text{in} \quad \mathbb{R},$$

where m_j is the mass per unit length of the jth tube, $k_j > 0$ is the stiffness constant of the jth tube and $p_0(x, t)$ is the pressure of the fluid at the point $x \in \Omega$ and at time t. Next, let us consider *Euler equation* in order to link p_0 and ϕ_0:

$$(1.5) \qquad \frac{\partial \mathbf{u}_0}{\partial t} + (\mathbf{u}_0 \cdot \nabla)\mathbf{u}_0 + \frac{1}{\rho}\nabla p_0 = \mathbf{0},$$

where ρ is the density of the fluid. The fluid being homogeneous, ρ is independent of x; neither it depends on t because the fluid is incompressible. We put $\rho = \rho_0 > 0$. Given that only small oscillations are being considered, linearizing Euler equation, we obtain the following relation:

$$(1.6) \qquad p_0 = -\rho_0 \frac{\partial \phi_0}{\partial t} + c(t),$$

where $c(t)$ is an arbitrary constant. This closes system (1.1), (1.4). Let us remark that the coupling between the fluid and the tubes occurs in (1.3) and (1.4).

As is usual in vibration models, we seek sinusoidal solutions of the form

$$(1.7) \qquad \phi_0(x, t) = \phi(x)e^{i\omega t}, \quad \mathbf{r}_{0j}(t) = \mathbf{r}_j e^{i\omega t},$$

where ω is the unknown (resonant) *vibration frequency* of the coupled system. When one search for solutions of the form (1.7) with the same frequency ω

---

† Moreover, the tube displacements are supposed to be small enough so that the geometrical variations of Ω induced by the motion of the tubes are neglected.

for ϕ_0 (fluid) and $\mathbf{r}_{0j}$ (tube), we see that both the left and the right sides of equation (1.4) have the same frequency and this justifies calling ω the resonant frequency of the system (see Paragraph 2.1 of Chapter I).

Substituting (1.6) into the right side of (1.4) and using (1.7) we can solve explicitly (1.4) and get

$$(1.8) \qquad\qquad \mathbf{r}_{0j}(t) = -\frac{i\rho_0\omega e^{i\omega t}}{k_j - m_j\omega^2} \int_{\Gamma_j} \phi \mathbf{n} d\gamma.$$

Note that the constant $c(t)$ in (1.6) does not contribute in any way because

$$\int_{\Gamma_j} \mathbf{n} d\gamma = \mathbf{0} \quad \forall j = 1, ..., K.$$

Using the above explicit formula (1.8) in the boundary condition (1.3) and the form (1.7) for ϕ_0 we see that (ω, ϕ) has to be a solution of the following spectral problem in Ω:

$$(1.9) \qquad \begin{cases} \Delta\phi = 0 \quad \text{in} \quad \Omega, \\[2mm] \dfrac{\partial\phi}{\partial n} = \dfrac{\rho_0\omega^2}{k_j - m_j\omega^2}\Big(\displaystyle\int_{\Gamma_j} \phi \mathbf{n} d\gamma\Big) \cdot \mathbf{n} \quad \text{on} \quad \Gamma_j, \ j = 1, ..., K, \\[4mm] \dfrac{\partial\phi}{\partial n} = 0 \quad \text{on} \quad \Gamma_0. \end{cases}$$

We refer to this model as the *incompressible case model* or *Laplace model*. Observe that the frequency parameter ω appears only on the boundary condition on Γ_j which is non-local. Note also that this is not of the same type as the Steklov eigenvalue problem seen in Paragraph 2.3 of Chapter I. Furthermore, there appears a rational function of ω^2 in the above problem rather than a simple linear function as in classical cases.

1.2. The Case of a Slightly Compressible Perfect Fluid

Let us now assume, dropping the incompressibility condition, that the density undergoes small variations around an equilibrium value ρ_0: $\rho = \rho_0 + \rho'$ with $\rho' \ll \rho_0$. Now, consider the *continuity equation*

$$(1.10) \qquad\qquad \frac{\partial\rho}{\partial t} + \text{div}\,(\rho\mathbf{u}_0) = 0.$$

Neglecting second order terms, we arrive at

$$(1.11) \qquad\qquad \frac{\partial\rho'}{\partial t} + \rho_0\text{div}\,\mathbf{u}_0 = 0.$$

Assuming a state equation in the form $p = p(\rho)$, one finds, in the first order approximation, that $p_0 = p(\rho_0)$ and $p' = c^2\rho'$, where $c^2 = (dp/d\rho)_{\rho=\rho_0}$†. Equation (1.11) can be rewritten in terms of p' as follows:

(1.12)
$$\frac{\partial p'}{\partial t} + \rho_0 c^2 \operatorname{div} \mathbf{u}_0 = 0.$$

Assuming again that the velocity field $\mathbf{u}_0$ admits a potential ϕ_0 and using the linearized Euler equation:

$$p = -\rho_0 \frac{\partial \phi_0}{\partial t} + c(t)$$

we conclude that

(1.13)
$$p' = -\rho_0 \frac{\partial \phi_0}{\partial t}$$

because the undetermined constant $c(t)$ can be absorbed in p_0. Substituting (1.13) in (1.12), we see that ϕ_0 satisfies the classical wave equation

$$\frac{\partial^2 \phi_0}{\partial t^2} - c^2 \Delta \phi_0 = 0 \quad \text{in} \quad \Omega \times \mathbb{R}.$$

The boundary conditions satisfied by ϕ_0 are the same as in the preceding case, i.e., (1.2) and (1.3). Looking once again for sinusoidal solutions of the form (1.7), we are led to the following eigenvalue problem in Ω:

(1.14)
$$\begin{cases} c^2 \Delta \phi + \omega^2 \phi = 0 \quad \text{in} \quad \Omega, \\ \dfrac{\partial \phi}{\partial n} = \dfrac{\rho_0 \omega^2}{k_j - m_j \omega^2}\left(\displaystyle\int_{\Gamma_j} \phi n d\gamma\right) \cdot \mathbf{n} \quad \text{on} \quad \Gamma_j, \ j = 1, ..., K, \\ \dfrac{\partial \phi}{\partial n} = 0 \quad \text{on} \quad \Gamma_0. \end{cases}$$

We call this problem the *compressible case model* or *Helmholtz model*. The only difference between (1.9) and (1.14) which is apparent is that in (1.14) the frequency parameter ω occurs in the equation in Ω as well as in the non-local boundary condition on Γ_j whereas in (1.9) it occurred only in the boundary condition on Γ_j.

---

† c is the speed of sound in the fluid; it is a strictly positive constant.

1.3. The Case of an Incompressible Viscous Fluid

In this model, we assume incompressibility but we consider a fluid which is viscous. We continue to assume that the tubes are rigid and the fluid is homogeneous. It is also supposed that the fluid motion is governed by the non-stationary Stokes system†:

$$(1.15) \qquad \begin{cases} \dfrac{\partial \mathbf{u}_0}{\partial t} = \operatorname{div} \sigma(\mathbf{u}_0, p_0) & \text{in} \quad \Omega \times \mathbb{R}, \\[2mm] \operatorname{div} \mathbf{u}_0 = 0 & \text{in} \quad \Omega \times \mathbb{R}, \end{cases}$$

where σ is the *stress-tensor* obeying *Stokes Law*:

$$(1.16) \qquad \sigma(\mathbf{u}_0, p_0) = -p_0 I + 2\nu e(\mathbf{u}_0).$$

Here I is the identity matrix of order 2, $\nu > 0$ is a given constant which represents the *kinematic viscosity* of the fluid and $e(\mathbf{u}_0)$ is the linear *strain tensor*, defined by

$$(1.17) \qquad 2e(\mathbf{u}_0) = \nabla \mathbf{u}_0 + (\nabla \mathbf{u}_0)^T.$$

The presence of viscosity results in the modification of the boundary conditions (1.2), (1.3) as well as the forcing term in (1.4). Indeed, it is common to impose the so-called *no-slip condition* on Γ_0 and Γ_j:

$$(1.18) \qquad \mathbf{u}_0 = \mathbf{0} \quad \text{on} \quad \Gamma_0 \times \mathbb{R},$$

$$(1.19) \qquad \mathbf{u}_0 = \frac{d\mathbf{r}_{0j}}{dt} \quad \text{on} \quad \Gamma_j \times \mathbb{R}, \ j = 1, ..., K.$$

On the other hand, in the forcing term the pressure p_0 should be replaced by the stress-tensor σ. Thus $\mathbf{r}_{0j}$ satisfies the following second order ode:

$$(1.20) \qquad m_j \frac{d^2 \mathbf{r}_{0j}}{dt^2} + k_j \mathbf{r}_{0j} = - \int_{\Gamma_j} \sigma \mathbf{n} \, d\gamma \quad \text{in} \quad \mathbb{R}.$$

As usual, we seek sinusoidal solutions of the form

$$(1.21) \qquad \mathbf{u}_0(x, t) = \mathbf{u}(x)e^{\omega t}, \quad p_0(x, t) = p(x)e^{\omega t}, \quad \mathbf{r}_{0j}(t) = \mathbf{r}_j e^{\omega t}$$

where ω is the unknown resonant frequency of the system. This time ω is taken to be a complex parameter because we expect dissipation of waves due to the presence of the viscosity.

---

† This is the case if viscosity is high or if the fluid moves slowly. For simplicity, we take $\rho_0 = 1$.

Substituting (1.21) into (1.20) and solving explicitly the ode, we get

$$(1.22) \qquad \mathbf{r}_{0j}(t) = \frac{-e^{\omega t}}{k_j + m_j \omega^2} \int_{\Gamma_j} \sigma(\mathbf{u}, p) \mathbf{n} d\gamma.$$

Following the procedure indicated in previous paragraphs, we see that $(\omega, \mathbf{u}, p)$ ought to be solution of the following spectral problem in Ω:

$$(1.23) \qquad \begin{cases} -\operatorname{div} \sigma(\mathbf{u}, p) + \omega \mathbf{u} = \mathbf{0} & \text{in} \quad \Omega, \\ \operatorname{div} \mathbf{u} = 0 & \text{in} \quad \Omega, \\ \mathbf{u} = \dfrac{-\omega}{k_j + m_j \omega^2} \displaystyle\int_{\Gamma_j} \sigma(\mathbf{u}, p) \mathbf{n} d\gamma & \text{on} \quad \Gamma_j, \ j = 1, ..., K. \\ \mathbf{u} = \mathbf{0} & \text{on} \quad \Gamma_0, \end{cases}$$

The above system is referred to as *Stokes model*. We point out that contrary to the earlier models the present one is a system in which both ω and ω^2 appear in the equation and in the boundary condition.

1.4. The Case of a Slightly Compressible Viscous Fluid

The modelling procedure in this case is a mixture of what we have done in the previous two paragraphs. We shall therefore not go into the details except to point out the modifications to be incorporated. Since the fluid is now slightly compressible and viscous, there are two changes to be done: one concerns the continuity equation in (1.15) and the other one is at the level of Stokes Law. Since $\mathbf{u}_0$ is no more divergence-free, we take (1.12) as the continuity equation as has already been done in Paragraph 1.2. Thus (1.15) is replaced by

$$(1.24) \qquad \begin{cases} \dfrac{\partial \mathbf{u}_0}{\partial t} = \operatorname{div} \sigma(\mathbf{u}_0, p') & \text{in} \quad \Omega \times \mathbb{R}, \\ \dfrac{\partial p'}{\partial t} + c^2 \operatorname{div} \mathbf{u}_0 = 0 & \text{in} \quad \Omega \times \mathbb{R}, \end{cases}$$

where now σ obeys the *modified Stokes Law* (see LANDAU and LIFCHITZ [1989] pp. 68–69):

$$(1.25) \qquad \sigma(\mathbf{u}_0, p') = -(p' + \frac{2}{3}\nu \operatorname{div} \mathbf{u}_0)I + 2\nu e(\mathbf{u}_0).$$

Following exactly the procedure presented in the previous paragraph, it can be checked that $(\omega, \mathbf{u}, p')$ must satisfy the following spectral problem in Ω:

$$(1.26) \qquad \begin{cases} -\operatorname{div} \sigma(\mathbf{u}, p') + \omega \mathbf{u} = \mathbf{0} & \text{in} \quad \Omega, \\ \omega p' + c^2 \operatorname{div} \mathbf{u} = 0 & \text{in} \quad \Omega, \\ \mathbf{u} = \dfrac{-\omega}{k_j + m_j \omega^2} \displaystyle\int_{\Gamma_j} \sigma(\mathbf{u}, p') \mathbf{n} d\gamma & \text{on} \quad \Gamma_j, \ j = 1, ..., K, \\ \mathbf{u} = \mathbf{0} & \text{on} \quad \Gamma_0, \end{cases}$$

where ω is, we recall, a complex parameter as in (1.23). We baptize this system as the *compressible Stokes model*. It is worth to point out that the unknown pressure p' can be eliminated from system (1.26) using the second equation in (1.26).

1.5. Other Cases

Many other models have been proposed by engineers in order to deal with the physical problem which we mentioned at the beginning of this section. Most of these, however, have not yet been studied from a rigourous mathematical point of view. It should be pointed out in this regard that these involve different aspects which are not taken into account in the models examined in the above paragraphs. These are, for example, cases where the tubes can undergo elastic deformations (non-rigid bodies), tubes can move around apart from oscillating around a state of equilibrium, fluid can obey a non-linear constitutive law (Navier-Stokes model) and so on. Some references for the engineering studies on this subject can be obtained in PLANCHARD [1988] or in CONCA, PLANCHARD, THOMAS and VANNINATHAN [1994]. As mentioned in Preface, fluid-solid interactions can be tackled from different points of view in both mathematical and physical contents. In this connection, we feel appropriate to cite the works of GIBERT [1988], MORAND and OHAYON [1992] and LANZA DE CRISTOFORIS and ANTMAN [1991]. The last article, in particular, proposes and treats sophisticated models to study large deformations of non-linear elastic tubes in two-dimensional flows.

§2. Existence Results

The study of the models presented in §1 commences with proving existence of eigenvalues and eigenvectors for which we use the methods of functional analysis introduced in Chapter I. The strategy is to employ *Green's operator technique* which consists of reducing the differential eigenvalue problem to finding the spectrum of a suitable operator acting on a Hilbert space. The construction of the Green's operator is based on the *variational formulation* of the corresponding stationary problem which we present in each and every case. In carrying out this program, we will see that each model possesses difficulties of different nature and we will overcome them by means of suitable techniques. For instance, the associated Green's operator are not self-adjoint and one has to introduce a convenient symmetrization procedure in each case.

2.1. The Laplace Model

Let us recall that this model is formally described by (1.9). Even though it is a spectral problem involving Laplace operator, we will see that it admits only finitely many eigenvalues. This property might seem strange at first sight, however, this will be explained very naturally by the presence of the coupling boundary condition on Γ_j. In fact, thanks to this condition, problem (1.9) reduces to an eigenvalue problem of an operator in a finite dimensional space.

It was implicit in our derivation of the model equation (1.9) that ω^2 is not in the set NF, where

$$(2.1) \qquad NF \overset{\text{def}}{=} \left\{ \frac{k_j}{m_j} \mid j = 1, ..., K \right\}.$$

Observe that (k_j/m_j) corresponds to the natural frequency of vibration of the jth tube in the absence of any fluid (i.e., in vacuum). It is intuitively clear that when one immerses such a tube in a fluid, dissipation of energy of vibration is expected and so the condition that $\omega^2 \notin NF$ is physically reasonable (see Exercise 2.1 below). This is why model (1.9) was presented with $(k_j - m_j\omega^2)$ occurring in the denominator. However, from a mathematical point of view, we cannot neglect apriori the frequency values in the set NF and so we will try to reformulate (1.9) in such a way that the new model is valid for all values of ω^2.

With this in mind, let us go back to the method of establishing the Laplace model (see Paragraph 1.1) and define $\mathbf{s} \in \mathbf{C}^{2K}$ by $\mathbf{s} = i\omega\mathbf{r}$, i.e.,

$$e^{i\omega t}\mathbf{s}_j = e^{i\omega t}i\omega\mathbf{r}_j = \frac{d}{dt}\mathbf{r}_{0j} \quad \forall j = 1, ..., K.$$

Observe that $\mathbf{s}_j$ represents the stationary component of the velocity of the jth tube. Now, we see that ω^2 corresponds to a vibration frequency of the system iff there exist $\phi \in H^1(\Omega)$ (not identically constant) and $\mathbf{s} \in \mathbf{C}^{2K}$ such that the triplet $(\omega, \phi, \mathbf{s})$ satisfies the following spectral problem in Ω:

$$(2.2) \qquad \begin{cases} \Delta\phi = 0 \quad \text{in} \quad \Omega, \\ \dfrac{\partial\phi}{\partial n} = \mathbf{s}_j \cdot \mathbf{n} \quad \text{on} \quad \Gamma_j, \ j = 1, ..., K, \\ \dfrac{\partial\phi}{\partial n} = 0 \quad \text{on} \quad \Gamma_0, \\ (k_j - m_j\omega^2)\mathbf{s}_j = \rho_0\omega^2 \displaystyle\int_{\Gamma_j} \phi\mathbf{n}d\gamma \quad \forall j = 1, ..., K. \end{cases}$$

We remark immediately that if $\omega^2 \notin NF$ then (2.2) coincides with (1.9). Moreover, the last condition can be split into an equivalent set of two conditions of different type. To see this, let us associate, to all $\omega \in \mathbf{C}$, the set of indices

$$(2.3) \qquad J(\omega^2) = \begin{cases} \emptyset & \text{if } \omega^2 \notin NF, \\ \left\{ j \mid 1 \le j \le K, \ \dfrac{k_j}{m_j} = \omega^2 \right\} & \text{if } \omega^2 \in NF. \end{cases}$$

Then clearly the last condition in (2.2) is equivalent to

$$(2.4) \qquad \begin{cases} \displaystyle\int_{\Gamma_j} \phi \mathbf{n} d\gamma = 0 & \forall j \in J(\omega^2), \\[2mm] s_j = \dfrac{\rho_0 \omega^2}{k_j - m_j \omega^2} \left(\displaystyle\int_{\Gamma_j} \phi \mathbf{n} d\gamma \right) & \text{on } \Gamma_j \ \forall j \notin J(\omega^2). \end{cases}$$

From now on, we take system (2.2) to represent the *Laplace model*. In passing from (1.9) to (2.2), we have added the unknowns (s_j) apart from ϕ and ω^2.

To solve the above system (2.2) we follow the Green's operator technique and introduce two finite dimensional operators S and T acting on $\mathbb{C}^{2K}$. To this end, for a given vector $\mathbf{a} \in \mathbb{C}^{2K}$, we shall group the components of $\mathbf{a}$ in subvectors $\mathbf{a}_j \in \mathbb{C}^2$ as follows:

$$\mathbf{a} = (\mathbf{a}_1, \mathbf{a}_2, ..., \mathbf{a}_K) = (\mathbf{a}_j);$$

$\mathbf{a}_j$ being associated with the jth tube. The usual inner product in the space $\mathbb{C}^{2K}$ and the corresponding norm are denoted by $(\cdot, \cdot)_K$ and $\|\cdot\|_K$ respectively. We now define $S: \mathbb{C}^{2K} \longrightarrow \mathbb{C}^{2K}$ by the following rule:

$$(2.5) \qquad S\mathbf{a} = \left(\int_{\Gamma_1} \psi \mathbf{n} d\gamma, \ ..., \ \int_{\Gamma_K} \psi \mathbf{n} d\gamma \right) \in \mathbb{C}^{2K}$$

where ψ is the (unique) solution defined up to an additive constant of the following Neumann problem:

$$(2.6) \qquad \begin{cases} \Delta \psi = 0 & \text{in } \Omega, \\[2mm] \dfrac{\partial \psi}{\partial n} = \mathbf{a}_j \cdot \mathbf{n} & \text{on } \Gamma_j, \ j = 1, ..., K, \\[2mm] \dfrac{\partial \psi}{\partial n} = 0 & \text{on } \Gamma_0. \end{cases}$$

A standard variational formulation of the above problem is

$$(2.7) \qquad \begin{cases} \dot\psi \in \dot H^1(\Omega), \\[2mm] \displaystyle\int_\Omega \nabla\psi \cdot \nabla\bar\varphi dx = \sum_{j=1}^{K} \mathbf{a}_j \cdot \int_{\Gamma_j} \bar\varphi \mathbf{n} d\gamma \quad \forall \dot\varphi \in \dot H^1(\Omega), \end{cases}$$

where $\dot{H}^1(\Omega)$ is the *quotient space* $(H^1(\Omega)/\mathbb{C})$. To simplify matters, we will use the same notation for a function φ in $H^1(\Omega)$ and also for the *equivalence class* $\{\varphi+c\}$ in $\dot{H}^1(\Omega)$ determined by it. Observe that the definition (2.5) of S depends only on the equivalence class of ψ. The existence and uniqueness of ψ follows easily from Lax-Milgram Lemma since the left side of (2.7) defines an elliptic sesquilinear form on $\dot{H}^1(\Omega)$ and the right side is a continuous linear form. Thus S is well-defined. We are now ready to introduce the operator T in terms of S. Indeed let $T\colon\mathbb{C}^{2K} \longrightarrow \mathbb{C}^{2K}$ be defined by

$$(2.8) \qquad T\mathbf{a} = \rho_0 \mathrm{ddiag}(k_j^{-1})S\mathbf{a} + \mathrm{ddiag}(m_j k_j^{-1})\mathbf{a} \quad \forall \mathbf{a} \in \mathbb{C}^{2K}$$

where, in a general manner, $\mathrm{ddiag}(\ell_j)$ will denote the diagonal matrix of order $2K$ whose diagonal entries are $\{\ell_1, \ell_1, ..., \ell_K, \ell_K\}$. Our interest in the operator T is due to the fact that its spectrum describes the frequencies of the Laplace model (2.2). More precisely, we have

Theorem 2.1. *Let* $(\omega, \phi, \mathbf{s}) \in \mathbb{C} \times \dot{H}^1(\Omega) \times \mathbb{C}^{2K}$ *be a solution of problem* (2.2) *such that* $\mathbf{s} \neq 0$. *Then* $\omega \neq 0$, $(1/\omega^2)$ *is an eigenvalue of* T *and* $\mathbf{s}$ *is it corresponding eigenvector. Conversely, if* $\mathbf{s} \in \mathbb{C}^{2K}$ *is an eigenvector of* T *corresponding to an eigenvalue* μ, *then* $\mu \neq 0$ *and there exists* $\phi \in \dot{H}^1(\Omega)$, *not identically constant, such that* $(1/\mu, \phi, \mathbf{s})$ *is a solution of* (2.2).

Proof. The first part of the theorem is an immediate consequence of the definition of T. For the converse part, it suffices to use also Proposition 2.2 below. To get the corresponding velocity potential ϕ, we solve (2.6) with **a=s**. ∎

Thus we see that from the spectrum of T, we recover not only the frequencies ω but also the corresponding velocity potential ϕ and the velocity vector **s** of the tubes. That is why, by Laplace model, we will understand the spectral problem for T. The following result summarizes some of the properties of S:

Proposition 2.2. *The operator* S *is a linear self-adjoint operator on* $\mathbb{C}^{2K}$ *and it is positive definite in the sense that there exists* $\alpha > 0$, *which in general depends on* K, *such that*

$$(2.9) \qquad (S\mathbf{a}, \mathbf{a})_K \geq \alpha \|\mathbf{a}\|_K^2 \quad \forall \mathbf{a} \in \mathbb{C}^{2K}.$$

Proof. Let $\mathbf{a}$ and $\tilde{\mathbf{a}}$ be given in $\mathbb{C}^{2K}$. Suppose that ψ and $\tilde{\psi}$ be the velocity potentials associated via (2.6) with $\mathbf{a}$ and $\tilde{\mathbf{a}}$ respectively. Taking $\tilde{\psi}$ as test function in (2.7), we get

$$(2.10) \qquad \int_{\Omega} \nabla\psi \cdot \nabla\tilde{\psi} dx = \sum_{j=1}^{K} \mathbf{a}_j \cdot \int_{\Gamma_j} \tilde{\psi} \mathbf{n} d\gamma = (\mathbf{a}, S\tilde{\mathbf{a}})_K.$$

Analogously, we also have

$$\int_{\Omega} \nabla\tilde{\psi} \cdot \nabla\bar{\psi} dx = \sum_{j=1}^{K} \tilde{\mathbf{a}}_j \cdot \int_{\Gamma_j} \bar{\psi} \mathbf{n} d\gamma = (\tilde{\mathbf{a}}, S\mathbf{a})_K.$$

The self-adjointness of S follows easily from these two relations. To prove that S is positive definite, it suffices to show that $(S\mathbf{a}, \mathbf{a})_K > 0 \; \forall \mathbf{a} \neq \mathbf{0}$. But from (2.10), we have

$$(2.11) \qquad (S\mathbf{a}, \mathbf{a})_K = \int_{\Omega} |\nabla\psi|^2 dx.$$

Hence $(S\mathbf{a}, \mathbf{a})_K$ vanishes iff ψ is constant, i.e., $\mathbf{a} = \mathbf{0}$. ∎

Regarding the operator T, we first remark that it is not self-adjoint because the matrices S and $\mathrm{ddiag}(k_j^{-1})$ do not in general commute except if S is diagonal or if $k_1 = \cdots = k_K$. To overcome this little difficulty, we symmetrize T by means of the change of variables $\mathbf{b} = \mathrm{ddiag}(k_j^{1/2})\,\mathbf{a}$. Under this transformation the eigenvalue equation $T\mathbf{a} = \mu\mathbf{a}$ becomes $\tilde{T}\mathbf{b} = \mu\mathbf{b}$, where

$$(2.12) \qquad \tilde{T} = \rho_0 \mathrm{ddiag}(k_j^{-1/2}) S \mathrm{ddiag}(k_j^{-1/2}) + \mathrm{ddiag}(m_j k_j^{-1}).$$

Since S is self-adjoint and positive definite, so is $\tilde{T}$. Therefore, it follows that the spectrum of T consists of strictly positive numbers which we arrange as follows:

$$(2.13) \qquad \mu_1 \geq \cdots \geq \mu_{2K} > 0.$$

Here and in what follows, the eigenvalues will always be numbered so that the same value is repeated according to its multiplicity. The corresponding eigenvectors of T are denoted as $\mathbf{s}_1, ..., \mathbf{s}_{2K}$. They automatically form an orthonormal basis for $\mathbb{C}^{2K}$ with respect to the weighted scalar product $(\mathrm{ddiag}(k_j)\mathbf{a}, \mathbf{b})_K$.

We are now in a position to describe the solutions of problem (2.2). This is done by means of the spectrum of T. For each $\ell = 1, ..., 2K$, let us define ω_ℓ^2 by

$$(2.14) \qquad \omega_\ell^2 = \frac{1}{\mu_\ell}$$

and denote by ϕ_ℓ the corresponding velocity potential defined by (2.6) with $\mathbf{a} = \mathbf{s}_\ell$. Then we have

Theorem 2.3. *The triplets* $\{(\omega_\ell^2, \phi_\ell, \mathbf{s}_\ell)\}_\ell$ *satisfy:*

(i) *For each* $\ell = 1, ..., 2K$, $(\omega_\ell^2, \phi_\ell, \mathbf{s}_\ell)$ *is a solution of* (2.2).

(ii) *The set* $\{\mathbf{s}_\ell\}_\ell$ *forms a basis for* $\mathbb{C}^{2K}$ *which is orthonormal in the following sense:*

$$(2.15) \qquad \left(\mathrm{ddiag}(k_j)\mathbf{s}_\ell, \mathbf{s}_m\right)_K = \delta_{\ell m} \quad \forall \ell, m = 1, ..., 2K.$$

(iii) *The triplets* $\{(\omega_\ell^2, \phi_\ell, \mathbf{s})\}_\ell$ *are all the solutions of problem* (2.2), *i.e., if* $(\omega_\ell^2, \phi_\ell, \mathbf{s}_\ell) \in \mathbb{C} \times \dot{H}^1(\Omega) \times \mathbb{C}^{2K}$ *satisfies* (2.2), *then there exists* $\ell \geq 1$ *such that* $\omega^2 = \omega_\ell^2$ *and* $(\phi, \mathbf{s})$ *can be written as a linear combination of all* $(\phi_m, \mathbf{s}_m)$'s *for which* ω_m^2 *is equal to* ω^2. ∎

The above result describes completely the solutions of (2.2). In particular, it shows that there exist only finitely many eigenvalues, the total number (including their multiplicities) being $2K$; two for each tube. The Laplace equation in the fluid part signifies the fact that the fluid does not vibrate before the immersion of the tube bundle. After this, however, the whole system vibrates with the frequencies $\{\omega_\ell^2\}$ described by the above theorem.

◇ **Exercise 2.1.** (Case of identical tubes) Assume that all the tubes are identical so that $m_1 = \cdots = m_K = m > 0$ (say) and $k_1 = \cdots = k_K = k > 0$ (say) and analyze the spectrum of the Laplace model in this case. In particular, show that $T\mathbf{s} = (1/\omega^2)\mathbf{s}$ with $\omega^2 = (k/m)$ iff there exists $\phi \in \dot{H}^1(\Omega)$ such that

$$(2.16) \qquad \begin{cases} \Delta\phi = 0 & \text{in} \quad \Omega, \\ \dfrac{\partial\phi}{\partial n} = \mathbf{s}_j \cdot \mathbf{n} & \text{on} \quad \Gamma_j, \; j = 1, ..., K, \\ \dfrac{\partial\phi}{\partial n} = 0 & \text{on} \quad \Gamma_0, \quad \displaystyle\int_{\Gamma_j} \phi \mathbf{n} d\gamma = 0 \quad \forall j = 1, ..., K. \end{cases}$$

Show further that (2.16) does not possess non-constant solutions ϕ. Conclude that (m/k) is not an eigenvalue of T, i.e., the fluid-solid structure cannot vibrate at the natural frequency of the tubes. ∎

Comments

The content of Theorem 2.1 is essentially due to PLANCHARD [1982] (see also IBNOU-ZAHIR [1984]). In this paper, he has used the technique of *added mass matrix* which reflects the interaction of tubes via the fluid. The approach presented in the text which uses the Green's operator is inspired from AGUIRRE and CONCA [1988] where an analogous problem with periodic boundary condition is treated. ∎

2.2. The Helmholtz Model

Here, we adapt the procedure of Paragraph 2.1 to prove existence of eigenvalues for the model formally described by (1.14). In this case, we have the vibrations of the fluid apart from those studied in the Laplace model. We must therefore work in a product of two spaces, one to capture the vibrations of the fluid and the other to capture those of the tubes. We expect a countable number of elements in the spectrum of the Helmholtz model.

As done in the previous paragraph, we start by reformulating problem (1.14) so that the values in the set NF should also be allowed to be admissible frequencies. We first introduce a new variable $\mathbf{s} \in \mathbb{C}^{2K}$ which, this time, is defined by

$$\mathbf{s} = c^2 i \omega \mathbf{r}.$$

Observe that the same variable was considered also in the incompressible case except for the constant c^2. Now, a closer examination of the method of arriving at the Helmholtz model (1.14) shows that ω^2 corresponds to an admissible vibration frequency of the system iff there exists $\phi \in H^1(\Omega)$ (not identically zero) and $\mathbf{s} \in \mathbb{C}^{2K}$ such that the triplet $(\omega, \phi, \mathbf{s})$ satisfies the following spectral problem in Ω:

(2.17)
$$
\begin{cases}
c^2 \Delta \phi + \omega^2 \phi = 0 & \text{in} \quad \Omega, \\[2mm]
c^2 \dfrac{\partial \phi}{\partial n} = \mathbf{s}_j \cdot \mathbf{n} & \text{on} \quad \Gamma_j, \ j = 1, ..., K, \\[2mm]
\dfrac{\partial \phi}{\partial n} = 0 & \text{on} \quad \Gamma_0, \\[2mm]
(k_j - m_j \omega^2) \mathbf{s}_j = \rho_0 c^2 \omega^2 \displaystyle\int_{\Gamma_j} \phi \mathbf{n} d\gamma & \forall j = 1, ..., K.
\end{cases}
$$

Once again, the last condition in (2.17) can be replaced by the two conditions in (2.4) wherein ρ_0 is replaced by $\rho_0 c^2$.

Before going into the details of the study of problem (2.17), let us remark that this has a trivial solution which is $\phi \equiv$ constant, $\omega = 0$ and $\mathbf{s} = \mathbf{0}$. The eigenvalue $\omega = 0$ has obviously multiplicity one. It is also worthwhile to notice via a simple integration by parts that all solutions of (2.17) satisfy

(2.18)
$$\omega^2 \int_{\Omega} \phi dx = 0.$$

Since the zero eigenvalue has already been eliminated, we always look for velocity potentials ϕ with zero average in Ω. Let us therefore introduce the space $L_0^2(\Omega)$:

(2.19)
$$L_0^2(\Omega) = \left\{ \psi \in L^2(\Omega) \ \Big| \ \int_{\Omega} \psi dx = 0 \right\},$$

endowed with the standard L^2-norm. We also need the space $V \overset{\text{def}}{=} H^1(\Omega) \cap L_0^2(\Omega)$. The semi-norm $\psi \to |\psi|_{1,\Omega}$ defines a norm on V equivalent to the norm induced by $H^1(\Omega)$. This is an immediate consequence of Poincaré-Wirtinger Inequality (1.10) of Chapter I.

In order to carry out the Green's operator technique to solve the above system, we introduce the operator $R: L_0^2(\Omega) \times \mathbb{C}^{2K} \longrightarrow L_0^2(\Omega) \times \mathbb{C}^{2K}$ defined by

$$(2.20) \quad R(f, \mathbf{a}) = (u, (\mathbf{b}_j)), \quad \mathbf{b}_j = \rho_0 c^2 k_j^{-1} \int_{\Gamma_j} u \, n d\gamma + m_j k_j^{-1} \mathbf{a}_j \quad \forall j = 1, ..., K,$$

and u is the unique weak solution in V of

$$(2.21) \quad \begin{cases} c^2 \Delta u + f = 0 & \text{in} \quad \Omega, \\ c^2 \dfrac{\partial u}{\partial n} = \mathbf{a}_j \cdot \mathbf{n} & \text{on} \quad \Gamma_j, \ j = 1, ..., K, \\ \dfrac{\partial u}{\partial n} = 0 & \text{on} \quad \Gamma_0, \quad \displaystyle\int_\Omega u dx = 0. \end{cases}$$

The motivation to consider the operator R is that its spectrum describes the vibration frequencies of the Helmholtz model in the following sense:

Theorem 2.4. *Let $(\omega, \phi, \mathbf{s}) \in \mathbb{C} \times V \times \mathbb{C}^{2K}$ be a solution of problem (2.17) such that $\phi \not\equiv 0$. Then ω is non-zero, $(1/\omega^2)$ is an eigenvalue of R and the couple $(\phi, \mathbf{s}/\omega^2)$ is a corresponding eigenvector. Conversely, if $(\phi, \mathbf{s})$ in $L_0^2(\Omega) \times \mathbb{C}^{2K}$ is an eigenvector of R with eigenvalue μ, then $\mu \neq 0$ and $((1/\mu), \phi, \mathbf{s}/\mu)$ is a solution of problem (2.17).*

Proof. Let $(\omega, \phi, \mathbf{s}) \in \mathbb{C} \times V \times \mathbb{C}^{2K}$ be a solution of (2.17). If $\omega = 0$, a simple integration by parts shows that $\phi \equiv 0$. Therefore $\omega \neq 0$. From the last condition in (2.17), it follows that

$$(2.22) \quad \frac{1}{\omega^2} \mathbf{s}_j = \rho_0 c^2 k_j^{-1} \int_{\Gamma_j} \phi n d\gamma + m_j k_j^{-1} \mathbf{s}_j.$$

Comparing this with the definition of R (see (2.20) and (2.21)), we get immediately that $R(\omega^2 \phi, \mathbf{s}) = (\phi, \mathbf{s}/\omega^2)$, i.e.,

$$(2.23) \quad R\left(\phi, \frac{\mathbf{s}}{\omega^2}\right) = \frac{1}{\omega^2}\left(\phi, \frac{\mathbf{s}}{\omega^2}\right)$$

which proves that $(1/\omega^2)$ is an eigenvalue of R. The proof of the converse part amounts to retracing the above steps provided $0 \notin \sigma(R)$. This will be proved below (see Proposition 2.7). ■

The above result shows that from the spectrum of R, we recover not only the frequencies ω but also the corresponding velocity potential ϕ and the velocity vector $\mathbf{s}$ of the tubes. That is why, by *Helmholtz model*, we will understand the spectral problem associated with R. Let us therefore proceed to analyze the structure of R and study its spectrum. Since R is a linear operator defined in the product space $L_0^2(\Omega) \times \mathbb{C}^{2K}$, it can formally be represented by a 2×2 matrix whose entries are linear operators. More precisely, we can rewrite R in the form

$$(2.24) \qquad R = \begin{bmatrix} R_{11} & R_{12} \\ \rho_0 c^2 \mathrm{ddiag}(k_j^{-1}) R_{21} & R_{22} \end{bmatrix},$$

where R_{11}, R_{12}, R_{21} and R_{22} are linear operators defined by the following rules:

$$(2.25) \qquad R_{11} \colon L_0^2(\Omega) \longrightarrow L_0^2(\Omega), \quad R_{11} f = u_1,$$

$$(2.26) \qquad R_{12} \colon \mathbb{C}^{2K} \longrightarrow L_0^2(\Omega), \quad R_{12} \mathbf{a} = u_2,$$

$$(2.27) \qquad R_{21} \colon L_0^2(\Omega) \longrightarrow \mathbb{C}^{2K}, \quad R_{21} f = \left(\int_{\Gamma_1} un d\gamma, \cdots, \int_{\Gamma_K} un d\gamma \right),$$

$$(2.28) \qquad \begin{cases} R_{22} \colon \mathbb{C}^{2K} \longrightarrow \mathbb{C}^{2K}, \quad R_{22} \mathbf{a} = (\mathbf{b}_j), \text{ where} \\ \\ \mathbf{b}_j = \rho_0 c^2 k_j^{-1} \displaystyle\int_{\Gamma_j} u_2 n d\gamma + m_j k_j^{-1} \mathbf{a}_j \quad \forall j = 1, ..., K. \end{cases}$$

In these definitions, u_1 and u_2 are the unique solutions of the following variational problems:

$$(2.29) \qquad \begin{cases} u_1 \in V, \\ \\ c^2 \displaystyle\int_\Omega \nabla u_1 \cdot \nabla \bar\varphi dx = \int_\Omega f \bar\varphi dx \quad \forall \varphi \in V, \end{cases}$$

$$(2.30) \qquad \begin{cases} u_2 \in V, \\ \\ c^2 \displaystyle\int_\Omega \nabla u_2 \cdot \nabla \bar\varphi dx = \sum_{j=1}^K \mathbf{a}_j \cdot \int_{\Gamma_j} \bar\varphi n d\gamma \quad \forall \varphi \in V. \end{cases}$$

Let us observe that $u_1 + u_2 = u$; u being the function which was used in the definition of R (see (2.20) and (2.21)). The operator R_{11} is very familiar in the study of vibration problems. It is the inverse, in the space $L_0^2(\Omega)$, of the operator $(-\Delta)$ with homogeneous Neumann boundary conditions on $\Gamma_0 \cup \Gamma_1 \cup \cdots \cup \Gamma_K$. The non-zero Neumann eigenvalues on Ω are nothing but the inverse of the eigenvalues of R_{11}. This situation has been studied in Paragraph 2.2, Chapter I (see also Example 3.5 and Exercise 3.13). In particular, we know that the spectrum of R_{11} consists of a countably infinite set of positive real numbers which converges to zero and each has a finite multiplicity. The corresponding eigenfunctions can be chosen in such a way that they form an orthonormal basis for $L_0^2(\Omega)$. We refer to R_{11} as the Neumann operator associated with problem (2.17).

Concerning the operator R_{22}, let us note that it is the operator T of the incompressible case where ρ_0 is to be replaced by $\rho_0 c^2$. Recall that a symmetrization procedure was used to study the spectrum of T and the same can be repeated here. We introduce therefore an operator $\tilde{R}_{22}$ which is self-adjoint, similar to R_{22} and defined by

$$(2.31) \qquad \tilde{R}_{22} = \rho_0 c^2 \operatorname{ddiag}(k_j^{-1/2}) S \operatorname{ddiag}(k_j^{-1/2}) + \operatorname{ddiag}(m_j k_j^{-1}),$$

where S is defined by (2.5) and (2.6).

Let us now proceed to study the operators R_{12} and R_{21}. They are clearly continuous and linear. Let R_{12}^* be the adjoint operator of R_{12}. Identifying the spaces $L_0^2(\Omega)$ and $\mathbb{C}^{2K}$ with their duals, R_{12}^* is defined by

$$(2.32) \qquad R_{12}^*: L_0^2(\Omega) \longrightarrow \mathbb{C}^{2K}, \quad \left(R_{12}^* f, \mathbf{a}\right)_K = \int_\Omega f\overline{(R_{12}\mathbf{a})} dx,$$

for all $f \in L_0^2(\Omega)$, $\mathbf{a} \in \mathbb{C}^{2K}$.

Proposition 2.5. *The operator R_{21} is the operator adjoint to R_{12}, i.e., $R_{21} = R_{12}^*$ and conversely.*

Proof. Let $f \in L_0^2(\Omega)$ and $\mathbf{a} \in \mathbb{C}^{2K}$ be given. Taking $\varphi = R_{12}\mathbf{a} = u_2$ in (2.29) and using the definition of $R_{12}^* f$, we obtain

$$\left(R_{12}^* f, \mathbf{a}\right)_K = \int_\Omega f\bar{u}_2 dx = c^2 \int_\Omega \nabla u_1 \cdot \nabla \bar{u}_2 dx,$$

where u_1 is the solution of (2.29). On the other hand, taking $\varphi = u_1$ in (2.30) and using the definition of $R_{21} f$ (see (2.27)), we get

$$c^2 \int_\Omega \nabla u_2 \cdot \nabla \bar{u}_1 dx = \sum_{j=1}^K \mathbf{a}_j \cdot \int_{\Gamma_j} \bar{u}_1 \mathbf{n} d\gamma = (\mathbf{a}, R_{21} f)_K.$$

The desired results follow immediately from the two relations above. ∎

From the above result, we conclude that R is not self-adjoint. However, as it was said in the introduction of this section, it is not hard to see that

R can be symmetrized in the sense that one can find a self-adjoint operator $\tilde{R}$ having the same spectrum as R. This is done using the symmetrization procedure introduced in Paragraph 2.1. It is easily seen that the above procedure produces the operator $\tilde{R}$ which, in its matrix representation, can be written as follows:

(2.33)
$$\begin{cases} \tilde{R}: L_0^2(\Omega) \times \mathbb{C}^{2K} \longrightarrow L_0^2(\Omega) \times \mathbb{C}^{2K}, \\ \tilde{R} = \begin{bmatrix} R_{11} & \rho_0^{1/2} c\, R_{12} \mathrm{ddiag}(k_j^{-1/2}) \\ \rho_0^{1/2} c\, \mathrm{ddiag}(k_j^{-1/2}) R_{21} & \tilde{R}_{22} \end{bmatrix}, \end{cases}$$

where $\tilde{R}_{22}$ was defined in (2.31).

Proposition 2.6. *A complex number μ is an eigenvalue of R with corresponding eigenvector $(\phi, \mathbf{s}) \in L_0^2(\Omega) \times \mathbb{C}^{2K}$ iff μ is an eigenvalue of $\tilde{R}$ with corresponding eigenvector $(\phi, \rho_0^{-1/2} c^{-1} \mathrm{ddiag}(k_j^{1/2})\mathbf{s})$.*

Proof. Let $\mu \in \mathbb{C}$ be an eigenvalue of R with eigenvector $(\phi, \mathbf{s}) \in L_0^2(\Omega) \times \mathbb{C}^{2K}$. We have then

(2.34)
$$\begin{cases} R_{11}\phi + R_{12}\mathbf{s} = \mu\phi, \\ \rho_0 c^2 \mathrm{ddiag}(k_j^{-1}) R_{21}\phi + R_{22}\mathbf{s} = \mu\mathbf{s}. \end{cases}$$

If we define $\mathbf{r} = \rho_0^{-1/2} c^{-1} \mathrm{ddiag}(k_j^{1/2})\mathbf{s}$, then the first equation in (2.34) can be rewritten as

(2.35)
$$R_{11}\phi + \rho_0^{1/2} c R_{12} \mathrm{ddiag}(k_j^{-1/2})\mathbf{r} = \mu\phi.$$

On the other hand, multiplying the second equation by $\rho_0^{-1/2} c^{-1} \mathrm{ddiag}(k_j^{1/2})$ and using the definition of $\mathbf{r}$, we have

$$\rho_0^{1/2} c\, \mathrm{ddiag}(k_j^{-1/2}) R_{21}\phi + \mathrm{ddiag}(k_j^{1/2}) R_{22} \mathrm{ddiag}(k_j^{-1/2})\mathbf{r} = \mu\mathbf{r}.$$

Using the definition of $\tilde{R}_{22}$, we can write

(2.36)
$$\rho_0^{1/2} c\, \mathrm{ddiag}(k_j^{-1/2}) R_{21}\phi + \tilde{R}_{22}\mathbf{r} = \mu\mathbf{r}.$$

The identities (2.35) and (2.36) prove that μ is an eigenvalue of $\tilde{R}$ with eigenvector $(\phi, \mathbf{r})$. This terminates the first part of Proposition 2.6. The proof of the converse is similar and so we omit it. ∎

Thus we are reduced to study the spectrum of $\tilde{R}$. The following proposition summarizes its main properties which are essential to describe its spectrum:

Proposition 2.7. *The operator $\tilde{R}$ is a compact self-adjoint operator on $L_0^2(\Omega) \times \mathbb{C}^{2K}$. Moreover, it is positive definite in the sense that the following inequality holds:*

$$(2.37) \quad \begin{cases} \displaystyle\int_\Omega \left[R_{11}f + \rho_0^{1/2}c\,\mathrm{ddiag}(k_j^{-1/2})\mathbf{a} \right] \bar{f}\,dx + \\ \qquad\qquad + \left(\rho_0^{1/2}c\,\mathrm{ddiag}(k_j^{-1/2})R_{21}f + \tilde{R}_{22}\mathbf{a}, \mathbf{a} \right)_K > 0 \end{cases}$$

for all $(f, \mathbf{a}) \neq (0, \mathbf{0})$, i.e., $\left(\tilde{R}(f, \mathbf{a}), (f, \mathbf{a}) \right) > 0 \ \forall (f, \mathbf{a}) \neq (0, \mathbf{0})$.

Proof. The self-adjointness of $\tilde{R}$ is obvious. Its compactness is an immediate consequence of the fact the R_{11} is compact. Thus it remains to prove (2.37). In order to do this, let $f \in L_0^2(\Omega)$ and $\mathbf{a} \in \mathbb{C}^{2K}$ be given. Let A stand for the quantity of the left side of (2.37). Since $R_{12}^* = R_{21}$, A can be rewritten as

$$(2.38) \quad A = \int_\Omega (R_{11}f)\bar{f}\,dx + 2\rho_0^{1/2}c\,\mathfrak{Re}\left(R_{21}f, \mathrm{ddiag}(k_j^{-1/2})\mathbf{a} \right)_K + \left(\tilde{R}_{22}\mathbf{a}, \mathbf{a} \right)_K.$$

Taking $\varphi = R_{11}f = u_1$ as test function in (2.29), we have

$$(2.39) \quad \int_\Omega (R_{11}f)\bar{f}\,dx = \int_\Omega u_1\bar{f}\,dx = c^2 \int_\Omega |\nabla u_1|^2\,dx.$$

We next define $u \in V$ as $R_{12}\mathrm{ddiag}(k_j^{-1/2})\mathbf{a}$. Using the definition of R_{12} (see (2.26) and (2.30)), this means that u is the unique solution of the following variational problem:

$$(2.40) \quad \begin{cases} u \in V, \\ \displaystyle c^2 \int_\Omega \nabla u \cdot \nabla\bar{\varphi}\,dx = \sum_{j=1}^K k_j^{-1/2}\mathbf{a}_j \cdot \int_{\Gamma_j} \bar{\varphi}n\,dx \quad \forall \varphi \in V. \end{cases}$$

Taking $\varphi = u_1$ as a test function in (2.40) and using the definition of $R_{21}f$ (see (2.27)), we obtain

$$(2.41) \quad \left(\mathrm{ddiag}(k_j^{-1/2})\mathbf{a}, R_{21}f \right)_K = c^2 \int_\Omega \nabla u \cdot \nabla\bar{u}_1\,dx.$$

We take now $\varphi = u$ in (2.40). The resulting equation can be rewritten in terms of S as follows:

$$\left(\mathbf{a}, \mathrm{ddiag}(k_j^{-1/2})S\,\mathrm{ddiag}(k_j^{-1/2})\mathbf{a} \right)_K = c^2 \int_\Omega |\nabla u|^2\,dx.$$

But it is clear from the definition of $\tilde{R}_{22}$ that for all $a \in \mathbb{C}^{2K}$, we have

$$\left(\tilde{R}_{22}\mathbf{a}, \mathbf{a}\right)_K \geq \rho_0 c^2 \left(\text{ddiag}(k_j^{-1/2})S\text{ddiag}(k_j^{-1/2})\mathbf{a}, \mathbf{a}\right)_K$$

and so we conclude that

$$(2.42) \qquad \left(\tilde{R}_{22}\mathbf{a}, \mathbf{a}\right)_K \geq \rho_0 c^4 \int_\Omega |\nabla u|^2 dx.$$

Using (2.39), (2.41) and (2.42) in the definition (2.38) of A, we get

$$(2.43) \qquad A \geq c^2 \int_\Omega |\nabla(u_1 + \rho_0^{1/2} c\, u)|^2 dx = c^2 |u_1 + \rho_0^{1/2} c\, u|_{1,\Omega}^2.$$

The announced result in Proposition 2.7 is now an easy consequence of the above inequality. ∎

From Propositions 2.6 and 2.7 it follows that the spectrum of R (and of $\tilde{R}$ also) consists of a countably infinite sequence of strictly positive numbers that converges to zero. We arrange the eigenvalues of R as follows:

$$(2.44) \qquad \mu_1 \geq \cdots \geq \mu_\ell \geq \cdots \longrightarrow 0.$$

The corresponding eigenvectors are denoted as $(\phi_1, \mathbf{s}_1), ..., (\phi_\ell, \mathbf{s}_\ell), ...$. They automatically form a basis of $L_0^2(\Omega) \times \mathbb{C}^{2K}$. Since the eigenvectors of $\tilde{R}$ form an orthonormal basis in $L_0^2(\Omega) \times \mathbb{C}^{2K}$, those of R also form an orthonormal basis in $L_0^2(\Omega) \times \mathbb{C}^{2K}$ but with respect to the scalar product in $L_0^2(\Omega) \times \mathbb{C}^{2K}$ modified according to Proposition 2.6. More precisely, we have

$$(2.45) \qquad \int_\Omega \phi_\ell \bar{\phi}_m dx + \rho_0^{-1} c^{-2} \left(\text{ddiag}(k_j)\mathbf{s}_\ell, \mathbf{s}_m\right)_K = \delta_{\ell m} \quad \forall \ell, m \geq 1.$$

The following theorem describes completely the spectrum of the vibration problem (2.17). We pose

$$(2.46) \qquad \omega_\ell^2 = \frac{1}{\mu_\ell} \quad \forall \ell \geq 1.$$

Theorem 2.8. *The triplets* $\{(\omega_\ell^2, \phi_\ell, \mathbf{s}_\ell)\}_\ell$ *satisfy:*

(i) *For each* $\ell \geq 1$, $(\omega_\ell^2, \phi_\ell, \omega_\ell^2 \mathbf{s}_\ell)$ *is a solution of* (2.17).

(ii) *The set* $\{(\phi_\ell, \mathbf{s}_\ell)\}_\ell$ *forms a basis in* $L_0^2(\Omega) \times \mathbb{C}^{2K}$ *which is orthonormal in the sense of* (2.45).

(iii) *The triplets* $\{(\omega_\ell^2, \phi_\ell, \omega_\ell^2 \mathbf{s}_\ell)\}_\ell$ *are all the solutions of problem* (2.17), *i.e., if* $(\omega^2, \phi, \mathbf{s})$ *in* $\mathbb{C} \times V \times \mathbb{C}^{2K}$ *satisfies* (2.17), *then there exists* $\ell \geq 1$ *such that* $\omega^2 = \omega_\ell^2$ *and* $(\phi, \mathbf{s})$ *can be written as a linear combination of all* $(\phi_n, \omega_n^2 \mathbf{s}_n)$*'s for which* $\omega_n^2 = \omega^2$. ∎

Among other things, the above result shows that there exists a sequence of vibration frequencies which tend to infinity in the compressible case. This property is one of the main differences between the compressible and the incompressible cases. Recall that the latter problem had only finitely many frequencies.

Comments

The material of this paragraph is taken from the article CONCA, PLANCHARD and VANNINATHAN [1989b]. A different approach using "balayage" in the frequency domain can be found for instance in IBNOU-ZAHIR [1984] and in PLANCHARD [1988] (see also CONCA, PLANCHARD, THOMAS and VANNINATHAN [1994]). ∎

2.3. The Stokes Model

This paragraph is devoted to proving existence of vibration frequencies for the Stokes model (1.23). We follow a pattern similar to that of the two previous paragraphs where we used Green's operator technique. The problem is non-standard because, as we shall see, it is reduced either to a *generalized eigenvalue problem* or to a *quadratic eigenvalue problem*. The end result shows that this model admits a countable infinite set of complex eigenvalues which converge to infinity. The question of existence is settled by transforming the problem to finding the spectrum of a compact operator acting on a Hilbert space. This operator is not self-adjoint and this is why the set of eigenvalues can contain elements with non-zero imaginary parts.

Once again, as in the previous paragraphs, a reformulation of the problem is called for. While deriving the model equation (1.23), it was implicitly assumed that the frequency parameter ω is such that $(-\omega^2) \notin NF$. This means that $\omega \notin INF$ where

$$(2.47) \qquad INF \overset{\text{def}}{=} \left\{ \pm i \sqrt{\frac{k_j}{m_j}} \;\middle|\; j = 1, ..., K \right\}.$$

In order to allow the values in INF to be admissible, we go back to the derivation of the model and examine it a little closely. As illustrated earlier in the case of Laplace and Helmholtz models, we can formulate the Stokes model in the following way by introducing the velocities (s_j) of the tubes: Observe that $\omega \in \mathbb{C}$ is a vibration frequency of the coupled system iff there exists a non-trivial pair $(\mathbf{u}, p) \in H^1(\Omega)^2 \times L^2(\Omega)$ and $\mathbf{s} \in \mathbb{C}^{2K}$ such that $(\omega, \mathbf{u}, p, \mathbf{s})$ satisfies

(2.48)
$$\begin{cases} -\operatorname{div} \sigma(\mathbf{u}, p) + \omega \mathbf{u} = 0 & \text{in } \Omega, \\ \operatorname{div} \mathbf{u} = 0 & \text{in } \Omega, \\ \mathbf{u} = \mathbf{s}_j & \text{on } \Gamma_j, \ j = 1, ..., K, \\ \mathbf{u} = 0 & \text{on } \Gamma_0, \\ (k_j + m_j \omega^2) \mathbf{s}_j = -\omega \displaystyle\int_{\Gamma_j} \sigma(\mathbf{u}, p) \mathbf{n} d\gamma & \forall j = 1, ..., K, \end{cases}$$

where $\sigma(\mathbf{u}, p)$ is the stress-tensor given by (1.16). We note immediately that (2.48) coincides with (1.23) if $\omega \notin INF$. The last condition splits into two conditions:

(2.49)
$$\begin{cases} \displaystyle\int_{\Gamma_j} \sigma(\mathbf{u}, p) \mathbf{n} d\gamma = 0 & \forall j \in J(-\omega^2), \\ \mathbf{s}_j = -\dfrac{\omega}{k_j + m_j \omega^2} \displaystyle\int_{\Gamma_j} \sigma(\mathbf{u}, p) \mathbf{n} d\gamma & \text{on } \Gamma_j \ \forall j \notin J(-\omega^2), \end{cases}$$

where $J(-\omega^2)$ was defined in (2.3). Observe also that $\omega = 0$ cannot be a vibration frequency in (2.48).

In the case of Stokes operator, it is a standard practice to look for the velocity $\mathbf{u}$ in the space of divergence-free fields. More precisely, let us introduce the space

(2.50) $\mathcal{H} = \left\{ \mathbf{v} \in H^1(\Omega)^2 \ \middle| \ \begin{array}{l} \operatorname{div} \mathbf{v} = 0 \quad \text{in } \Omega, \ \mathbf{v} = 0 \quad \text{on } \Gamma_0 \text{ and} \\ \mathbf{v} \text{ is a constant vector on } \Gamma_j, \ j = 1, ..., K \end{array} \right\}$.

This space is equipped with the usual inner product in $H^1(\Omega)^2$. It is then a closed subspace of $H^1(\Omega)^2$. *Korn's Inequality* says that the map

$$\mathbf{v} \longmapsto \|e(\mathbf{v})\|_{0, \Omega} = \left(\int_\Omega e(\mathbf{v}) : e(\bar{\mathbf{v}}) dx \right)^{1/2},$$

where $e(\mathbf{v})$ is the strain-tensor defined by (1.17)†, becomes a norm in $\mathcal{H}$ equivalent to the one induced by $H^1(\Omega)^2$. Regarding Korn's Inequality, we refer the reader to RAVIART and THOMAS [1983] pp. 50–53.

---

† By the expression $e(\mathbf{v}) : e(\bar{\mathbf{v}})$ we mean the standard scalar product between tensors.

Once the function spaces have been introduced, we proceed to give a variational formulation of (2.48). Multiplying the first equation in (2.48) by $\bar{\mathbf{v}} \in \mathcal{H}$ and integrating by parts in Ω using the boundary conditions on $\mathbf{v}$, it follows that any solution $(\omega, \mathbf{u}, p)$ of (2.48) satisfies

$$2\nu \int_\Omega e(\mathbf{u}) : e(\bar{\mathbf{v}})dx + \omega \int_\Omega \mathbf{u} \cdot \bar{\mathbf{v}}dx - \sum_{j=1}^K \int_{\Gamma_j} \sigma(\mathbf{u}, p)\mathbf{n} \cdot \bar{\mathbf{v}}d\gamma = 0.$$

Using now the last condition in (2.48) and the fact that the test function $\mathbf{v}$ is constant on each Γ_j, we see that (2.48) reduces to: Find $\omega \in \mathbb{C}, \mathbf{u} \in \mathcal{H}$, $\mathbf{u} \neq \mathbf{0}$ such that

(2.51)
$$\begin{cases} 2\nu \int_\Omega e(\mathbf{u}) : e(\bar{\mathbf{v}})dx + \omega \left[\int_\Omega \mathbf{u} \cdot \bar{\mathbf{v}}dx + \sum_{j=1}^K m_j \gamma_j(\mathbf{u}) \cdot \gamma_j(\bar{\mathbf{v}}) \right] + \\ \qquad\qquad\qquad + \sum_{j=1}^K \frac{k_j}{\omega} \gamma_j(\mathbf{u}) \cdot \gamma_j(\bar{\mathbf{v}}) = 0 \end{cases}$$

for all $\mathbf{v} \in \mathcal{H}$. Here $\gamma_j(\mathbf{u})$ denotes the trace of $\mathbf{u}$ on Γ_j. In the following we solve equation (2.51) for the variables $(\omega, \mathbf{u})$. One $\mathbf{u}$ is found, the pressure (up to an additive constant), is obtained by a standard application of *De Rham's Theorem* (see TEMAM [1987] Propositions 1.1 and 1.2) and $\mathbf{s}$ is recovered by means of the boundary condition on Γ_j in (2.48). Thus, in contrast to Laplace model, we are working with a formulation which eliminates the velocity vector of the tubes.

It can be remarked that problem (2.51) is a non-standard eigenvalue problem since the way the frequency parameter ω appears is not quite common; both ω and ω^{-1} occur and so it is in fact a *quadratic eigenvalue problem*.

In order to carry out the Green's operator technique, we introduce three bounded linear operators U_1, U_2 and Q defined by the following rules:

(2.52)
$$U_1 : \mathcal{H} \longrightarrow \mathcal{H}, \quad U_1\mathbf{f} = \mathbf{u}_1,$$

(2.53)
$$U_2 : \mathbb{C}^{2K} \longrightarrow \mathcal{H}, \quad U_2\mathbf{s} = \mathbf{u}_2,$$

(2.54)
$$Q : \mathcal{H} \longrightarrow \mathbb{C}^{2K}, \quad Q\mathbf{v} = (\gamma_1(\mathbf{v}), ..., \gamma_K(\mathbf{v})).$$

In these definitions, the vector functions $\mathbf{u}_1$ and $\mathbf{u}_2$ are the unique solutions of the following variational problems:

(2.55)
$$\begin{cases} \mathbf{u}_1 \in \mathcal{H}, \\ 2\nu \int_\Omega e(\mathbf{u}_1) : e(\bar{\mathbf{v}})dx + \int_\Omega \mathbf{f} \cdot \bar{\mathbf{v}}dx = 0 \quad \forall \mathbf{v} \in \mathcal{H}, \end{cases}$$

$$(2.56) \quad \begin{cases} \mathbf{u}_2 \in \mathcal{H}, \\ 2\nu \int_\Omega e(\mathbf{u}_2) : e(\bar{\mathbf{v}}) dx + \sum_{j=1}^{K} \mathbf{s}_j \cdot \gamma_j(\bar{\mathbf{v}}) = 0 \quad \forall \mathbf{v} \in \mathcal{H}. \end{cases}$$

Lax-Milgram Lemma guarantees the existence and the uniqueness of these solutions and therefore the operators U_1 and U_2 are well-defined. Using these operators, equation (2.51) can be recast in the form

$$(2.57) \qquad [U_1 + U_2 \mathrm{ddiag}(m_j)Q]\mathbf{u} + \frac{1}{\omega^2} U_2 \mathrm{ddiag}(k_j)Q\mathbf{u} = \frac{1}{\omega}\mathbf{u},$$

where, as in the previous paragraph, ddiag(ℓ_j), in general, denotes the diagonal matrix of order $2K$ whose entries are $\{\ell_1, \ell_1, ..., \ell_K, \ell_K\}$.

Conversely, if $(\omega, \mathbf{u}) \in \mathbb{C} \times \mathcal{H}$ satisfies (2.57) then a brief computation shows that it is a solution of problem (2.51). Now, we make the transformation

$$(2.58) \qquad\qquad\qquad \lambda = \frac{1}{\omega},$$

then we can write (2.57) in the form of the following eigenvalue problem in Ω: Find $\lambda \in \mathbb{C}$ for which there exists $\mathbf{u} \in \mathcal{H}, \mathbf{u} \neq \mathbf{0}$ such that

$$(2.59) \qquad [U_1 + U_2 \mathrm{ddiag}(m_j)Q]\mathbf{u} - \lambda\mathbf{u} + \lambda^2 U_2 \mathrm{ddiag}(k_j)Q\mathbf{u} = \mathbf{0}.$$

Thus we have transformed our original problem of this paragraph to (2.51) or (2.59) on which we concentrate our attention in the sequel. The equations (2.51) or (2.59) are obviously a quadratic eigenvalue problem and will be referred to as the *Stokes model*.

It is now time to study some properties of various operators appearing in (2.59). We begin by observing that U_1 is a very familiar operator in the study of Stokes equations. It is nothing but the inverse of the operator $(-2\nu \mathrm{div}\, e(\cdot))$ in the space $\mathcal{H}$. The following proposition summarizes the main properties of U_1 needed for our requirements. Its proof is classical and based on Rellich's Lemma.

Proposition 2.9. *The operator U_1 is a compact self-adjoint linear operator in $\mathcal{H}$ which is negative definite in the sense that*

$$(2.60) \qquad 2\nu \int_\Omega e(U_1\mathbf{u}) : e(\bar{\mathbf{u}}) dx = -\|\mathbf{u}\|_{0,\Omega}^2 < 0 \quad \forall \mathbf{u} \in \mathcal{H}, \ \mathbf{u} \neq \mathbf{0}. \ \blacksquare$$

Let us now proceed to study the operators U_1 and Q. They are clearly bounded linear and both are of finite rank. Identifying the Hilbert spaces

$\mathcal{H}$ and $\mathbb{C}^{2K}$ with their duals, $U_2^*: \mathcal{H} \longrightarrow \mathbb{C}^{2K}$, the operator adjoint to U_2, is defined by

(2.61) $\qquad \left(U_2^*(\mathbf{v}), \mathbf{s}\right)_K = \int_\Omega e(\mathbf{v}) : e(\overline{U_2 \mathbf{s}}) dx \quad \forall \mathbf{s} \in \mathbb{C}^{2K}, \, \forall \mathbf{v} \in \mathcal{H}.$

Similarly, $Q^*: \mathbb{C}^{2K} \longrightarrow \mathcal{H}$ is defined by

(2.62) $\qquad \int_\Omega e(Q^*\mathbf{s}) : e(\bar{\mathbf{v}}) dx = \left(\mathbf{s}, Q\mathbf{v}\right)_K \quad \forall \mathbf{s} \in \mathbb{C}^{2K}, \, \forall \mathbf{v} \in \mathcal{H}.$

Proposition 2.10. *The operators* U_2, Q *and their adjoints are related by*

$$U_2^* = -\frac{1}{2\nu} Q, \quad Q^* = -2\nu U_2.$$

Proof. Let $\mathbf{v} \in \mathcal{H}$ and $\mathbf{s} \in \mathbb{C}^{2K}$ be given. From (2.56), we have

$$\left(U_2^* \mathbf{v}, \mathbf{s}\right)_K = -\frac{1}{2\nu} \sum_{j=1}^{K} \bar{s}_j \cdot \gamma_j(\mathbf{v})$$

which, according to the definition of Q, can be rewritten as

$$\left(U_2^* \mathbf{v}, \mathbf{s}\right)_K = -\frac{1}{2\nu}\left(Q\mathbf{v}, \mathbf{s}\right)_K.$$

This proves the relation $U_2^* = (-1/2\nu)Q$. The other one follows from the fact that $A = A^{**}$ for any bounded linear operator A. ∎

It is an immediate corollary of the above result that U_2 is injective since Q is obviously onto $\mathbb{C}^{2K}$. Hence the image F of U_2 is of dimension exactly $2K$. Next, we examine the operators $U_2 \mathrm{ddiag}(m_j)Q$ and $U_2 \mathrm{ddiag}(k_j)Q$ which occur in our eigenvalue problem (2.59). The following result highlights their properties:

Proposition 2.11. *The operators* $U_2 \mathrm{ddiag}(m_j)Q$ *and* $U_2 \mathrm{ddiag}(k_j)Q$ *are both self-adjoint operators from* $\mathcal{H}$ *into* $\mathcal{H}$. *They have rank equal to* $2K$ *and more precisely, their images coincide with that of* U_2:

(2.63) $\qquad \mathcal{R}\left(U_2 \mathrm{ddiag}(m_j)Q\right) = \mathcal{R}\left(U_2 \mathrm{ddiag}(k_j)Q\right) = \mathcal{R}(U_2) = F.$

These operators are non-positive in the sense that the following inequalities hold:

(2.64) $\qquad \begin{cases} 2\nu \displaystyle\int_\Omega e\left(U_2 \mathrm{ddiag}(m_j)Q\mathbf{v}\right) : e(\bar{\mathbf{v}}) dx = -\sum_{j=1}^{K} m_j |\gamma_j(\mathbf{v})|^2 \leq 0, \\[4mm] 2\nu \displaystyle\int_\Omega e\left(U_2 \mathrm{ddiag}(k_j)Q\mathbf{v}\right) : e(\bar{\mathbf{v}}) dx = -\sum_{j=1}^{K} k_j |\gamma_j(\mathbf{v})|^2 \leq 0 \end{cases}$

for all $\mathbf{v}$ *in* $\mathcal{H}$.

Proof. The self-adjointness of these operators is an obvious consequence of Proposition 2.10. Since the matrices $\mathrm{ddiag}(m_j)$ and $\mathrm{ddiag}(k_j)$ are non-singular, we get

$$\mathcal{R}\big(\mathrm{ddiag}(m_j)Q\big) = \mathcal{R}\big(\mathrm{ddiag}(k_j)Q\big) = \mathcal{R}(Q) = \mathbb{C}^{2K},$$

which clearly implies (2.63). Finally, to prove (2.64), let us set

$$\mathbf{w} = U_2\mathrm{ddiag}(m_j)Q\mathbf{v}.$$

By definition, it satisfies

$$2\nu \int_\Omega e(\mathbf{w}) : e(\mathbf{z})dx = -\sum_{j=1}^{K} m_j\gamma_j(\mathbf{v}) \cdot \gamma_j(\bar{\mathbf{z}}) \quad \forall \mathbf{z} \in \mathcal{H}.$$

This gives the first inequality if we take $\mathbf{z} = \mathbf{v}$. The second one is similar. This completes the proof of Proposition 2.11. ∎

With the preparatory material out of our way, we are now ready to introduce the Green's operator U associated with the Stokes model which will allow us to describe the spectrum of problem (2.59). To this end, we begin by putting (2.59) in the form of a generalized eigenvalue problem corresponding to two linear operators A and B on the product space $\mathcal{H} \times F$. Let $(\lambda, \mathbf{u})$ be a solution of (2.59). Define $\mathbf{v}$ by

(2.65) $$\mathbf{v} = \mathbf{u} - \frac{1}{\lambda}\big[U_1 + U_2\mathrm{ddiag}(m_j)Q\big]\mathbf{u}.$$

Then (2.59) can be written as

(2.66) $$\mathbf{v} = \lambda U_2\mathrm{ddiag}(k_j)Q\mathbf{u}.$$

This shows that $\mathbf{v}$ belongs to the finite dimensional space F (see (2.63)). Now evidently, (2.65) and (2.66) can be clubbed into one vector equation:

(2.67) $$\begin{bmatrix} U_1 + U_2\mathrm{ddiag}(m_j)Q & 0 \\ 0 & I_F \end{bmatrix} \begin{bmatrix} \mathbf{u} \\ \mathbf{v} \end{bmatrix} = \lambda \begin{bmatrix} I_H & -i_F \\ U_2\mathrm{ddiag}(k_j)Q & 0 \end{bmatrix} \begin{bmatrix} \mathbf{u} \\ \mathbf{v} \end{bmatrix},$$

where I_H and I_F are the identity operators in H and F respectively and i_F denotes the canonical embedding of F into $\mathcal{H}$.

Conversely, it is straightforward to see that if $(\lambda, \mathbf{u}, \mathbf{v}) \in \mathbb{C} \times \mathcal{H} \times F$ satisfies (2.67), then $(\lambda, \mathbf{u})$ is a solution of (2.59) and $\mathbf{v}$ satisfies (2.65). Thus (2.59) and (2.67) are equivalent.

Let us next define the operators $A, B : \mathcal{H} \times F \longrightarrow \mathcal{H} \times F$ by

(2.68) $$A[\mathbf{u}, \mathbf{v}] = \begin{bmatrix} U_1 + U_2\mathrm{ddiag}(m_j)Q & 0 \\ 0 & I_F \end{bmatrix} \begin{bmatrix} \mathbf{u} \\ \mathbf{v} \end{bmatrix},$$

$$(2.69) \qquad B[\mathbf{u}, \mathbf{v}] = \begin{bmatrix} I_H & -i_F \\ U_2 \mathrm{ddiag}(k_j)Q & 0 \end{bmatrix} \begin{bmatrix} \mathbf{u} \\ \mathbf{v} \end{bmatrix}.$$

Using these notations, (2.67) can be recast in the form of the following generalized eigenvalue problem associated with the operators A and B: Find
$\lambda \in \mathbb{C}$ for which there exists $[\mathbf{u}, \mathbf{v}] \in \mathcal{H} \times F, [\mathbf{u}, \mathbf{v}] \neq [0, 0]$ such that

$$(2.70) \qquad A[\mathbf{u}, \mathbf{v}] = \lambda B[\mathbf{u}, \mathbf{v}].$$

We are now going to prove the following proposition which allows us to reduce
(2.70) into a *standard eigenvalue problem*:

Proposition 2.12. *The operator B has a linear continuous inverse B^{-1}.*

Proof. Let us write the operator B as a perturbation of the identity:

$$(2.71) \qquad B = I_{\mathcal{H} \times F} - D,$$

where the operator $D \colon \mathcal{H} \times F \longrightarrow \mathcal{H} \times F$ is given by

$$(2.72) \qquad D[\mathbf{u}, \mathbf{v}] = \begin{bmatrix} 0 & i_F \\ -T_2 \mathrm{ddiag}(k_j)Q & I_F \end{bmatrix} \begin{bmatrix} \mathbf{u} \\ \mathbf{v} \end{bmatrix}.$$

Since the image of D is $F \times F$, D has finite rank and the dimension of its
image is $4K$. Thus D is compact and so B can be regarded as a compact
perturbation of the identity operator in $\mathcal{H} \times F$. By virtue of the Fredholm's
Alternative (see Theorem 3.13 of Chapter I), to prove that B is bijective it
is enough to show that B is injective.

To this end, let $[\mathbf{u}, \mathbf{v}] \in \mathcal{H} \times F$ be such that $B[\mathbf{u}, \mathbf{v}] = [0, 0]$. According
to the definition of B, this means that

$$(2.73) \qquad \begin{cases} \mathbf{u} - \mathbf{v} = 0, \\ U_2 \mathrm{ddiag}(k_j)Q\mathbf{u} = 0. \end{cases}$$

Since U_2 and $\mathrm{ddiag}(k_j)$ are injective, the second condition in (2.73) implies
that $Q\mathbf{u} = 0$; hence $\mathbf{u} \in H_0^1(\Omega)^2$. This shows indeed that

$$\mathcal{N}[U_2 \mathrm{ddiag}(k_j)Q] \text{ (kernel of } U_2 \mathrm{ddiag}(k_j)Q) = H_0^1(\Omega)^2.$$

Since this operator is self-adjoint, the orthogonal complement of $H_0^1(\Omega)^2$ in
$\mathcal{H}$ is nothing but its image, i.e., F (see Proposition 2.11). Hence necessarily
$H_0^1(\Omega)^2 \cap F = \{0\}$. On the other hand, the first condition in (2.73) implies
that $\mathbf{u} \in F$ and therefore $\mathbf{u} = 0$. This finishes the proof of the injectivity as
well as the bijectivity of B.

Now, to conclude that B^{-1} is continuous, it is enough to apply the classical Open Mapping Theorem. ∎

By using the foregoing result, it follows that problem (2.70) is in fact equivalent to the following eigenvalue problem: Find $\lambda \in \mathbb{C}$ for which there exists $[\mathbf{u}, \mathbf{v}] \in \mathcal{H} \times F, [\mathbf{u}, \mathbf{v}] \neq [\mathbf{0}, \mathbf{0}]$ such that

$$(2.74) \qquad\qquad U[\mathbf{u}, \mathbf{v}] = \lambda[\mathbf{u}, \mathbf{v}],$$

where the operator U is defined by

$$(2.75) \qquad\qquad U = B^{-1}A.$$

Since U_1 and U_2ddiag$(m_j)Q$ are compact (see Propositions 2.9 and 2.11), A is compact. Thus U is compact too since B^{-1} is continuous (see Proposition 3.10 of Chapter I).

Applying the spectral theory of compact operators (see Paragraph 3.3 of Chapter I), we see that the spectrum of U consists of a countable sequence of complex numbers whose only possible accumulation point is zero. It is a consequence that 0 is not an eigenvalue of U due to the inequalities (2.60) and (2.64). Thus all eigenvalues of U have finite multiplicities and we denote them by

$$(2.76) \qquad\qquad \lambda_1, ..., \lambda_n, ...,$$

where, as usual, the eigenvalues are repeated according to their multiplicities. We remark that $\bar{\lambda}$ is an eigenvalue if λ is so. The corresponding eigenvectors of U are denoted by

$$(2.77) \qquad\qquad [\mathbf{u}_1, \mathbf{v}_1], ..., [\mathbf{u}_n, \mathbf{v}_n], ...$$

We conclude this paragraph with the main result which describes the solutions of the Stokes model. We pose

$$(2.78) \qquad\qquad \omega_n = \frac{1}{\lambda_n} \quad n \geq 1.$$

Theorem 2.13. *The pairs* $\{(\omega_n, \mathbf{u}_n)\}_n$ *satisfy:*
(i) *For each* $n \geq 1$, $(\omega_n, \mathbf{u}_n)$ *is a solution of* (2.51).
(ii) *These are all the solutions of* (2.51), *i.e., if* $(\omega, \mathbf{u}) \in \mathbb{C} \times \mathcal{H}$ *is any solution of* (2.51), *then there exists at least one* $n \geq 1$ *such that* $\omega = \omega_n$ *and* $\mathbf{u}$ *is a linear combination of all* $\mathbf{u}_\ell$'s *for which* $\omega_\ell = \omega$.
(iii) *There exists a subsequence* $\{\omega_{n'}\}$ *of the set of eigenvalues* $\{\omega_n\}$ *such that*

$$(2.79) \qquad\qquad \omega_{n'} \in \mathbb{R} \quad and \quad \lim_{n' \to \infty} \omega_{n'} = -\infty.$$

Proof. It remains only to prove (iii). To this end, let ε be a non-negative parameter and define the operator $S_\varepsilon \colon \mathcal{H} \longrightarrow \mathcal{H}$ by

$$(2.80) \qquad S_\varepsilon = \left[U_1 + U_2\mathrm{ddiag}(m_j)Q\right] + \varepsilon U_2\mathrm{ddiag}(k_j)Q.$$

By virtue of Propositions 2.9 and 2.11, it is clear that S_ε is compact, self-adjoint and negative definite. Denote its eigenvalues by

$$\Lambda_1(\varepsilon) \le \cdots \le \Lambda_n(\varepsilon) \le \cdots \longrightarrow 0.$$

Since the dependence of S_ε is analytical on ε, using Corollary 4.2 of Chapter I we see that $\Lambda_\ell(\varepsilon)$ defines a continuous function of $\varepsilon \ge 0$. Let us regard S_ε as a perturbation of $[U_1 + U_2\mathrm{ddiag}(m_j)Q]$ which is compact, self-adjoint and negative definite. Its eigenvalues form then a sequence $\{\alpha_\ell\}$ of negative numbers converging to zero. They are, as usual, arranged such that

$$(2.81) \qquad \alpha_1 \le \cdots \le \alpha_\ell \le \cdots \longrightarrow 0.$$

Since S_ε is a perturbation of $[U_1 + U_2\mathrm{ddiag}(m_j)Q]$ by an operator which is non-positive and of rank $2K$, $\Lambda_\ell(\varepsilon)$'s are non-increasing functions of ε and they verify the following interlacing inequalities with the eigenvalues $\{\alpha_\ell\}$ (see Theorem 4.3 of Chapter I):

$$(2.82) \qquad \begin{cases} \alpha_{\ell-2K} \le \Lambda_\ell(\varepsilon) \le \alpha_\ell & \forall \ell \ge 2K+1, \\ \Lambda_\ell(\varepsilon) \le \alpha_1 & \forall \ell = 1,...,2K, \end{cases}$$

for all $\varepsilon \ge 0$.

Now let us fix $\ell \ge 2K + 1$. From (2.82), it follows that the graphs of the functions $\varepsilon \longmapsto \Lambda_\ell(\varepsilon)$ and $\varepsilon \longmapsto -\varepsilon^{1/2}$ intersect at least at one point, the first of which is denoted as ε_ℓ, i.e.,

$$(2.83) \qquad \varepsilon_\ell = \min\{\varepsilon \ge 0 \mid \Lambda_\ell(\varepsilon) = -\varepsilon^{1/2}\}.$$

Next we define $\lambda'_\ell = \Lambda_\ell(\varepsilon_\ell) = -\varepsilon_\ell^{1/2}$ and let $\mathbf{u}'_\ell$ be an eigenvector of S_{ε_ℓ} corresponding to the eigenvalue $\Lambda_\ell(\varepsilon_\ell)$. It is obvious that $(\lambda'_\ell, \mathbf{u}'_\ell)$ is a solution of (2.59). The above process provides then an infinite sequence $\{(\lambda'_\ell, \mathbf{u}'_\ell)\}_{\ell \ge 2K+1}$ of solutions of (2.59) such that λ'_ℓ is real and they satisfy

$$\alpha_{\ell-2K} \le \lambda'_\ell \le \alpha_\ell \quad \forall \ell \ge 2K + 1.$$

Since $\alpha_\ell \longrightarrow 0$, so does λ'_ℓ. Their inverses ω'_ℓ obviously satisfy the requirement (iii) of Theorem 2.13. This finishes the proof. ∎

The foregoing result describes the vibration frequencies of the Stokes model. In particular, they admit a subsequence of real values tending to $-\infty$ apart from containing other values which may have non-zero imaginary parts. The presence of infinitely many different frequencies is due to viscosity of the fluid. This is one of the main differences with respect to the Laplace model.

Comments

The properties of the Stokes model described is this paragraph were obtained by CONCA, DURAN and PLANCHARD [1992]. We could have applied the general abstract theory of quadratic eigenvalue problems developed in Paragraph 4.3 of Chapter I to reach the conclusions of the above result. However, for the sake of clarity, we preferred to go through the details of the construction of real solutions. ∎

◇ **Exercise 2.2.** (Green's operator for the compressible Stokes model) The goal of this exercise is to transform (in a formal manner) system (1.26) of the compressible Stokes model to a generalized eigenvalue problem. To this end, we adapt a procedure similar to the one followed in Paragraph 2.3. As usual, we first reformulate (1.26) to accommodate the frequency values in the set INF (see (2.47)).

(i) Show that ω is a vibration frequency of the compressible Stokes model iff there exists a non-trivial pair $(\mathbf{u}, p) \in H^1(\Omega)^2 \times L^2(\Omega)$ and $\mathbf{s} \in \mathbb{C}^{2K}$ such that $(\omega, \mathbf{u}, p, \mathbf{s})$ satisfies

(2.84)
$$
\begin{cases}
-\operatorname{div} \sigma(\mathbf{u}, p) + \omega \mathbf{u} = 0 & \text{in } \Omega, \\
\omega p + c^2 \operatorname{div} \mathbf{u} = 0 & \text{in } \Omega, \\
\mathbf{u} = \mathbf{s}_j & \text{on } \Gamma_j, \ j = 1, ..., K, \\
\mathbf{u} = 0 & \text{on } \Gamma_0, \\
(k_j + m_j \omega^2) \mathbf{s}_j = -\omega \displaystyle\int_{\Gamma_j} \sigma(\mathbf{u}, p) \mathbf{n} d\gamma & \forall j = 1, ..., K,
\end{cases}
$$

where $\sigma(\mathbf{u}, p)$ is the stress-tensor which is now given by (1.25). To obtain the variational formulation of the problem, introduce the space

(2.85) $\quad \mathcal{G} = \{\mathbf{v} \in H^1(\Omega)^2 \mid \mathbf{v} = 0 \text{ on } \Gamma_0, \ \mathbf{v} \text{ is constant on } \Gamma_j, \ j = 1, ..., K\}$

and consider the bilinear form on $\mathcal{G} \times \mathcal{G}$

(2.86) $\qquad a(\mathbf{u}, \mathbf{v}) = 2\nu \displaystyle\int_\Omega e(\mathbf{u}) : e(\bar{\mathbf{v}}) dx - \frac{2\nu}{3} \displaystyle\int_\Omega \operatorname{div} \mathbf{u} \operatorname{div} \bar{\mathbf{v}} dx.$

(ii) Show that there is a constant $\alpha > 0$ such that

(2.87) $\qquad\qquad\qquad a(\mathbf{u}, \mathbf{v}) \geq \alpha \|\mathbf{u}\|_{1,\Omega}^2 \quad \forall \mathbf{u} \in \mathcal{G}.$

(iii) Prove that $\omega = 0$ cannot be a vibration frequency admitted by (2.84) and that a variational formulation of (2.84) can be given as follows: Find $\omega \in \mathbb{C}, \mathbf{u} \in \mathcal{G}, \mathbf{u} \neq 0$ such that

$$
(2.88) \quad
\begin{cases}
a(\mathbf{u}, \mathbf{v}) + \omega \left[\int_\Omega \mathbf{u} \cdot \bar{\mathbf{v}} dx + \sum_{j=1}^{K} m_j \gamma_j(\mathbf{u}) \cdot \gamma_j(\bar{\mathbf{v}}) \right] + \\
\qquad + \frac{1}{\omega} \left[c^2 \int_\Omega \operatorname{div} \mathbf{u} \operatorname{div} \bar{\mathbf{v}} dx + \sum_{j=1}^{K} k_j \gamma_j(\mathbf{u}) \cdot \gamma_j(\bar{\mathbf{v}}) \right] = 0
\end{cases}
$$

for all $\mathbf{v} \in \mathcal{G}$. Once $(\omega, \mathbf{u})$ is found from (2.88), the pressure p is recovered from the second equation in (2.84) and s_j is obtained from the boundary condition on Γ_j in (2.84). Observe that (2.88) is a quadratic eigenvalue problem in ω.

(iv) With the idea of writing (2.88) in the form of a generalized eigenvalue problem, find suitable bounded linear operators $U_1 \in \mathcal{L}(\mathcal{G}, \mathcal{G})$, $U_2 \in \mathcal{L}(\mathbb{C}^{2K}, \mathcal{G})$, $U_3 \in \mathcal{L}(\mathcal{G}, \mathcal{G})$ and $Q \in \mathcal{L}(\mathcal{G}, \mathbb{C}^{2K})$ such that (2.88) can be recast in the following form:

$$
(2.89) \quad \left[U_1 + U_2 \operatorname{ddiag}(m_j)Q \right] \mathbf{u} + \frac{1}{\omega^2} \left[c^2 U_3 + U_2 \operatorname{ddiag}(k_j)Q \right] \mathbf{u} = \frac{1}{\omega} \mathbf{u}.
$$

Notice that the compressibility of the fluid is reflected by the presence of an additional term involving U_3 in (2.89) (compare with (2.59)).

(v) Make the transformation $\lambda = (1/\omega)$ and define operators $A, B \in \mathcal{L}(\mathcal{G} \times \mathcal{G}, \mathcal{G} \times \mathcal{G})$ in a such a way that (2.89) is transformed to the following generalized eigenvalue problem: Find $\lambda \in \mathbb{C}$ for which there exists $[\mathbf{u}, \mathbf{v}] \in \mathcal{G} \times \mathcal{G}, [\mathbf{u}, \mathbf{v}] \neq [0, 0]$ such that

$$
(2.90) \qquad\qquad A[\mathbf{u}, \mathbf{v}] = \lambda B[\mathbf{u}, \mathbf{v}].
$$

Is the operator A obtained compact? Is B invertible? Can one apply the method followed in the case of Stokes model to solve (2.90)? ∎

§3. Bounds on Eigenvalues

Having established in §2 existence of eigenvalues for various models of fluid-solid structures, our aim in this section is to study some of their qualitative properties. We will focus our attention on obtaining sharp upper and lower bounds for the eigenvalues which give precise information on their location in the real axis or in the complex plane as the case may be. Of course, we are interested in realistic bounds which can be easily computed. Most of the bounds are nothing but *interlacing inequalities* which compare the vibration frequencies of the coupled fluid-solid structures with those of fluid or solid independently. We go further in our study and present some refined results on the location of eigenvalues. They give the exact number of eigenvalues in certain special intervals of $\mathbb{R}$. The main techniques to achieve this goal are based on mini-max characterization of eigenvalues (see Paragraph 3.3 of Chapter I), the general theory developed in §4 of Chapter I on the effects of perturbations and on some homotopy arguments to be presented below. The results of this section will be exploited in Chapter III to carry out the study of the asymptotic behaviour of vibrations of fluid-solid structures.

3.1. Bounds for the Laplace Model

Let us begin by recalling that the eigenvalues of the Laplace model (see (2.2)) are recovered from the characteristic values of the operator $\tilde{T} \in \mathcal{L}(\mathbb{C}^{2K})$ defined by (2.12) (see Theorems 2.1 and 2.3). The definition of $\tilde{T}$ involves another operator denoted by S which was introduced in (2.5). This operator S depends only on the geometry of the problem, namely on the domain Ω. According to Proposition 2.2, S is self-adjoint and positive definite and so the standard quantities $m_- = m_-(S)$ (smallest eigenvalue of S) and $m_+ = m_+(S)$ (largest eigenvalue of S) are both positive. We announce now the first result providing some estimates on the eigenvalues of the Laplace model.

Theorem 3.1. *Let $\{\omega_\ell^2\}_{\ell=1}^{2K}$ be the eigenvalues of the Laplace model (2.2) given by Theorem 2.3. Then for all $\ell = 1, ..., 2K$, we have*

$$(3.1) \qquad 0 < \frac{k_\ell}{m_\ell + \rho_0 m_+} \leq \omega_\ell^2 \leq \frac{k_\ell}{m_\ell + \rho_0 m_-} < \frac{k_\ell}{m_\ell}.$$

Proof. Since $m_- I \leq S \leq m_+ I$, we have the following estimate for $\tilde{T}$ via (2.12):

$$\mathrm{ddiag}(\rho_0 k_\ell^{-1} m_- + m_\ell k_\ell^{-1}) \leq \tilde{T} \leq \mathrm{ddiag}(\rho_0 k_\ell^{-1} m_+ + m_\ell k_\ell^{-1}).$$

Therefore the eigenvalues $\{\mu_\ell\}_{\ell=1}^{2K}$ of $\tilde{T}$ satisfy

$$k_\ell^{-1}(m_\ell + \rho_0 m_-) \leq \mu_\ell \leq k_\ell^{-1}(m_\ell + \rho_0 m_+) \quad \forall \ell = 1, ..., 2K.$$

These inequalities are nothing but (3.1) if we recall that ω_ℓ^2 is the inverse of μ_ℓ. ∎

Remembering that k_ℓ is the vibration frequency of the ℓth tube, we see that the upper bound in (3.1) signifies that the ℓth vibration frequency of the fluid-solid structure cannot exceed that of the ℓth tube in the vacuum.

This conclusion seems physically reasonable. In the case of identical tubes (see Exercise 2.1), i.e., $k_1 = \cdots = k_K = k$ and $m_1 = \cdots = m_K = m$, the inequalities (3.1) become

$$(3.2) \qquad 0 < \frac{k}{m + \rho_0 m_+} \leq \omega_\ell^2 \leq \frac{k}{m + \rho_0 m_-} < \frac{k}{m} \quad \forall \ell = 1, ..., 2K.$$

In the case of an arbitrary geometry of Ω, the quantity $m_+ = m_+(S)$ can be as large as we please and therefore the lower bound in (3.1) can approach zero. For special geometries, however, one can find a uniform lower bound which is bounded away from zero. We will see examples in Chapter III.

We now examine the monotonicity of the largest vibration frequency of the Laplace model with respect to the number of tubes. Consider the case of identical tubes with parameters k and m in the cavity Ω_0. Remember that the fluid occupies the domain Ω defined by

$$(3.3) \qquad \Omega = \Omega_0 \backslash \bigcup_{j=1}^{K} \bar{\mathcal{O}}_j,$$

where the $\mathcal{O}_j$'s represent sections of the tubes. Let us now introduce a new tube with cross section denoted by $\mathcal{O}_{K+1}$ such that $\bar{\mathcal{O}}_{K+1} \subset \Omega$. Call $\tilde{\Omega}$ the new region occupied by the fluid:

$$(3.4) \qquad \tilde{\Omega} = \Omega_0 \backslash \bigcup_{j=1}^{K+1} \bar{\mathcal{O}}_j.$$

It is assumed that $\{\mathcal{O}_j\}_{j=1}^{K+1}$ are all connected bounded open sets, pair-wise disjoint, and that have locally Lipschitz boundaries.

Theorem 3.2. *Let $\omega_{\max}^2$ (respectively $\tilde{\omega}_{\max}^2$) be the greatest eigenvalue of the Laplace model considered in Ω (respectively in $\tilde{\Omega}$). Then*

$$(3.5) \qquad \omega_{\max}^2 \leq \tilde{\omega}_{\max}^2,$$

i.e., the greatest eigenvalue in the Laplace model increases with K provided Ω_0 remains unchanged.

Proof. By the Courant-Fischer principle (see Theorem 3.18 of Chapter I), we have

$$(3.6) \qquad \tilde{\lambda}_{\max} = \max_{\tilde{s} \in \mathbf{C}^{2(K+1)}} \frac{(\tilde{S}\tilde{s}, \tilde{s})_{K+1}}{\|\tilde{s}\|_{K+1}^2},$$

where $\tilde{S}$ is the operator associated with the Laplace model on $\tilde{\Omega}$ (see (2.5) for its definition) and $\tilde{\lambda}_{\max}$ is its greatest eigenvalue. For each $\tilde{s} \in \mathbb{C}^{2(K+1)}$, let $\tilde{\phi}$ be the unique solution of

$$(3.7) \qquad \begin{cases} \tilde{\phi} \in \dot{H}^1(\tilde{\Omega}), \\ \displaystyle\int_{\tilde{\Omega}} \nabla\tilde{\phi} \cdot \nabla\bar{\psi} dx = \sum_{j=1}^{K+1} \tilde{s}_j \cdot \int_{\Gamma_j} \bar{\psi} n d\gamma \quad \forall \psi \in \dot{H}^1(\tilde{\Omega}). \end{cases}$$

By the definition of $\tilde{S}$, if we take $\psi = \tilde{\phi}$ in (3.7) it follows that

$$\left(\tilde{S}\tilde{s}, \tilde{s}\right)_{K+1} = \int_{\tilde{\Omega}} |\nabla\tilde{\phi}|^2 dx.$$

Hence, choosing $\tilde{s} \in \mathbb{C}^{2(K+1)}$ in the form $(s, 0)$ with $s \in \mathbb{C}^{2K}$, we have

$$(3.8) \qquad \tilde{\lambda}_{\max} \geq \max_{s \in \mathbb{C}^{2K}} \frac{1}{\|s\|_K^2} \int_{\tilde{\Omega}} |\nabla\tilde{\phi}|^2 dx.$$

On the other hand, for each $s \in \mathbb{C}^{2K}$, let ϕ be the unique solution of

$$(3.9) \qquad \begin{cases} \phi \in \dot{H}^1(\Omega), \\ \displaystyle\int_{\Omega} \nabla\phi \cdot \nabla\bar{\psi} dx = \sum_{j=1}^{K} s_j \cdot \int_{\Gamma_j} \bar{\psi} n d\gamma \quad \forall \psi \in \dot{H}^1(\Omega). \end{cases}$$

Taking ϕ as a test function in (3.7), which is possible because $\tilde{\Omega} \subset \Omega$, and using that $s_{K+1} = 0$, we get

$$\int_{\tilde{\Omega}} \nabla\tilde{\phi} \cdot \nabla\bar{\phi} dx = \sum_{j=1}^{K} s_j \cdot \int_{\Gamma_j} \bar{\phi} n d\gamma = \int_{\Omega} |\nabla\phi|^2 dx$$

and by Cauchy-Schwarz's Inequality, we conclude that

$$\int_{\Omega} |\nabla\phi|^2 dx \leq \int_{\tilde{\Omega}} |\nabla\tilde{\phi}|^2 dx.$$

Combining the above inequality with (3.8), we obtain

$$\tilde{\lambda}_{\max} \geq \max_{s \in \mathbb{C}^{2K}} \frac{1}{\|s\|_K^2} \int_{\Omega} |\nabla\phi|^2 dx = \max_{s \in \mathbb{C}^{2K}} \frac{(Ss, s)_K}{\|s\|_K^2} = \lambda_{\max}$$

where $\lambda_{\max}$ obviously denotes the greatest eigenvalue of S.

To conclude, it suffices to observe that the following relation connecting the eigenvalues λ's of S and the vibration frequencies ω's of the Laplace model in Ω holds in case of identical tubes:

$$\lambda = \frac{\rho_0 \omega^2}{k - m\omega^2}. \tag{3.10}$$

Of course, a similar relation holds between $\tilde{\lambda}$'s and $\tilde{\omega}$'s. They are valid because ω^2 is never equal to (k/m) (see Exercise 2.1 or Theorem 3.1). ∎

If we increase the number of tubes in the bundle immersed in the fluid, it is reasonable to expect that the dominant frequency of the coupled system increases. This is reflected in (3.5).

Let us now present a comparison result for the first eigenvalue of the Laplace model. More precisely, we compare the first eigenvalue with a single tube and that corresponding to multiple tubes. We do this only in the particular configuration of tubes described below.

Let $Y =]0,1[^2$ be the reference cell in $\mathbb{R}^2$ and $\mathcal{O}$ be a connected open set with a locally Lipschitz boundary Γ, such that $\bar{\mathcal{O}} \subset Y$. We put $Y^* = Y\backslash\bar{\mathcal{O}}$ and let $\{Y_m^*\}_{m\in I}$ be a finite family of cells which are translates of Y^*(i.e., $Y_m^* = Y^* + \mathbf{a}_m$ for some $\mathbf{a}_m \in \mathbb{R}^2$) such that they are mutually disjoint. We denote by K the cardinality of I and define (see Figure 3.1)

$$\tilde{\Omega} = \text{int}\left(\bigcup_{m\in I} \bar{Y}_m^* \right).$$

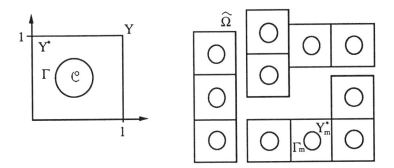

Figure 3.1: *The reference cell Y^* and the domain $\tilde{\Omega}$.*

Theorem 3.3. *Let $\tilde{\omega}^2_{\min}$ (respectively $\omega_{\min}$) be the lowest eigenvalue of the Laplace model considered in Ω (respectively in Y^*). Then*

$$(3.11) \qquad\qquad\qquad \tilde{\omega}^2_{\min} \le \omega^2_{\min}.$$

Proof. By the Courant-Fischer characterization of eigenvalues, we have

$$(3.12) \qquad\qquad \lambda_{\min} = \min_{s \in \mathbb{C}^2} \frac{1}{\|s\|^2_1} \int_{Y^*} |\nabla \psi|^2 dx,$$

where ψ is the unique solution of

$$(3.13) \qquad\qquad \begin{cases} \psi \in \dot{H}^1(Y^*), \\ \displaystyle\int_{Y^*} \nabla \psi \cdot \nabla \bar{\varphi} dx = s \cdot \int_\Gamma \bar{\varphi} n d\gamma \quad \forall \varphi \in \dot{H}^1(Y^*). \end{cases}$$

Here, $\lambda_{\min}$ is the first eigenvalue of the operator S associated with the Laplace model on Y^*. Analogously, we have

$$(3.14) \qquad\qquad \tilde{\lambda}_{\min} = \min_{t \in \mathbb{C}^{2K}} \frac{1}{\|t\|^2_K} \int_{\tilde{\Omega}} |\nabla \phi|^2 dx,$$

where ϕ is the unique solution of

$$(3.15) \qquad\qquad \begin{cases} \phi \in \dot{H}^1(\tilde{\Omega}), \\ \displaystyle\int_\Omega \nabla \phi \cdot \nabla \bar{\varphi} dx = \sum_{m \in I} t_m \cdot \int_{\Gamma_m} \bar{\varphi} n d\gamma \quad \forall \varphi \in \dot{H}^1(\tilde{\Omega}). \end{cases}$$

Here $\tilde{\lambda}_{\min}$ is the first eigenvalue of the operator $\tilde{S}$ associated with the Laplace model on $\tilde{\Omega}$. Let $s_{\min} \in \mathbb{C}^2$ be the point at which the minimum of (3.12) is attained and $\psi_{\min}$ be the corresponding solution of (3.13). It is clear that $\lambda_{\min}$ is also the first eigenvalue of the Laplace model in Y^*_m for all $m \in I$. The corresponding eigenvector is $s_{\min}$. The associated function ψ_m, solution of (3.13) posed on Y^*_m, is just the translate of $\psi_{\min}$ by a_m:

$$\psi_m(\cdot) = \psi(\cdot + a_m).$$

Choose now $t = (s_{\min}, ..., s_{\min}) \in \mathbb{C}^{2K}$ in (3.14). The corresponding solution ϕ of (3.15) is taken as a test function in the variational problem defining ψ_m. We then have

$$\int_{Y^*_m} \nabla \psi_m \cdot \nabla \bar{\phi} dx = s_{\min} \cdot \int_{\Gamma_m} \bar{\phi} n d\gamma \quad \forall m \in I.$$

Summing up over $m \in I$ and using (3.15), we obtain

$$\sum_{m \in I} \int_{Y_m^*} \nabla \psi_m \cdot \nabla \bar{\phi} dx = \sum_{m \in I} \mathbf{s}_{\min} \cdot \int_{\Gamma_m} \bar{\phi} \mathbf{n} d\gamma = \sum_{m \in I} \mathbf{t}_m \cdot \int_{\Gamma_m} \bar{\phi} \mathbf{n} d\gamma = \int_{\tilde{\Omega}} |\nabla \phi|^2 dx.$$

By Cauchy-Schwarz's Inequality, we conclude

$$\int_{\tilde{\Omega}} |\nabla \phi|^2 dx \leq \sum_{m \in I} \int_{Y_m^*} |\nabla \psi_m|^2 dx = K \int_{Y^*} |\nabla \psi_{\min}|^2 dx =$$

$$= K \lambda_{\min} \|\mathbf{s}_{\min}\|_1^2 = \lambda_{\min} \|\mathbf{t}\|_K^2.$$

Thus, using (3.14), we conclude that

$$\tilde{\lambda}_{\min} \leq \lambda_{\min}.$$

To finish the proof, it is enough to use (3.10) which links the eigenvalues of S with the vibration frequencies of the Laplace model. ■

3.2. Interlacing Inequalities for the Helmholtz Model

The object of this paragraph is to present two kinds of bounds for the eigenvalues of the Helmholtz model, the existence of which is known from Theorem 2.8. The idea consists in comparing the set of eigenvalues of the Helmholtz model with the set consisting of those of the Laplace model and the Neumann eigenvalues associated with Laplace operator in Ω. The first result shows that there is a shift between both sets of eigenvalues which is of at most $2K$ positions; K being the number of tubes in Ω and the shift being understood in the sense of interlacing inequalities. The second one estimates the difference between the values of the eigenvalues in both sets.

We have seen in Paragraph 2.2 that the eigenvalue problem (2.17) of the Helmholtz model includes, in a certain sense, two other eigenvalue problems mentioned above, namely the Neumann eigenvalue problem defined by R_{11} and that of the Laplace model defined by R_{22} (see in particular (2.33)). It is therefore natural to compare the eigenvalues of these two sets. This is precisely the aim of this paragraph.

The main technique used to achieve this goal is the tool developed in §4 of Chapter I on perturbations. The main results of this paragraph are, as we shall see, easy consequences of Theorems 4.1 and 4.3 of Chapter I.

Let us begin by recalling that the eigenvalues in the Helmholtz model are obtained as characteristic values of the operator $\tilde{R}$ defined in (2.33) (see Theorem 2.4 and Proposition 2.6):

$$\tilde{R} = \begin{bmatrix} R_{11} & \rho_0^{1/2} c \, R_{12} \text{ddiag}(k_j^{-1/2}) \\ \rho_0^{1/2} c \, \text{ddiag}(k_j^{-1/2}) R_{21} & \tilde{R}_{22} \end{bmatrix},$$

where the operators R_{11}, R_{12}, R_{21} and $\tilde{R}_{22}$ are defined in (2.25), (2.26), (2.27) and (2.31) respectively. In order to apply Theorem 4.3 of Chapter I on the interlacing inequalities concerning perturbations by operators of finite rank, we make the following choice:

$$
(3.16) \quad
\begin{cases}
H = L_0^2(\Omega) \times \mathbb{C}^{2K}, \\[2mm]
A = \begin{bmatrix} R_{11} & 0 \\ 0 & \tilde{R}_{22} \end{bmatrix}, \\[4mm]
B = \begin{bmatrix} 0 & \rho_0^{1/2} c\, R_{12}\mathrm{ddiag}(k_j^{-1/2}) \\ \rho_0^{1/2} c\,\mathrm{ddiag}(k_j^{-1/2})R_{21} & 0 \end{bmatrix}.
\end{cases}
$$

Since the operators R_{11} and $\tilde{R}_{22}$ are self-adjoint and compact (in fact, $\tilde{R}_{22}$ is of finite rank), we see that A is compact and self-adjoint. Let us denote its characteristic values by

$$
(3.17) \qquad 0 < \nu_1^2 \le \cdots \le \nu_\ell^2 \le \cdots \longrightarrow \infty.
$$

They contain the Neumann eigenvalues of Ω (i.e., characteristic values of R_{11}) and those corresponding to the Laplace model (i.e., characteristic values of $\tilde{R}_{22}$). Concerning the operator B, we must verify that B satisfies the hypotheses of Theorem 4.3 of Chapter I, namely that it is a self-adjoint operator of finite rank. The purpose of the following proposition is to establish these facts:

Proposition 3.4. *Let B be defined as in* (3.16). *Then B is a self-adjoint operator and its rank is $4K$. Moreover, there exists a set of $2K$ strictly positive numbers $0 < \xi_1^2 \le \cdots \le \xi_{2K}^2$ such that ξ_i and $(-\xi_i)$ are precisely the characteristic values of B.*

Proof. The self-adjointness of B is an immediate consequence of the fact that $R_{12} = R_{21}^*$ (see Proposition 2.5). Let us now proceed to prove that its rank is $4K$. Let $\xi \in \mathbb{C}$ be a characteristic value of B with corresponding eigenvector $(\phi, s) \in L_0^2(\Omega) \times \mathbb{C}^{2K}$. By the definition of B, this means that

$$
\begin{cases}
\rho_0^{1/2} c\, R_{12}\mathrm{ddiag}(k_j^{-1/2})s = \xi^{-1}\phi, \\
\rho_0^{1/2} c\,\mathrm{ddiag}(k_j^{-1/2})R_{21}\phi = \xi^{-1}s.
\end{cases}
$$

Applying the operator $\rho_0^{1/2} c\,\mathrm{ddiag}(k_j^{-1/2})R_{21}$ to the first equation above, we get

$$
(3.18) \qquad\qquad C s = \xi^{-2} s,
$$

where the operator $C: \mathbb{C}^{2K} \longrightarrow \mathbb{C}^{2K}$ is defined as follows:

$$
(3.19) \qquad C = \rho_0 c^2 \mathrm{ddiag}(k_j^{-1/2})R_{21}R_{12}\mathrm{ddiag}(k_j^{-1/2}).
$$

Thus ξ^2 is a characteristic value of C and s is a corresponding eigenvector.

Conversely, let $\xi^2 \in \mathbb{C}$ be a characteristic value of C with corresponding eigenvector $s \in \mathbb{C}^{2K}$. Since $R_{12}^* = R_{21}$, we see that C is self-adjoint and so $\xi^2 \in \mathbb{R}$. Moreover, C is easily seen to be positive definite. Thus $\xi^2 > 0$. Taking

(3.20)
$$\begin{cases} \xi = +\sqrt{\xi^2}, \\ \phi = \xi \rho_0^{1/2} R_{12} \mathrm{ddiag}(k_j^{-1/2})s, \end{cases}$$

we verify that (ϕ, s) provides an eigenvector of B with characteristic value ξ. In a similar manner, we see that $(-\xi)$ is also a characteristic value of B with eigenvector $(-\phi, s)$. Thus the characteristic values of B are nothing but the square roots of those of C. Denoting by $0 < \xi_1^2 \leq \cdots \leq \xi_{2K}^2$ the characteristic values of C, we conclude the proof of Proposition 3.4. ∎

$\Diamond$ **Exercise 3.1.** (Interpretation of the spectral equation of C) Show that $(\xi^2, s) \in \mathbb{R}^+ \times \mathbb{C}^{2K}$ satisfies (3.18) iff

$$s_j = \frac{\rho_0 c^2}{k_j - m_j \xi^2} \int_{\Gamma_j} u \, n \, d\gamma \quad \forall j = 1, \dots, K,$$

where u is the unique solution of

$$\begin{cases} \Delta^2 u = 0 \quad \text{in} \quad \Omega, \\ \dfrac{\partial u}{\partial n} = 0 \quad \text{on} \quad \Gamma_0 \cup \Gamma_1 \cup \cdots \cup \Gamma_K, \\ c^2 \dfrac{\partial \Delta u}{\partial n} = \dfrac{\rho_0 \xi^2}{k_j - m_j \xi^2} \left(\int_{\Gamma_j} u \, n \, d\gamma \right) \cdot n \quad \text{on} \quad \Gamma_j, \ j = 1, \dots, K, \\ \dfrac{\partial \Delta u}{\partial n} = 0 \quad \text{on} \quad \Gamma_0. \end{cases} \ \blacksquare$$

Thus, knowing that there are $2K$ positive and $2K$ negative eigenvalues for B, we are in a position to apply Theorem 4.3 of Chapter I. A word of caution is however not bad. Theorem 4.3 was phrased in terms of eigenvalues of the operators in question and we now apply it to their characteristic values.

Theorem 3.5. *Let* $\{\omega_\ell^2\}$ *be the eigenvalues of the Helmholtz model* (2.17). *They satisfy the following interlacing inequalities with respect to the eigenvalues* $\{\nu_\ell^2\}$:

(3.21)
$$\begin{cases} \nu_{\ell-2K}^2 \leq \omega_\ell^2 \leq \nu_{\ell+2K}^2 \quad \forall \ell \geq 2K+1, \\ \omega_\ell^2 \leq \nu_{\ell+2K}^2 \quad \forall \ell = 1, \dots, 2K. \ \blacksquare \end{cases}$$

Let us denote by (k/m) the following quantity:

$$\frac{k}{m} \overset{\text{def}}{=} \max\left\{\frac{k_j}{m_j} \mid 1 \le j \le K\right\}.$$

Of course $(k/m) > 0$. Let N be the total number of eigenvalues ν_ℓ^2 that lie in the interval $]0, (k/m)[$. By the result of Theorem 3.1, we have the following lower bound for N:

$$(3.22) \qquad\qquad\qquad N \ge 2K.$$

Combining this inequality with Theorem 3.5, we obtain immediately that

Corollary 3.6. *Let N_H be the number of eigenvalues of the Helmholtz model (2.17) that lie in the interval* $]0, (k/m)[$. *Then*

$$(3.23) \qquad\qquad N - 2K \le N_H \le N + 2K. \; \blacksquare$$

This is a first result estimating the number of eigenvalues present in the interval $]0, (k/m)[$. A finer result in this direction will be seen in the next paragraph. To obtain a second result on estimating the difference between the eigenvalues $\{\omega_\ell^2\}$ and $\{\nu_\ell^2\}$, we can apply directly Theorem 4.1 of Chapter I with the same choice (3.16). Recalling that the greatest eigenvalue of B is ξ_1^{-1} according to Proposition 3.4, we then have

Theorem 3.7. *The following estimates are satisfied by the eigenvalues $\{\omega_\ell^2\}$ of the Helmholtz model and those given by (3.17):*

$$(3.24) \qquad\qquad \left|\frac{1}{\omega_\ell^2} - \frac{1}{\nu_\ell^2}\right| \le \frac{1}{\xi_1} \quad \forall \ell \ge 1. \; \blacksquare$$

$\diamond$ **Exercise 3.2.** Show, for each $\ell \ge 1$, that the ℓth Neumann eigenvalue in Ω cannot exceed the ℓth Neumann eigenvalue in Ω_0. $\blacksquare$

3.3. More on the Eigenvalues of the Helmholtz Model

We have already seen, in the earlier paragraph, two kinds of bounds on the eigenvalues of the Helmholtz model. These were deduced from the general results in the theory of perturbations. In particular, we could deduce bounds on the number of eigenvalues found in the interval $]0, (k/m)[$ (see Corollary 3.6.). In this paragraph, we go further in this direction and derive some refined results on the position of the eigenvalues.

Contrary to the nature of the results of the previous paragraph, those of this one will be rather special to the problem under consideration. The main result here (see Theorem 3.17) asserts that there are exactly $2K$ eigenvalues of the Helmholtz model in the interval $]0, (k/m)[$, where we recall that K is the number of tubes immersed in the fluid. While the earlier results on the Helmholtz eigenvalues were valid in general, the main theorem of this paragraph will be derived under a suitable geometric condition which we formulate below (see condition (3.52)). This calls for the introduction of a new eigenvalue problem in the domain Ω (see problem (3.25)).

Let us recall, from Paragraph 2.2, that the eigenvalues, if any, in the set NF (defined by (2.1)) have to be given special treatment. That is why we had to reformulate the initial problem in the form (2.17) to include the values in the set NF. One question which arises naturally in this context is the following: When can one say that there is no characteristic values of the associated operator R in the set NF? The answer to this question is related to the geometric condition referred to above. This will be explored in Theorem 3.9.

As we shall see below, the geometric condition will also enable us to pass from the spectral problem of the Helmholtz model to that of the Laplace model. This passage is achieved by introducing a positive parameter ε and by constructing a homotopy between these two cases (see Subparagraph 3.3.2). The passage requires a priori bounds on the eigenvalues with respect to the parameter ε. These are deduced in 3.3.3. Of course, many results remain valid without the geometric condition which is formulated in (3.52). Finally, we apply the various results derived in the earlier part of this paragraph to the spectrum of the Helmholtz model in Subparagraph 3.3.4.

3.3.1. A New Eigenvalue Problem

We start by introducing a new eigenvalue problem in Ω: Find $\beta \in \mathbb{C}$ for which there is a non-zero $\phi \in V$ and $\mathbf{s} \in \mathbb{C}^{2K}$ such that

(3.25)
$$\begin{cases} c^2 \Delta\phi + \beta^2 \phi = 0 & \text{in} \quad \Omega, \\ c^2 \dfrac{\partial\phi}{\partial n} = \mathbf{s}_j \cdot \mathbf{n} & \text{on} \quad \Gamma_j, \ j = 1, ..., K, \\ \dfrac{\partial\phi}{\partial n} = 0 & \text{on} \quad \Gamma_0, \\ \displaystyle\int_{\Gamma_j} \phi \mathbf{n} d\gamma = 0 & \forall j = 1, ..., K, \end{cases}$$

where as in Paragraph 2.2., we define $V = H^1(\Omega) \cap L_0^2(\Omega)$. In order to give a variational formulation of the above problem, we introduce the following space:

(3.26)
$$W = \left\{ \psi \in V \ \middle| \ \int_{\Gamma_j} \psi \mathbf{n} d\gamma = 0 \quad \forall j = 1, ..., K \right\}.$$

It will be convenient to recast the second condition in (3.25) in the following form:

$$(3.27) \qquad \int_{\Gamma_j} \frac{\partial \phi}{\partial n} \bar{\psi} d\gamma = 0 \quad \forall \psi \in W, \ j = 1, ..., K.$$

For $f \in L_0^2(\Omega)$, the following variational problem is readily solved uniquely using Lax-Milgram Lemma:

$$(3.28) \qquad \begin{cases} u \in W, \\ c^2 \int_\Omega \nabla u \cdot \nabla \bar{\psi} dx = \int_\Omega f \bar{\psi} dx \quad \forall \psi \in W. \end{cases}$$

As usual, the above variational equation can be interpreted as the following boundary value problem:

$$(3.29) \qquad \begin{cases} -c^2 \Delta u = f \quad \text{in} \quad \Omega, \\ \int_{\Gamma_j} \frac{\partial u}{\partial n} \bar{\psi} d\gamma = 0 \quad \forall \psi \in W, \ j = 1, ..., K, \\ \frac{\partial u}{\partial n} = 0 \quad \text{on} \quad \Gamma_0, \\ \int_{\Gamma_j} u n d\gamma = 0 \quad \forall j = 1, ..., K, \quad \int_\Omega u dx = 0. \end{cases}$$

Once again, we observe that the second condition can be replaced by the following equivalent one: There exists $s \in \mathbb{C}^{2K}$ such that

$$(3.30) \qquad \frac{\partial u}{\partial n} = s_j \cdot n \quad \text{on} \quad \Gamma_j, \ j = 1, ..., K.$$

Associated with problem (3.28) there is an operator $G: L_0^2(\Omega) \longrightarrow L_0^2(\Omega)$, defined by:

$$(3.31) \qquad G f = u,$$

where u is the unique solution of (3.28). It is readily seen that G is a compact self-adjoint operator which is positive definite. It then follows from the abstract spectral theory developed in Chapter I that the spectrum consists of non-zero eigenvalues together with zero. The important point for us is to observe that the eigenvalues of problem (3.25) are precisely the characteristic values of the operator G and thus we obtain the following result:

Theorem 3.8. *There is a sequence of pairs* $\{(\beta_\ell^2, \chi_\ell)\}_\ell$ *in* $\mathbb{R} \times W$ *such that*
(i) *They are solutions of problem* (3.25).
(ii) $0 < \beta_1^2 \leq \cdots \leq \beta_\ell^2 \leq \cdots \longrightarrow \infty$. *Each eigenvalue* β_ℓ^2 *has finite multiplicity and it is repeated as many times as its multiplicity.*
(iii) *The set* $\{\chi_\ell\}_\ell$ *forms an orthonormal basis for* $L_0^2(\Omega)$.
(iv) *They provide all the solutions of problem* (3.25), *i.e., if* $(\beta^2, \phi) \in \mathbb{R} \times W$ *is any solution of* (3.25) *then there exists* $\ell \geq 1$ *such that* $\beta^2 = \beta_\ell^2$ *and* ϕ *can be written as a linear combination of all* $\chi_{\ell'}$ *'s for which* $\beta_{\ell'}^2 = \beta^2$. ∎

There is a relationship between the spectrum of R (see (2.24) for its definition) and the eigenvalues of problem (3.25). In fact, this served as the main motivation to consider problem (3.25). To formulate this property, we make the hypothesis that

$$(3.32) \qquad k_j = k, \quad m_j = m \quad \forall j = 1, ..., K$$

for some positive constants k and m. We would like to know whether (k/m) is a characteristic value of the operator R. As a reply, we have

Theorem 3.9. *We suppose that* (3.32) *holds. Then* (k/m) *is a characteristic value of the operator* R *iff* (k/m) *is an eigenvalue of problem* (3.25).

Proof. Let us recall the definition of R: $L_0^2(\Omega) \times \mathbb{C}^{2K} \longrightarrow L_0^2(\Omega) \times \mathbb{C}^{2K}$,

$$R(f, \mathbf{a}) = (u, (\mathbf{b}_j)), \quad \mathbf{b}_j = \rho_0 c^2 k^{-1} \int_{\Gamma_j} u n d\gamma + m k^{-1} \mathbf{a}_j \quad \forall j = 1, ..., K,$$

where u is the unique solution of (2.21). Using this, we simply notice that (k/m) is a characteristic value of R, i.e., there is a non-zero $(\phi, \mathbf{s}) \in L_0^2(\Omega) \times \mathbb{C}^{2K}$ such that $(k/m)R(\phi, \mathbf{s}) = (\phi, \mathbf{s})$ iff $(\phi, \mathbf{s})$ satisfies system (3.25) with $\beta^2 = (k/m)$. This is to say that (k/m) is an eigenvalue of problem (3.25). ∎

In order to deduce more information on the location of the eigenvalues of the Helmholtz model, we will make the following hypothesis in the sequel:

$$(3.33) \qquad \frac{k}{m} \text{ is not an eigenvalue of problem } (3.25).$$

We remark that the above condition is on the geometry of the tubes and the domain Ω_0.

3.3.2. A Homotopy between Laplace and Helmholtz Models

We will now define a homotopy by means of a positive parameter ε which links continuously Helmholtz model with Laplace model extensively studied in Paragraphs 2.1 and 3.1. For ε in the interval $[0, 1]$, we consider the following one parameter family of eigenvalue problems: Find $\omega^2(\varepsilon) \in \mathbb{C}$ for which there exists a non-zero $\phi^\varepsilon \in V$ such that

(3.34)
$$\begin{cases} c^2 \Delta \phi^\varepsilon + \varepsilon \omega^2(\varepsilon) \phi^\varepsilon = 0 \quad \text{in} \quad \Omega, \\[2mm] \dfrac{\partial \phi^\varepsilon}{\partial n} = \dfrac{\rho_0 \omega^2(\varepsilon)}{k - m \omega^2(\varepsilon)} \Big(\int_{\Gamma_j} \phi^\varepsilon \mathbf{n} d\gamma \Big) \cdot \mathbf{n} \quad \text{on} \quad \Gamma_j, \; j = 1, ..., K, \\[2mm] \dfrac{\partial \phi^\varepsilon}{\partial n} = 0 \quad \text{on} \quad \Gamma. \end{cases}$$

Observe that when $\varepsilon = 0$, problem (3.34) coincides with Laplace model and when $\varepsilon = 1$, it corresponds to Helmholtz model. In this sense, the family of problems (3.34) as ε varies ($0 \leq \varepsilon \leq 1$) defines a homotopy between both models.

As in Paragraph 2.2, one can set about studying problem (3.34) for each fixed $\varepsilon > 0$. In particular, one introduces the self-adjoint operator $\tilde{R}^\varepsilon \colon L_0^2(\Omega) \times \mathbb{C}^{2K} \longrightarrow L_0^2(\Omega) \times \mathbb{C}^{2K}$ corresponding to problem (3.34) by the following rule:

(3.35)
$$\begin{cases} \tilde{R}^\varepsilon(f, \mathbf{a}) = (u_1^\varepsilon + \rho_0^{1/2} c k^{-1/2}, (\mathbf{b}_j^\varepsilon)), \\[2mm] \mathbf{b}_j^\varepsilon = \rho_0^{1/2} c k^{-1/2} \displaystyle\int_{\Gamma_j} u_1^\varepsilon \mathbf{n} d\gamma + \rho_0 k^{-1} \int_{\Gamma_j} u_2 \mathbf{n} d\gamma + m k^{-1} \mathbf{a}_j \quad \forall j \end{cases}$$

where u_1^ε and u_2 are the unique solutions of

(3.36)
$$\begin{cases} u_1^\varepsilon \in V, \\[2mm] c^2 \displaystyle\int_\Omega \nabla u_1^\varepsilon \cdot \nabla \bar{\varphi} dx = \varepsilon \int_\Omega f \bar{\varphi} dx \quad \forall \varphi \in V, \end{cases}$$

(3.37)
$$\begin{cases} u_2 \in V, \\[2mm] c^2 \displaystyle\int_\Omega \nabla u_2 \cdot \nabla \bar{\varphi} dx = \sum_{j=1}^K \mathbf{a}_j \cdot \int_{\Gamma_j} \bar{\varphi} \mathbf{n} d\gamma \quad \forall \varphi \in V. \end{cases}$$

Notice that problem (3.37) does not depend on ε. As in Paragraph 2.2, we reach the following conclusions regarding problem (3.34):

Theorem 3.10. *The following properties hold good for $0 < \varepsilon \leq 1$:*
(i) *$\tilde{R}^\varepsilon$ is a compact self-adjoint operator on $L_0^2(\Omega) \times \mathbb{C}^{2K}$.*
(ii) *We have the relation*

$$(3.38) \quad (\tilde{R}^\varepsilon(f, \mathbf{a}), (f, \mathbf{a})) = \int_\Omega |\nabla(u_1^\varepsilon + \rho_0^{1/2} c k^{-1/2} u_2)|^2 dx + m k^{-1} \|\mathbf{a}\|^2$$

for all $(f, \mathbf{a}) \in L_0^2(\Omega) \times \mathbb{C}^{2K}$. In particular, zero is not an eigenvalue of $\tilde{R}^\varepsilon$.
(iii) *The characteristic values of $\tilde{R}^\varepsilon$ form a sequence $\{\omega_\ell^2(\varepsilon)\}$ which can be arranged as follows:*

$$0 < \omega_1^2(\varepsilon) \leq \cdots \leq \omega_\ell^2(\varepsilon) \leq \cdots \longrightarrow \infty.$$

The corresponding eigenvectors $(\phi_\ell^\varepsilon, \mathbf{s}_\ell^\varepsilon)$ form a basis for $L_0^2(\Omega) \times \mathbb{C}^{2K}$ which is orthonormal in the sense (2.45).
(iv) *If $\omega^2(\varepsilon)$ is an eigenvalue of problem (3.34) different from (k/m), then $\omega^2(\varepsilon)$ is a characteristic value of $\tilde{R}^\varepsilon$. Conversely, if $\omega^2(\varepsilon)$ is a characteristic value of $\tilde{R}(\varepsilon)$ different from (k/m), then $\omega^2(\varepsilon)$ is an eigenvalue of problem (3.34).* ∎

We also have the following result which is analogous to Theorem 3.9:

Theorem 3.11. *(k/m) is a characteristic value of $\tilde{R}^\varepsilon$ iff $\varepsilon(k/m)$ is an eigenvalue of problem (3.25).* ∎

Later on, we will take into account these results and make suitable hypothesis on the geometry of the domain under which we will be able to deduce some results on the position of the eigenvalues of the Helmholtz model.

3.3.3. Continuity Properties of the Homotopy

Here we study the behaviour of the eigenvalues $\omega_\ell^2(\varepsilon)$ of problem (3.34) as $\varepsilon \to 0$. For the first result in this direction, we apply Corollary 4.2 of Chapter I. We remark that it is sufficient to use the properties of $\tilde{R}^\varepsilon$ described in Theorem 3.10 and use the fact that u_1^ε can be written as $u_1^\varepsilon = \varepsilon u_1$ where u_1 is the unique solution of (3.36) with $\varepsilon = 1$. We thus obtain

Theorem 3.12. *$(1/\omega_\ell^2(\varepsilon))$ is a Lipschitz function of ε in $]0,1]$ for each $\ell \geq 1$. In particular, $\omega_\ell^2(\varepsilon)$ is a continuous function of ε in $]0,1]$.* ∎

The above result implies, in addition, that for each $\ell \geq 1$, $\omega_\ell^2(\varepsilon)$ is bounded as a function of ε in $[\delta, 1]$ with $\delta > 0$. As we shall see below, they need not all be bounded on the entire interval $]0,1]$. Observe first that when we put $\varepsilon = 0$ in (3.34) we obtain formally the Laplace model. Since there are exactly $2K$ eigenvalues in this model, we expect that the first $2K$ characteristic values $\{\omega_\ell^2(\varepsilon)\}_{\ell=1}^{2K}$ remain bounded as $\varepsilon \to 0$. We also expect that the other characteristic values $\omega_\ell^2(\varepsilon) \longrightarrow \infty$ as $\varepsilon \to 0$ for all $\ell \geq 2K+1$. This is exactly what we plan to prove now. The proof relies on the Courant-Fischer characterization of eigenvalues.

Theorem 3.13. *We have the following estimates on $\omega_\ell^2(\varepsilon)$:*
(i) *There exists a constant $\alpha_0 > 0$, independent of $\varepsilon > 0$ and $\ell \geq 1$, such that*

$$\omega_\ell^2(\varepsilon) \geq \alpha_0 \quad \forall \varepsilon > 0, \ \forall \ell \geq 1.$$

(ii) *For all $\ell = 1, ..., 2K$,*

$$\omega_\ell^2(\varepsilon) \leq \frac{k}{m + \rho_0 c^2 m_-} \quad \forall \varepsilon > 0,$$

where $m_- = m_-(S)$ is the lowest eigenvalues of the operator S associated with the Laplace model.
(iii) *For all $\ell \geq 2K+1$, $\omega_\ell^2(\varepsilon) \longrightarrow \infty$ as $\varepsilon \to 0$.*

Proof. In the first place, let us derive bounds on the solution u_1^ε of problem (3.36). This will be useful throughout our subsequent analysis. Taking $\varphi = u_1^\varepsilon$ in (3.36) and using Poincaré Inequality, we easily deduce the following estimates on u_1^ε:

$$(3.39) \qquad \int_\Omega |\nabla u_1^\varepsilon|^2 \, dx \leq c_1 \varepsilon^2 \int_\Omega |f|^2 dx,$$

$$(3.40) \qquad \int_\Omega |u_1^\varepsilon|^2 \, dx \leq c_1 \varepsilon^2 \int_\Omega |f|^2 dx,$$

for some constant c_1 independent of ε.

We now prove (i). It is obviously sufficient to show the inequality for $\ell = 1$. By the Courant-Fischer principle (see Theorem 3.18 of Chapter I), we have

$$\frac{1}{\omega_1^2(\varepsilon)} = \max_{(f,\mathbf{a}) \in L_0^2(\Omega) \times \mathbb{C}^{2K}} \frac{\left(\tilde{R}^\varepsilon(f, \mathbf{a}), (f, \mathbf{a})\right)}{\|(f, \mathbf{a})\|^2}.$$

We have already given another expression for the numerator of the Rayleigh quotient in (3.38). The numerator is thus bounded from above by

$$2 \int_{\Omega} |\nabla u_1^\varepsilon|^2 dx + \rho_0 c^2 k^{-1} \int_{\Omega} |\nabla u_2|^2 dx + mk^{-1} \|\mathbf{a}\|^2$$

which is equal to

$$2 \int_{\Omega} |\nabla u_1^\varepsilon| dx + \rho_0 c^2 k^{-1}(S\mathbf{a}, \mathbf{a}) + mk^{-1} \|\mathbf{a}\|^2,$$

where S is the operator associated to the Laplace model. We can now use (3.39) to deduce the inequality

$$\frac{1}{\omega_1^2(\varepsilon)} \le \alpha_0^{-1},$$

where we define α_0 by

$$\alpha_0^{-1} = \max\{2c_1 \varepsilon^2, \rho_0 k^{-1} \|S\|, mk^{-1}\}.$$

This proves (i).

To prove (ii), it is convenient to use the mini-max principle in the following form:

$$\frac{1}{\omega_{2K}^2(\varepsilon)} = \max_{F_{2K}} \min_{(f,\mathbf{a}) \in F_{2K}} \frac{(\tilde{R}^\varepsilon(f,\mathbf{a}), (f,\mathbf{a}))}{\|(f,\mathbf{a})\|^2},$$

where the maximum is taken over all subspaces F_{2K} of $L_0^2(\Omega) \times \mathbb{C}^{2K}$ such that the dimension of F_{2K} is equal to $2K$. We choose $F_{2K} = \{0\} \times \mathbb{C}^{2K}$. Then we have

$$\frac{1}{\omega_{2K}^2(\varepsilon)} \ge \min_{\mathbf{a} \in \mathbb{C}^{2K}} \frac{(\tilde{R}^\varepsilon(0,\mathbf{a}), (0,\mathbf{a}))}{\|\mathbf{a}\|^2}.$$

Apart from making the above choice, we also use (3.38) with $f = 0$. Then the solution u_1^ε of (3.36) is obviously zero and so we obtain

$$(\tilde{R}^\varepsilon(0,\mathbf{a}),(0,\mathbf{a})) = \rho_0 c^2 k^{-1} \int_{\Omega} |\nabla u_2|^2 dx + mk^{-1} \|\mathbf{a}\|^2 =$$

$$= \rho_0 c^2 k^{-1}(S\mathbf{a}, \mathbf{a}) + mk^{-1} \|\mathbf{a}\|^2 \ge \rho_0 c^2 k^{-1} m_- \|\mathbf{a}\|^2 + mk^{-1} \|\mathbf{a}\|^2.$$

From these inequalities, the upper bound given in (ii) follows for $\ell = 2K$. Obviously, this is sufficient.

We now proceed with the proof of (iii). It is enough to take $\ell = 2K + 1$. We show below the following equivalent assertion:

(3.41)
$$\frac{1}{\omega_{2K+1}^2(\varepsilon)} \longrightarrow 0 \text{ as } \varepsilon \to 0.$$

From the Courant-Fischer principle, we have

$$\frac{1}{\omega_{2K+1}^2(\varepsilon)} = \min_{F_{2K}} \max_{(f,\mathbf{a})\in F_{2K}^{\perp}} \frac{\left(\tilde{R}^{\varepsilon}(f,\mathbf{a}),(f,\mathbf{a})\right)}{\|(f,\mathbf{a})\|^2},$$

where the minimum is taken over all subspaces F_{2K} of $L_0^2(\Omega) \times \mathbf{C}^{2K}$ of dimension $2K$. Once again, choosing $F_{2K} = \{0\} \times \mathbf{C}^{2K}$ we get the following inequality:

$$\frac{1}{\omega_{2K+1}^2(\varepsilon)} \leq \max_{f\in L_0^2(\Omega)} \frac{\left(\tilde{R}^{\varepsilon}(f,0),(f,0)\right)}{\|f\|^2}.$$

On the other hand, via (3.38), we see that

$$\left(\tilde{R}^{\varepsilon}(f,0),(f,0)\right) = \int_{\Omega} |\nabla u_1^{\varepsilon}|^2 dx,$$

where u_1^{ε} is the solution of problem (3.36). Applying now the estimate (3.39), we get

$$\frac{1}{\omega_{2K+1}^2(\varepsilon)} \leq c_1 \varepsilon^2.$$

This completes the proof of (3.41) and hence of the theorem. ∎

As a consequence of the bounds obtained in the previous theorem, we can now prove the following convergence result:

Theorem 3.14. *The first $2K$ characteristic values of the operator $\tilde{R}^{\varepsilon}$ converge to the corresponding eigenvalues of the Laplace model as $\varepsilon \to 0$, i.e.,*

$$\omega_{\ell}^2(\varepsilon) \longrightarrow \omega_{\ell}^2(0) \quad \forall \ell = 1,...,2K.$$

Proof. We proceed in two steps. In the first step, we show that one can pass to the limit in problem (3.34) as $\varepsilon \to 0$ and the limit is nothing but the eigenvalue problem of the Laplace model. More precisely, given a sequence $\varepsilon \to 0$, we show that there exist a subsequence (again denoted by ε) and pairs $(\omega_{0\ell}^2, \psi_{\ell}^{(0)})$, $\ell = 1,...,2K$ such that

(3.42) $$\omega_{\ell}^2(\varepsilon) \longrightarrow \omega_{0\ell}^2,$$

(3.43) $$\phi_{\ell}^{\varepsilon} \rightharpoonup \psi_{\ell}^{(0)} \quad \text{in} \quad V \quad \text{weakly},$$

(3.44) $$s_{\ell}^{\varepsilon} \longrightarrow s_{\ell}^{(0)} \quad \text{in} \quad \mathbf{C}^{2K},$$

(3.45) $$(\omega_{0\ell}^2, \psi_{\ell}^{(0)}) \quad \text{satisfy (1.9)}, \quad \psi_{\ell}^{(0)} \neq 0,$$

where

(3.46) $$s_{\ell}^{\varepsilon} = \frac{\rho_0 c^2 \omega_{\ell}^2(\varepsilon)}{k - m\omega_{\ell}^2(\varepsilon)} \left(\int_{\Gamma_1} \phi_{\ell}^{\varepsilon} n d\gamma, \cdots, \int_{\Gamma_K} \phi_{\ell}^{\varepsilon} n d\gamma \right),$$

(3.47) $$\mathbf{s}_\ell^{(0)} = \frac{\rho_0 c^2 \omega_{0\ell}^2}{k - m\omega_{0\ell}^2} \left(\int_{\Gamma_1} \psi_\ell^{(0)} \mathbf{n} d\gamma, \cdots, \int_{\Gamma_K} \psi_\ell^{(0)} \mathbf{n} d\gamma \right).$$

In the second step, we show that the limits obtained above provide all the solutions of the Laplace model.

First Step: From the bounds given by Theorem 3.13, we infer, in particular, that none of the values $\omega_\ell^2(\varepsilon)$, $\ell = 1, ..., 2K$ equals (k/m) and in fact they remain bounded away from (k/m) as ε varies. Consequently, they are eigenvalues of problem (3.34). For the moment, let us fix ℓ such that $1 \leq \ell \leq 2K$ and work with the eigenvalues $\omega_\ell^2(\varepsilon)$. For ease of writing, let us suppress the index ℓ: $\omega_\ell^2(\varepsilon) = \omega^2(\varepsilon)$, $\phi_\ell^\varepsilon = \phi^\varepsilon$, etc.

Multiplying (3.34) by ϕ^ε and integrating by parts, we arrive at the following energy identity:

(3.48) $$c^2 \int_\Omega |\nabla \phi^\varepsilon|^2 dx = \varepsilon \omega^2(\varepsilon) \int_\Omega |\phi^\varepsilon| dx + \frac{\rho_0 c^2 \omega^2(\varepsilon)}{k - m\omega^2(\varepsilon)} \sum_{j=1}^K |\int_{\Gamma_j} \phi^\varepsilon \mathbf{n} d\gamma|^2$$

Let us recall the normalization condition on the pairs $(\phi_{\ell'}^\varepsilon, \mathbf{s}_{\ell'}^\varepsilon)$ (see (2.45)):

(3.49) $$\int_\Omega \phi_j^\varepsilon \bar{\phi}_{\ell'}^\varepsilon dx + \rho_0 c^2 k^{-1} (\mathbf{s}_j^\varepsilon, \mathbf{s}_{\ell'}^\varepsilon) = \delta_{j\ell'} \quad \forall j, \ell' \geq 1.$$

Taking $j = \ell' = \ell$ in (3.49), we see that the right hand side of (3.48) remains bounded as $\varepsilon \to 0$. Therefore $\{\phi^\varepsilon\}$ is a bounded sequence in V and so one can extract a subsequence and choose $\psi^{(0)} \in V$ such that (3.43) holds. One can pass to a further subsequence (if necessary) to achieve (3.42) and (3.44). We thus have all the necessary ingredients to pass to the limit in problem (3.34) along the chosen subsequence. At the limit, we see that $(\omega_0^2, \psi^{(0)})$ satisfy (1.9). To complete the proof of the first step, it remains to establish $\psi^{(0)} \neq 0$. But this follows easily if we take $j = \ell' = \ell$ in the normalization condition and let $\varepsilon \to 0$. We get

(3.50) $$\int_\Omega |\psi^{(0)}|^2 dx + \rho_0 c^2 k^{-1} \|\mathbf{s}^{(0)}\|^2 = 1.$$

Second Step: Since the eigenvalues $\{\omega_\ell^2(\varepsilon)\}$ are arranged in the ascending fashion, we will have the following order at the limit:

$$\omega_{0,1}^2 \leq \cdots \leq \omega_{0,2K}^2.$$

Remark that the proof of Theorem 3.14 would be over if we could show that there is no other eigenvalue of the Laplace model except the limit points

$\{\omega_{0,\ell}^2\}_{\ell=1}^{2K}$ obtained in the first step. To this end, we first observe that the pairs $\{(\psi_\ell^{(0)}, s_\ell^{(0)})\}_j$ are linearly independent in $L_0^2(\Omega) \times \mathbf{C}^{2K}$. Indeed, they are orthogonal in the following sense:

$$(3.51) \qquad \int_\Omega \psi_\ell^{(0)} \bar{\psi}_{\ell'}^{(0)} \, dx + \rho_0 c^2 k^{-1} (s_\ell^{(0)}, s_{\ell'}^{(0)}) = \delta_{\ell\ell'} \quad \forall \ell, \ell' \geq 1.$$

This property is easily obtained if we pass to the limit in (3.49) as $\varepsilon \to 0$. Of course, this does not in general imply that $\{s_\ell^{(0)}\}_{\ell=1}^{2K}$ are linearly independent in $\mathbf{C}^{2K}$. But in this particular situation, it does and that can be seen as follows: Let

$$\sum_{\ell=1}^{2K} \alpha_\ell s_\ell^{(0)} = 0$$

be a relation expressing linear dependence of $\{s_\ell^{(0)}\}_{\ell=1}^{2K}$. Let us pose $\psi^{(0)} = \sum_\ell \alpha_\ell \psi_\ell^{(0)}$. Then it follows from the fact that $\psi_\ell^{(0)}$ satisfies (1.9):

$$\begin{cases} \Delta\psi^{(0)} = 0 \quad \text{in} \quad \Omega, \\ \dfrac{\partial\psi^{(0)}}{\partial n} = 0 \quad \text{on} \quad \Gamma_j, \ j = 1, ..., K, \\ \dfrac{\partial\psi^{(0)}}{\partial n} = 0 \quad \text{on} \quad \Gamma_0, \quad \int_\Omega \psi^{(0)} \, dx = 0. \end{cases}$$

We conclude therefore that $\psi^{(0)} = 0$. This would contradict the linear independence of the pairs $\{(\psi_\ell^{(0)}, s^{(0)})\}_{\ell=1}^{2K}$ unless $\alpha_\ell = 0$ for all $\ell = 1, ..., 2K$.

Thus we have found a basis $\{s_\ell^{(0)}\}_{\ell=1}^{2K}$ of $\mathbf{C}^{2K}$ consisting of eigenvectors of S associated with the Laplace model. This means that the entire set of eigenvalues is given by $\{\omega_{0,\ell}^2\}_{\ell=1}^{2K}$. This completes the proof of the second step and hence the theorem. ∎

We would like to make two remarks concerning the convergence of the eigenvectors $\{\phi_\ell^\varepsilon\}_{\varepsilon>0}$ and eigenvalues $\{\omega_\ell^2(\varepsilon)\}_{\varepsilon>0}$.

Remark. We have proved in Theorem 3.14 the weak convergence of a *subsequence* $\{\omega_\ell^2(\varepsilon)\}_{\varepsilon>0}$. Since its limit is the ℓth eigenvalue of the Laplace model, we conclude that the *entire sequence* $\{\omega_\ell^2(\varepsilon)\}_{\varepsilon>0}$ converges. On the other hand, one cannot, in general, say anything about the convergence of the *whole sequence* of eigenvectors $\{\phi_\ell^\varepsilon\}_{\varepsilon>0}$. This is due to the possible multiplicity of the eigenvalues. However, if an eigenvalue, say ℓ^{th} eigenvalue is simple, then we can arrange in such a way that the convergence (3.43) happens for the whole sequence. ∎

Remark. The weak convergence (3.43) of eigenvectors is in fact strong. To prove this, it is sufficient to prove the convergence of the norms. The later property is a consequence of (3.48). Indeed, letting $\varepsilon \to 0$ in (3.48), we obtain

$$\lim_{\varepsilon \to 0} c^2 \int_\Omega |\nabla \phi^\varepsilon|^2 dx = \frac{\rho_0 c^2 \omega^2(0)}{k - m\omega^2(0)} \sum_{j=1}^K |\int_{\Gamma_j} \psi^{(0)} \mathbf{n} d\gamma|^2 = c^2 \int_\Omega |\nabla \psi^{(0)}| dx,$$

since $(\omega^2(0), \psi^{(0)})$ satisfy (1.9). ∎

3.3.4. A Geometric Condition and a Consequence

In order to extract useful information from the analysis of the previous subparagraphs, we will impose the following condition:

$$(3.52) \qquad \frac{k}{m} < \beta_1^2,$$

where β_1^2 is the first eigenvalue of problem (3.25) given by Theorem 3.8. It is obviously a condition on the geometry of the domain Ω_0 and of the tubes Γ_j, $j = 1, ..., K$. We add to say that the above condition is satisfied if (k/m) is sufficiently small. Let us recall that (k/m) stands for the frequency of the system of tubes in the absence of any fluid.

The main motivation to introduce condition (3.52) (as for condition (3.33)) comes from our desire to have complete equivalence between the spectrum of $\tilde{R}^\varepsilon$ and that of problem (3.34) for all ε such that $0 < \varepsilon \leq 1$. This will be the case iff $\omega_\ell^2(\varepsilon)$ are all different from (k/m). We have indeed

Proposition 3.15. *If condition (3.52) holds then the eigenvalues* $\{\omega_\ell^2(\varepsilon)\}_\ell$ *are all different from* (k/m), *for all* ε *such that* $0 < \varepsilon \leq 1$.

Proof. Indeed, condition (3.52) can be rewritten as

$$\varepsilon \frac{k}{m} < \varepsilon \beta_1^2 \quad \forall \varepsilon \in]0, 1].$$

In particular, this implies that $\varepsilon(k/m)$ is never an eigenvalue of problem (3.25). This is in turn equivalent to saying that (k/m) is not a characteristic value of $\tilde{R}^\varepsilon$ via Theorem 3.11. Since $\{\omega_\ell^2(\varepsilon)\}_\ell$ are precisely the characteristic values of $\tilde{R}^\varepsilon$, this finishes the proof. ∎

We will apply the above proposition crucially in obtaining some precise information on the position of the eigenvalues of the Helmholtz model. We have already seen some results in this direction (see especially Theorem 3.5 and Corollary 3.6). We will now present results which improve upon them. The first one follows directly from Theorem 3.13.

Theorem 3.16. *There are at least $2K$ eigenvalues of the Helmholtz model in the interval $]0, (k/m)[$. More precisely, we have*

$$0 < \alpha_0 \leq \omega_\ell^2 \leq \frac{k}{m + \rho_0 c^2 m_-} < \frac{k}{m} \quad \forall \ell = 1, ..., 2K,$$

where α_0 and m_- are as in Theorem 3.13 and $\{\omega_\ell^2\}$ are the eigenvalues of the Helmholtz model. ∎

We emphasize the fact that the above result is generally valid and we have not used any geometric condition such as (3.52) in deducing it. On the other hand, if we assume (3.52) then we can deduce the following refinement using a continuity argument:

Theorem 3.17. *Assume that (3.52) holds. Then there are exactly $2K$ eigenvalues $\{\omega_\ell^2\}_{\ell=1}^{2K}$ of the Helmholtz model (multiplicities included) in the interval $]0, (k/m)[$.*

Proof. We already know from our previous result that the first $2K$ eigenvalues lie in the interval $]0, (k/m)[$. To conclude that the other eigenvalues lie outside this interval, we need the condition (3.52).

To this end, we infer from Theorem 3.13 that for each $\ell \geq 2K + 1$ we have for ε sufficiently small

$$\omega_\ell^2(\varepsilon) > \frac{k}{m}.$$

On the other hand, it follows from Proposition 3.15 that the eigenvalues $\omega_\ell^2(\varepsilon)$ are never equal to (k/m) for $\ell \geq 1$ and $0 < \varepsilon \leq 1$. Since $\omega_\ell^2(\varepsilon)$ is a continuous function of $\varepsilon \in]0, 1]$ (see Theorem 3.12), it follows that

$$\omega_\ell^2(\varepsilon) > \frac{k}{m} \quad \forall \ell \geq 1, \, 0 < \varepsilon \leq 1$$

Taking $\varepsilon = 1$, we get the required result. ∎

3.4. Bounds for the Stokes Model

In this paragraph, we pay attention to the Stokes model (2.48) introduced in Paragraph 2.3 and we concentrate on some qualitative properties of its eigenvalues provided by Theorem 2.13. Recall that Stokes model corresponds to a quadratic eigenvalue problem in which the zero order term is a bounded self-adjoint operator which is negative definite and the quadratic term is a bounded self-adjoint operator which is non-positive definite and of finite rank. Thus the abstract general theory developed in Paragraph 4.3 of Chapter I was useful in deducing the existence result and the same will be used now in obtaining some qualitative properties. In particular, we will deduce significant interlacing inequalities between some of the real eigenvalues of the Stokes model and the eigenvalues of a Stokes-type operator in Ω. This latter eigenvalue problem is of standard type (not quadratic) and its solutions are easily computed. From the abstract theory, we will also conclude that there are at most $4K$ non-real eigenvalues of the Stokes model, K being the number of tubes in Ω. Apart from these two estimates, there is a third type of sharp inequalities which show the regions of the complex plane where the eigenvalues are located.

Let us start by recalling that the original Stokes model was transferred to the quadratic eigenvalue problem (2.59): Find $\lambda \in \mathbb{C}$ for which there exists $\mathbf{u} \in \mathcal{H}, \mathbf{u} \neq \mathbf{0}$ such that

$$\left[U_1 + U_2 \,\mathrm{ddiag}(m_j)Q\right]\mathbf{u} - \lambda\mathbf{u} + \lambda^2 U_2\,\mathrm{ddiag}(k_j)Q\mathbf{u} = \mathbf{0},$$

where $\mathcal{H}$ is the space defined by (2.50) and the operators U_1, U_2 and Q were defined in (2.52)–(2.56). The eigenvalue parameter λ represents the inverse of the frequencies of the Stokes model. In order to apply the general theory of Paragraph 4.3 of Chapter I to the present case, we make the following choice:

(3.53)
$$\begin{cases} H = \mathcal{H}, \\ R = -\left[U_1 + U_2\,\mathrm{ddiag}(m_j)Q\right], \\ S = -U_2\,\mathrm{ddiag}(k_j)Q. \end{cases}$$

From Propositions 2.9 and 2.11, we know that R is a compact self-adjoint and positive definite operator. On the other hand, S is a self-adjoint non-negative operator of rank $2K$. The characteristic values of $(-R)$ are arranged in the descending order as indicated below:

(3.54)
$$0 > \nu_1 \geq \cdots \geq \nu_\ell \geq \cdots \longrightarrow -\infty.$$

Observe that R is the Green's operator associated with a Stokes-type system.

We are now in a position to apply Theorem 4.5 of Chapter I and thereby improve upon the existence result described by Theorem 2.13. However, we should exercise caution while applying Theorem 4.5 of Chapter I and while translating the result from eigenvalues to characteristic values. It is not out of place to note that λ^{-1} corresponds to the eigenvalue ω of the Stokes model (2.51).

Theorem 3.18. *Let $\{\omega_n\}$ be the eigenvalues of the Stokes model (2.51). Then the number of non-real elements in $\{\omega_n\}$ including multiplicities is at most $4K$. Moreover, if $\{\omega_\ell\}$ is the subsequence of $\{\omega_n\}$ consisting of real eigenvalues constructed in Theorem 2.13, then it satisfies the following interlacing inequalities with respect to the eigenvalues $\{\nu_\ell\}$:*

(3.55)
$$\nu_{\ell+2K} \leq \omega_\ell \leq \nu_\ell \quad \forall \ell \geq 1. \ \blacksquare$$

Remark. In the preceding theorem, if we group each eigenvalue with its conjugate, then we see that there exist at most $2K$ of such groups of non-real eigenvalues. ∎

To announce the next result, we introduce the quantity μ_{min} which is the smallest eigenvalue of the following spectral problem in Ω: Find $\mu \in \mathbb{C}$ for which there exists $\mathbf{u} \in \mathcal{H}, \mathbf{u} \neq 0$ such that

$$(3.56)\quad 2\nu \int_\Omega e(\mathbf{u}) : e(\bar{\mathbf{v}})dx = \mu\left[\int_\Omega \mathbf{u}\cdot\bar{\mathbf{v}}dx + \sum_{j=1}^{K} m_j\gamma_j(\mathbf{u})\cdot\gamma_j(\bar{\mathbf{v}})\right] \quad \forall \mathbf{v} \in \mathcal{H}.$$

This is an elliptic eigenvalue problem which enters the set-up of the theory of compact self-adjoint operators of Chapter I. It has a sequence of positive eigenvalues which admit the Courant-Fischer characterization. In the sequel, we need only the first eigenvalue μ_{min} which is characterized as

$$(3.57)\qquad\qquad \mu_{min} = \min_{\mathbf{v}\in\mathcal{H}} \frac{\nu\|e(\mathbf{v})\|_{0,\Omega}^2}{\|\mathbf{v}\|_{0,\Omega}^2 + \sum_{j=1}^{K} m_j|\gamma_j(\mathbf{v})|^2}.$$

We note that $\mu_{min} > 0$ and that is closely linked to the geometric shape of the domain Ω.

Theorem 3.19. *Let $(\omega, \mathbf{u})$ be any solution of the Stokes model (2.51). Then the following assertions hold:*
(i) $\mathfrak{Re}(\omega) < 0$.
(ii) *If $\mathfrak{Im}(\omega) \neq 0$, then*

$$(3.58)\qquad\qquad \begin{cases} |\omega|^2 < (k/m), \\ \mathfrak{Re}(\omega) \leq -\mu_{min}, \\ \displaystyle\sum_{j=1}^{K} k_j|\gamma_j(\mathbf{u})|^2 > 0, \end{cases}$$

where, we recall, (k/m) denotes the quantity $\max\{(k_j/m_j) \mid j = 1, ..., K\}$.

Proof. In the course of the proof, we will use the following notation for the inner product in $\mathcal{H}$:

$$(3.59)\qquad\qquad ((\mathbf{u}, \mathbf{v})) = \int_\Omega e(\mathbf{u}) : e(\bar{\mathbf{v}})dx.$$

Taking the scalar product of equation (2.57) with $\mathbf{u}$, we obtain

$$\omega^2\left(([U_1 + U_2\mathrm{ddiag}(m_j)Q]\mathbf{u}, \mathbf{u})\right) + ((U_2\mathrm{ddiag}(k_j)Q\mathbf{u}, \mathbf{u})) = \omega\|e(\mathbf{u})\|_{0,\Omega}^2.$$

We use $\mathbf{v} = \mathbf{u}$ as a test function in (2.55) and combine it with (2.64). We see that the above relation can be rewritten as

$$\omega^2 \left[\|u\|_{0,\Omega}^2 + \sum_{j=1}^{K} m_j |\gamma_j(\mathbf{u})|^2 \right] + \sum_{j=1}^{K} k_j |\gamma_j(\mathbf{u})|^2 + 2\nu\omega \|e(\mathbf{u})\|_{0,\Omega}^2 = 0.$$

Splitting the above identity into real and imaginary parts, we obtain

$$(3.60) \quad \begin{cases} \left[\Re\mathfrak{e}(\omega)^2 - \Im\mathfrak{m}(\omega)^2 \right] \left[\|u\|_{0,\Omega}^2 + \sum_{j=1}^{K} m_j |\gamma_j(\mathbf{u})|^2 \right] + \\[2ex] \qquad + \sum_{j=1}^{K} k_j |\gamma_j(\mathbf{u})|^2 + +2\nu\Re\mathfrak{e}(\omega) \|e(\mathbf{u})\|_{0,\Omega}^2 = 0 \end{cases}$$

and

$$(3.61) \quad 2\Im\mathfrak{m}(\omega) \left[\Re\mathfrak{e}(\omega) \left(\|u\|_{0,\Omega}^2 + \sum_{j=1}^{K} m_j |\gamma_j(\mathbf{u})|^2 \right) + \nu \|e(\mathbf{u})\|_{0,\Omega}^2 \right] = 0.$$

If $\Im\mathfrak{m}(\omega) = 0$, then (3.60) implies that $\Re\mathfrak{e}(\omega) \leq 0$. On the other hand, if $\Im\mathfrak{m}(\omega) \neq 0$, then (3.61) reduces to

$$(3.62) \quad \Re\mathfrak{e}(\omega) \left[\|u\|_{0,\Omega}^2 + \sum_{j=1}^{K} m_j |\gamma_j(\mathbf{u})|^2 \right] + \nu \|e(\mathbf{u})\|_{0,\Omega}^2 = 0,$$

which clearly implies that $\Re\mathfrak{e}(\omega) < 0$. Combining these results and using the fact that zero is not an eigenvalue of the Stokes model (2.51), we conclude that $\Re\mathfrak{e}(\omega) < 0$ always.

Let us now assume that $\Im\mathfrak{m}(\omega) \neq 0$. By virtue of (3.62), we get

$$-\Re\mathfrak{e}(\omega) = \frac{\nu \|e(\mathbf{u})\|_{0,\Omega}^2}{\|u\|_{0,\Omega}^2 + \sum\limits_{j=1}^{K} m_j |\gamma_j(\mathbf{u})|^2}$$

which clearly proves that $-\Re\mathfrak{e}(\omega) \geq \mu_{\min}$.

Let us now show $|\omega|^2 < (k/m)$. To this end, we introduce the vector functions $\mathbf{z}_1$ and $\mathbf{z}_2$ defined by

$$\begin{cases} \mathbf{z}_1 = U_1 \mathbf{u}, \\[2ex] \mathbf{z}_2 = U_2 \mathrm{ddiag} \left(\dfrac{k_j + m_j \omega^2}{\omega^2} \right) Q \mathbf{u}. \end{cases}$$

In terms of these functions, equation (2.57) can be rewritten as follows:

$$(3.63) \quad \mathbf{z}_1 + \mathbf{z}_2 = \frac{1}{\omega} \mathbf{u}.$$

Using the definitions of U_1 and U_2, we have

$$(3.64) \qquad 2\nu((\mathbf{z}_1, \mathbf{v})) = -\int_\Omega \mathbf{u} \cdot \bar{\mathbf{v}} dx \quad \forall \mathbf{v} \in \mathcal{H},$$

$$(3.65) \qquad 2\nu((\mathbf{z}_2, \mathbf{v})) = -\sum_{j=1}^K \left(\frac{k_j + m_j \omega^2}{\omega^2} \right) \gamma_j(\mathbf{u}) \cdot \gamma_j(\bar{\mathbf{v}}) \quad \forall \mathbf{v} \in \mathcal{H}.$$

Multiplying (3.63) by $\mathbf{z}_1$, we obtain

$$(3.66) \qquad \|e(\mathbf{z}_1)\|_{0,\Omega}^2 + ((\mathbf{z}_2, \mathbf{z}_1)) = \frac{1}{\omega}((\mathbf{u}, \mathbf{z}_1)).$$

Taking $\mathbf{v} = \mathbf{u}$ as test function in (3.64), we get

$$((\mathbf{u}, \mathbf{z}_1)) = -\frac{1}{2\nu} \|\mathbf{u}\|_{0,\Omega}^2.$$

Using this in (3.66), we can write (3.66) equivalently as

$$(3.67) \qquad \|e(\mathbf{z}_1)\|_{0,\Omega}^2 + \frac{1}{2\nu\omega} \|\mathbf{u}\|_{0,\Omega}^2 + ((\mathbf{z}_2, \mathbf{z}_1)) = 0.$$

On the other hand, multiplying (3.63) by $\mathbf{z}_2$ this time, we get

$$((\mathbf{z}_1, \mathbf{z}_2)) + \|e(\mathbf{z}_2)\|_{0,\Omega}^2 = \frac{1}{\omega}((\mathbf{u}, \mathbf{z}_2)).$$

Taking $\mathbf{v} = \mathbf{u}$ as test function in (3.65), we arrive at

$$((\mathbf{u}, \mathbf{z}_2)) = -\frac{1}{2\nu} \sum_{j=1}^K \left(\frac{k_j + m_j \bar{\omega}^2}{\bar{\omega}^2} \right) |\gamma_j(\mathbf{u})|^2$$

and so combining these two relations, we get

$$(3.68) \qquad ((\mathbf{z}_1, \mathbf{z}_2)) = -\|e(\mathbf{z}_2)\|_{0,\Omega}^2 - \frac{1}{2\nu\omega} \sum_{j=1}^K \left(\frac{k_j + m_j \bar{\omega}^2}{\bar{\omega}^2} \right) |\gamma_j(\mathbf{u})|^2.$$

Combining (3.67) with (3.68), we deduce

$$\|e(\mathbf{z}_1)\|_{0,\Omega}^2 - \|e(\mathbf{z}_2)\|_{0,\Omega}^2 + \frac{1}{2\nu\omega} \|\mathbf{u}\|_{0,\Omega}^2 - \frac{1}{2\nu\bar{\omega}} \sum_{j=1}^K \left(\frac{k_j}{\omega^2} + m_j \right) |\gamma_j(\mathbf{u})|^2 = 0.$$

Taking imaginary part of this equation, we get

$$\Im\mathfrak{m}(\frac{1}{\omega})\|\mathbf{u}\|^2_{0,\Omega} - \frac{1}{|\omega|^2}\Im\mathfrak{m}(\frac{1}{\omega})\sum_{j=1}^{K}k_j|\gamma_j(\mathbf{u})|^2 + \Im\mathfrak{m}(\frac{1}{\omega})\sum_{j=1}^{K}m_j|\gamma_j(\mathbf{u})|^2 = 0.$$

Dividing this by $\Im\mathfrak{m}(1/\omega)$, the last relation becomes

$$(3.69) \qquad \|\mathbf{u}\|^2_{0,\Omega} - \frac{1}{|\omega|^2}\sum_{j=1}^{K}k_j|\gamma_j(\mathbf{u})|^2 + \sum_{j=1}^{K}m_j|\gamma_j(\mathbf{u})|^2 = 0.$$

This clearly proves the third inequality in (3.58) since $\mathbf{u} \neq \mathbf{0}$. It also implies that

$$\sum_{j=1}^{K}m_j|\gamma_j(\mathbf{u})|^2\left(\frac{k_j}{m_j} - |\omega|^2\right) > 0$$

and so we obtain

$$\frac{k}{m} - |\omega|^2 > 0.$$

This concludes the proof of Theorem 3.19. ∎

The above result gives precise information of the zone in $\mathbb{C}$ where the eigenvalues are located. In Figure 3.2, we present a graphical view of this zone.

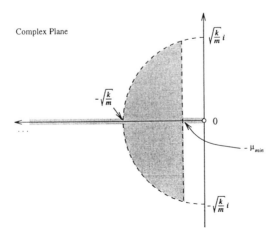

Figure 3.2: *Location in $\mathbb{C}$ of the eigenvalues of the Stokes model.*

The fact that the spectrum lies in the left half-plane is due to the viscosity of the fluid and it signifies damping of the vibrations. On the other hand, the estimate (3.58) on the non-real eigenvalues and the fact that there exist at most $2K$ of such eigenvalues (see Remark following Theorem 3.18) are very much in the spirit of the Laplace model. In fact, non-real eigenvalues of the Stokes model should be viewed as perturbations with respect to viscosity from those of the Laplace model. (A word of caution may not be out of place here. One should keep in mind that we sought the solutions of the form $e^{i\omega t}$ and $e^{\omega t}$ in the Laplace and Stokes model respectively). Regarding real eigenvalues of the Stokes model, one may conjecture that they go to $-\infty$ as the viscosity tends to 0.

§4. Numerical Methods in Fluid-Solid Structures

The aim of this section is to present some of the most favourite methods employed by engineers to compute the eigenvalues of various mathematical models introduced in §1. The presentation of the methods is followed by a thorough mathematical justification and examples of numerical calculations in specific problems. An interpretation of the results is also added.

We start by the Laplace model and present what is known as the technique of *added mass matrix* commonly used by engineers. This is formed by defining the *influence functions* which stand for the interactions between various tubes via the fluid. These functions depend only on the geometry of Ω and allow us to "eliminate" the fluid from the picture and formulate the problem exclusively in terms of the displacement of tubes. This is done in Paragraph 4.1 and a numerical example is illustrated in Paragraph 4.2. On the other hand, the Stokes model is examined from a numerical point of view in Paragraph 4.6 by using finite element method.

The remaining parts of this section are devoted to developing another original technique called *unitary operator method* or *frequency scanning method*. This technique was originally conceived to calculate the entire spectrum of elliptic boundary value problems in bounded domains. Following the original paper of GRÉGOIRE, NÉDELEC and PLANCHARD [1976], we introduce the method for the Laplace operator with Robin boundary condition. We then apply this to the Helmholtz model and then give guidelines to its application to the problem of the acoustic resonator. The idea behind this method consists of constructing a family of operators parametrized by the real frequency ω and analyzing the behaviour of their spectra as ω varies. Hence the name of the method. A numerical example is provided to illustrate its good performance.

4.1. Technique of Added Mass Matrix for the Laplace Model

The Laplace model was introduced in Paragraph 1.1 and the existence of its solutions was established in Paragraph 2.1 by following the Green's operator technique. In this approach, formulation of the problem was in terms of the velocity potential $\phi_0 = \phi_0(x, t)$ of the fluid and the displacement vector $\mathbf{r}_0 = (\mathbf{r}_{0j}(t))$ of the tubes did not figure explicitly. On the contrary, the method of added mass presented here consists in transforming Laplace model to a system of second order ode's for the displacement vector $\mathbf{r}_0$. This system involves the so-called added mass matrix that is constructed with the aid of influence functions which represent the fluid dynamic interaction with the tubes. They allow us to "eliminate" the potential ϕ_0 of the fluid and formulate the Laplace model exclusively in terms of $\mathbf{r}_0$.

It is clear that ϕ_0 depends linearly on the velocity $(d\mathbf{r}_{0j}/dt)$ of the tubes. We therefore express

$$(4.1) \qquad \phi_0(x, t) = \sum_{j=1}^{K} \frac{d\mathbf{r}_{0j}(t)}{dt} \cdot \mathbf{q}_j(x),$$

where $\mathbf{q}_j = [q_{j1}, q_{j2}], j = 1, ..., K$. Here each component $q_{j\ell}$ of $\mathbf{q}_j$ is the (unique) solution defined up to an additive constant of the following Neumann problem:

$$(4.2) \qquad \begin{cases} \Delta q_{j\ell} = 0 & \text{in} \quad \Omega, \\ \dfrac{\partial q_{j\ell}}{\partial n} = 0 & \text{on} \quad \Gamma_0, \\ \dfrac{\partial q_{j\ell}}{\partial n} = n_\ell \delta_{jj'} & \text{on} \quad \Gamma_{j'}, \ j' = 1, ..., K, \end{cases}$$

where n_ℓ is the ℓth component of the normal $\mathbf{n}$ on the boundary and $\delta_{jj'}$ is the Kronecker tensor. In order to fix $q_{j\ell}$ uniquely, one can, for example, impose the condition that its mean value is zero:

$$(4.3) \qquad \int_\Omega q_{j\ell} dx = 0 \quad \forall j = 1, ... K, \ \forall \ell = 1, 2.$$

The functions $q_{j\ell}$ are called the *influence functions*. In fact $q_{j\ell}$ signifies the influence of the jth tube on the other tubes because of its movement in the direction of the x_ℓ-axis. We point out the link that exists between problems (4.2) and (2.6) occurring in the definition of the operators S of the Laplace model. Indeed, $q_{j\ell}$ is nothing but the solution of (2.6) when $\mathbf{a}$ is an element of the canonical basis of $\mathbb{C}^{2K}$. We have the following property of functions $q_{j\ell}$ which is very easy to prove.

Proposition 4.1. *The functions $\{q_{j\ell}\}$ are linearly independent.*

Proof. Let there be a linear combination of these functions which vanishes, i.e., there exist scalars $\alpha_{j\ell}$ such that

$$\phi \overset{\text{def}}{=} \sum_{j=1}^{K} \sum_{\ell=1}^{2} \alpha_{j\ell} q_{j\ell} = 0.$$

Let us calculate the normal derivative of ϕ on $\Gamma_{j'}$ using the last condition in (4.2). We get

$$\frac{\partial \phi}{\partial n} = \sum_{\ell=1}^{2} \alpha_{j'\ell} n_\ell \quad \text{on} \quad \Gamma_{j'}, \ j' = 1, ..., K.$$

The fact that this vanishes imply automatically that $\alpha_{j'\ell} = 0 \ \forall j', \ell$ and hence the proposition. ∎

Substituting the expression (4.1) in (1.4) and (1.6), we get the equation

$$m_j \frac{d^2 \mathbf{r}_{0j}}{dt^2} + k_j \mathbf{r}_{0j} = -\rho_0 \sum_{j'=1}^{K} \int_{\Gamma_j} \left(\frac{d^2 \mathbf{r}_{0j'}}{dt^2} \cdot \mathbf{q}_{j'} \right) \mathbf{n} d\gamma, \quad j = 1, ..., K.$$

which can be rewritten in the following form:

(4.4) $$[\text{ddiag}(m_j) + \rho_0 S'] \frac{d^2 \mathbf{r}_0}{dt^2} + \text{ddiag}(k_j) \mathbf{r}_0 = \mathbf{0},$$

where S' is a matrix of order $2K$ whose entries are given by

(4.5) $$\int_{\Gamma_j} q_{j'\ell'} n_\ell d\gamma \quad \forall j, j' = 1, ..., K, \ \forall \ell, \ell' = 1, 2.$$

S' is called *added mass matrix*. It enjoys the following properties:

Proposition 4.2. *S' is self-adjoint and positive definite.*

Proof. It is sufficient to observe that the elements defined by (4.5) are nothing but the matrix entries of the operator S defined by (2.5) with respect to the canonical basis of $\mathbb{C}^{2K}$. Therefore, Proposition 4.2 is nothing but another version of Proposition 2.2. ∎

As in Paragraph 2.1, we seek sinusoidal solutions to (4.4) of the form $r_{0j}(t) = r_j e^{i\omega t}, j = 1, ..., K$. We thus arrive at the following eigenvalue problem: Find μ for which there exists a non-zero $\mathbf{r} \in \mathbb{C}^{2K}$ such that

$$(4.6) \qquad \left[\mathrm{ddiag}(m_j) + \rho_0 S'\right]\mathbf{r} = \mu\,\mathrm{ddiag}(k_j)\mathbf{r},$$

or equivalently

$$(4.7) \qquad T'\mathbf{r} = \mu\mathbf{r},$$

where T' is the matrix representation of the operator T in (2.8), i.e.,

$$(4.8) \qquad T' = \mathrm{ddiag}(m_j k_j^{-1}) + \rho_0\,\mathrm{ddiag}(k_j^{-1})S'.$$

As already noted in Paragraph 2.2, T' is not self-adjoint and we overcame this difficulty by introducing a symmetrized version of T'. More exactly, we defined

$$(4.9) \qquad \tilde{T}' = \mathrm{ddiag}(m_j k_j^{-1}) + \rho_0\,\mathrm{ddiag}(k_j^{-1/2})S'\mathrm{ddiag}(k_j^{-1/2}).$$

Then problem (4.7) is transformed to

$$(4.10) \qquad \tilde{T}'\mathbf{t} = \mu\mathbf{t}, \quad \mathbf{t} = \mathrm{ddiag}(k_j^{1/2})\mathbf{r}.$$

We observe that (4.10) is a standard eigenvalue problem associated with the matrix $\tilde{T}'$ which is now self-adjoint and positive definite.

To compute the influence functions, finite element method is very commonly used to discretize problem (4.2) and reduce it to a linear system corresponding to a self-adjoint positive definite matrix. Finally, we remark that there are several numerical methods and corresponding computer codes to resolve such linear systems and eigenvalue problems of the type (4.10).

4.2. A Numerical Example Concerning Laplace Model

Several numerical experiments were carried out by IBNOU-ZAHIR [1984] at Electricité de France in Clamart, FRANCE (see also IBNOU-ZAHIR and PLANCHARD [1983]). One of his examples will be presented below: He considered a sheaf of 100 identical cylindrical tubes arranged in 10 rows; 10 tubes in each row. They are distributed periodically with period ε inside a square cavity filled with water. Let r be the radius of the tubes. We give now some of the frequency values which correspond to the following data:

$$
(4.11) \quad
\begin{cases}
r = 0.5 \text{ cm}, \quad \varepsilon = 1.44 \text{ cm}, \quad \rho_0 = 1000 \text{ kg/m}^3, \\
\text{natural vibration frequency of tubes} = \sqrt{k/m} = 56.6 \text{ Hz}.
\end{cases}
$$

$$
\begin{aligned}
\omega_1 &= \omega_2 &&= 52.6570 \text{ Hz}, \\
\omega_3 &= \omega_4 &&= 52.6199 \text{ Hz}, \\
\omega_5 &= \omega_6 &&= 52.5620 \text{ Hz}, \\
\omega_7 &= \omega_8 &&= 52.4893 \text{ Hz}, \\
&\ \vdots \\
\omega_{99} &= \omega_{100} &&= 45.4883 \text{ Hz}, \\
\omega_{101} &= \omega_{102} &&= 45.4744 \text{ Hz}, \\
&\ \vdots \\
\omega_{197} &= \omega_{198} &&= 39.86630 \text{ Hz}, \\
\omega_{199} &= \omega_{200} &&= 39.84660 \text{ Hz}.
\end{aligned}
$$

Observe that the estimate (3.2) is verified in this numerical example. As shown by the numerical values listed above, the frequency values cluster to each other. This phenomenon is more pronounced as the number of tubes increases and this will be the subject of investigation in Chapter III.

The following Figs. 4.1 and 4.2 show some of the corresponding eigenvectors which represent the displacements of the tubes. Observe the different symmetries that the solution exhibits in this case.

4.3. Unitary Operator Method

The aim of this paragraph is to present a method due to GRÉGOIRE, NÉDELEC and PLANCHARD [1976] which is meant to calculate all the eigenvalues and eigenvectors of elliptic boundary value problems in bounded domains†. We illustrate the method in the case of the Laplace operator with Robin boundary condition (see Paragraph 2.2 of Chapter I). More precisely,

---

† This technique is based on an original idea due to P. Caseau (CASEAU [1972]).

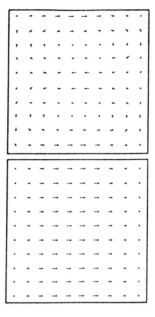

Figure 4.1.

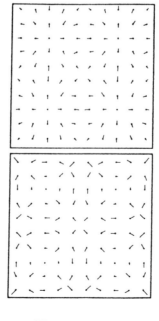

Figure 4.2.

let us consider the eigenvalue problem (2.16) of Chapter I with $c = 1$, i.e., find $\lambda \in \mathbb{C}$ for which there exists u not identically zero such that

$$(4.12) \qquad \begin{cases} \Delta u + \lambda u = 0 & \text{in} \quad \Omega, \\ \dfrac{\partial u}{\partial n} + bu = 0 & \text{on} \quad \Gamma. \end{cases}$$

Here we take b to be a constant to simplify the matters. It is assumed that

$$(4.13) \qquad 0 \le b \le \infty$$

with the understanding that $b = \infty$ corresponds to Dirichlet boundary condition on Γ.

The method essentially consists in constructing, for each real frequency $\omega \ne 0$, a unitary operator $B = B(\omega)$ and investigating the behaviour of its spectrum with respect to ω. We define $B \in \mathcal{L}(L^2(\Gamma))$ in the following way: We put

$$(4.14) \qquad Bg = g' \overset{\text{def}}{=} -\frac{\partial u}{\partial n} + i\omega u,$$

where $u \in H^1(\Omega)$ is the unique weak solution of the following boundary-value problem:

$$(4.15) \qquad \begin{cases} \Delta u + \omega^2 u = 0 & \text{in} \quad \Omega, \\ \dfrac{\partial u}{\partial n} + i\omega u = g & \text{on} \quad \Gamma. \end{cases}$$

Let us admit for the moment that (4.15) is well-posed for each $g \in L^2(\Gamma)$. We now see how the computation of the eigenvalues of B provides a novel method of finding the eigenvalues of (4.12). Let therefore $\mu \in \mathbb{C}$ be an eigenvalue of $B(\omega)$ with eigenvector g. It is then easily checked that the corresponding unique solution of (4.15) is an eigenfunction in (4.12) with

$$(4.16) \qquad \lambda = \omega^2, \quad b = i\omega \frac{\mu - 1}{\mu + 1}.$$

Observe that b defined by (4.16) is a real number. Indeed, b can be expressed alternatively as

$$(4.17) \qquad b = -\frac{2\omega \Im\mathrm{m}(\mu)}{|\mu + 1|^2}.$$

This is because $B(\omega)$ is unitary and hence $|\mu| = 1$. Therefore, since $b \ge 0$ and $\omega \ne 0$, the following equivalences hold true:

$$(4.18) \qquad \begin{cases} \Im\mathrm{m}(\mu) < 0 \iff \omega > 0, \\ \Im\mathrm{m}(\mu) > 0 \iff \omega < 0. \end{cases}$$

Thus, it is seen that one can recover eigenvalues of problem (4.12) at least for some values of b if ω varies over $\mathbb{R}\backslash\{0\}$. A refined analysis of the spectrum of the operator B and its variation with respect to ω will allow us to prove that for any fixed b all the eigenvalues of (4.12) are recovered by this method. For this reason, this is also known as the *method of frequency scanning*.

Conversely, let u be an eigenfunction of (4.12) corresponding to the eigenvalue λ. Then it is easily verified that

$$(4.19) \qquad \mu = \frac{i\omega + b}{i\omega - b} \quad (\text{with } \omega = +\sqrt{\lambda})$$

is an eigenvalue of $B = B(\omega)$ with eigenvector

$$(4.20) \qquad g = (i\omega - b)u.$$

Moreover, μ and λ have the same multiplicity as we shall see below. Obviously, a similar situation exists for the operator $B(-\omega)$.

Let us now proceed to study the operator B in a more detailed fashion. Our first aim is to prove the existence of a unique solution to (4.15) whose variational formulation is given by

$$(4.21) \quad \left\{ \begin{array}{l} u \in H^1(\Omega), \\[2mm] \displaystyle\int_\Omega \nabla u \cdot \nabla \bar{v}\,dx - \omega^2 \int_\Omega u\bar{v}\,dx + i\omega \int_\Gamma u\bar{v}\,d\gamma = \int_\Gamma g\bar{v}\,d\gamma \quad \forall v \in H^1(\Omega). \end{array} \right.$$

Theorem 4.3. *For any $g \in L^2(\Gamma)$ and for any real $\omega \neq 0$, problem (4.21) has a unique solution $u \in H^1(\Omega)$ and there exists a constant c depending on ω such that*

$$(4.22) \qquad \|u\|_{1,\Omega} \leq c\|g\|_{0,\Gamma}.$$

Proof. *Uniqueness:* If $g = 0$ then we have

$$\int_\Omega |\nabla u|^2\,dx - \omega^2 \int_\Omega |u|^2\,dx + i\omega \int_\Gamma |u|^2\,d\gamma = 0.$$

Taking imaginary part of the above relation, we obtain $u = 0$ on Γ. Thus u satisfies

$$(4.23) \quad \left\{ \begin{array}{l} \Delta u + \omega^2 u = 0 \quad \text{in} \quad \Omega, \\[2mm] u = \dfrac{\partial u}{\partial n} = 0 \quad \text{on} \quad \Gamma. \end{array} \right.$$

By the *Uniqueness Theorem of Holmgren* for the Cauchy problem for the operator $\Delta + \omega^2 I$, it follows that $u \equiv 0$ (for more modern versions of this result, the reader can consult LIONS [1988] p.56 for smooth domains and GRISVARD [1989] p.233 for Lipschitz domains).

Existence: We use a Galerkin procedure with a special basis to construct the solution. Let $\{v_n\}$ be a complete basis of $H^1(\Omega)$ formed by Neumann eigenfunctions of $(-\Delta)$ in Ω (see Paragraph 2.2 of Chapter I). For each $m \geq 1$, let us consider the following approximation of problem (4.21): Find u_m in the linear span V_m of $\{v_1, ..., v_m\}$ such that

$$(4.24) \quad \int_\Omega \nabla u_m \cdot \nabla \bar{v} dx - \omega^2 \int_\Omega u_m \bar{v} dx + i\omega \int_\Gamma u_m \bar{v} d\gamma = \int_\Gamma g \bar{v} d\gamma \quad \forall v \in V_m.$$

Since this is a finite dimensional problem, to prove the existence of u_m, it suffices to establish uniqueness. Suppose therefore that $g = 0$ in (4.24). We deduce then that

$$\int_\Omega |\nabla u_m|^2 dx - \omega^2 \int_\Omega |u_m|^2 dx + i\omega \int_\Gamma |u_m|^2 d\gamma = 0.$$

Taking imaginary part of this relation, we see that $u_m = 0$ on Γ. But u_m is a finite linear combination of Neumann eigenvectors. Therefore u_m satisfies (4.23). Appealing once again to the Uniqueness Theorem of Holmgren, we conclude that $u_m \equiv 0$.

Now, let us prove that $\{u_{m|\Gamma}\}$ is bounded in $L^2(\Gamma)$. Taking $v = u_m$ in (4.24) and separating the imaginary part of the resulting relation, one obtains

$$\omega \int_\Gamma |u_m|^2 d\gamma = \Im m \left(\int_\Gamma g \bar{u}_m d\gamma \right)$$

which easily implies, by Cauchy-Schwarz's Inequality, that $\{u_m\}$ is bounded in $L^2(\Gamma)$.

Our next step is to prove that $\{u_m\}$ is bounded in $L^2(\Omega)$ and also in $H^1(\Omega)$. To this end, let us suppose the contrary and assume that $\|u_m\|_{0,\Omega} \longrightarrow \infty$. Define $\hat{u}_m = u_m / \|u_m\|_{0,\Omega}$. Taking $v = u_m$ in (4.24), separating real parts and dividing by $\|u_m\|_{0,\Omega}^2$ we get

$$\int_\Omega |\nabla \hat{u}_m|^2 dx - \omega^2 = \frac{1}{\|u_m\|_{0,\Omega}} \Re e \left(\int_\Gamma g \hat{u}_m d\gamma \right).$$

As a consequence $\|\nabla \hat{u}_m\|_{0,\Omega}^2 \longrightarrow \omega^2$ and $\{\hat{u}_m\}$ is bounded in $H^1(\Omega)$. By extracting a subsequence (still denoted by m) we conclude that there exists $\hat{u} \in H^1(\Omega)$ such that $\hat{u}_m \rightharpoonup \hat{u}$ in $H^1(\Omega)$ weakly and in $L^2(\Omega)$ strongly. Thus $\|\hat{u}\|_{0,\Omega} = 1$ and $\hat{u} = 0$ on Γ. On the other hand, dividing (4.24) by $\|u_m\|_{0,\Omega}$ and passing to the limit, we get

$$\int_\Omega \nabla \hat{u} \cdot \nabla \bar{v} dx - \omega^2 \int_\Omega \hat{u} \bar{v} dx = 0 \quad \forall v \in H^1(\Omega).$$

Thus, using the Uniqueness Theorem of Holmgren once again we see that $\hat{u} \equiv 0$ which is clearly a contradiction. Thus $\{u_m\}$ is bounded in $L^2(\Omega)$ and aposteriori bounded in $H^1(\Omega)$. We can therefore extract a subsequence (still denoted by m) such that

$$u_m \rightharpoonup u \quad \text{in} \quad H^1(\Omega) \text{ weakly.}$$

A simple passage to the limit in (4.24) shows that u is a solution of (4.21). The whole sequence $\{u_m\}$ converges because u is unique. Moreover, it is easy to see that $u_m \longrightarrow u$ in $H^1(\Omega)$ strongly and that (4.22) holds. ∎

The foregoing result provides a unique solution $u \in H^1(\Omega)$ and so $u_{|\Gamma}$ is in $H^{1/2}(\Gamma)$. Therefore it follows from the boundary condition in (4.15) that $\frac{\partial u}{\partial n} \in L^2(\Gamma)$. Thus B defined by (4.14) is a well-defined bounded operator from $L^2(\Gamma)$ into itself.

Theorem 4.4. *The operator B is unitary in $L^2(\Gamma)$. Moreover, we have $B^*(\omega) = B(-\omega)$, where $B^*(\omega)$ is the adjoint of $B(\omega)$.*

Proof. To see that B is unitary, it is enough to prove that

$$\|g\|_{0,\Gamma} = \|g'\|_{0,\Gamma} \quad \forall g \in L^2(\Gamma).$$

Indeed, we have

$$\|g\|_{0,\Gamma}^2 = \int_\Gamma \left[|\frac{\partial u}{\partial n}|^2 + \omega^2 |u|^2 \right] d\gamma - i\omega \int_\Gamma \left[\frac{\partial u}{\partial n}\bar{u} - \frac{\partial \bar{u}}{\partial n}u \right] d\gamma.$$

By using Green's formula, we easily prove that the last integral is zero. Therefore

$$\|g\|_{0,\Gamma}^2 = \int_\Gamma \left[|\frac{\partial u}{\partial n}|^2 + \omega^2 |u|^2 \right] d\gamma.$$

In the same way, it is clear that

$$\|g'\|_{0,\Gamma}^2 = \int_\Gamma \left[|\frac{\partial u}{\partial n}|^2 + \omega^2 |u|^2 \right] d\gamma.$$

Thus B is unitary.

To prove the latter part of the theorem, we compute the adjoint $B^*(\omega)$. It is easily checked by Green's formula that $B^*(\omega)$ is given by

(4.25) $$B^*(\omega)h = \frac{\partial v}{\partial n} + i\omega v \quad \forall h \in L^2(\Gamma),$$

where v is the unique solution (which exists by Theorem 4.3) of

(4.26)
$$\begin{cases} \Delta v + \omega^2 v = 0 & \text{in } \Omega, \\ -\dfrac{\partial v}{\partial n} + i\omega v = h & \text{on } \Gamma. \end{cases}$$

A further checking easily yields $B^*(\omega) = B(-\omega)$. ∎

From the above result, it follows that $\sigma(B)$ is non-empty, it is contained in the unit circle and $\sigma_r(B)$ is empty (see Exercise 3.2 of Chapter I). It was already noted that the eigenvalues of B give rise to those of problem (4.12). We are now in a position to prove that the multiplicities are preserved in this correspondence.

Proposition 4.5. *Let λ be an eigenvalue of (4.12). Then μ given by (4.19) is an eigenvalue of $B(\omega)$ and both have the same multiplicity.*

Proof. Let $\{u_j\}$ be a basis of the eigenspace corresponding to λ. Assume that the corresponding functions $\{g_j\}$ given by (4.20) are not linearly independent. Then there exist scalars $\{\alpha_j\}$ such that $\sum_j \alpha_j g_j = 0$. From (4.20) it follows that

$$u \overset{\text{def}}{=} \sum_j \alpha_j u_j = 0 \quad \text{on} \quad \Gamma.$$

But u satisfies (4.12). Once again by Holmgren Theorem, it follows that $u \equiv 0$ in Ω and hence $\alpha_j = 0 \ \forall j$. This shows that the multiplicity of λ is lower than the multiplicity of μ. The other way inequality is similar; we need simply to use the result of Theorem 4.3. ∎

We now come to the crucial step of the method, namely the analysis of the spectrum $\sigma(B)$ of B.

Theorem 4.16. *For any non-zero real $\omega \neq 0$, -1 is an element of the spectrum $\sigma(B)$ of B and $\sigma(B)\backslash\{-1\}$ is an infinite sequence consisting of eigenvalues of B, each of finite multiplicity, which converges to -1. If -1 is not an eigenvalue of B then $\sigma_c(B) = \{-1\}$. Finally, if $\omega > 0$ (respectively $\omega < 0$) then there are at most a finite number of eigenvalues with negative (respectively positive) imaginary parts.*

Proof. *First Step*: We show that all elements of $\sigma(B)$ different from -1 are eigenvalues of B. Let then $-1 \neq \mu \in \sigma(B)$ be given. By virtue of Weyl's criterion (see Theorem 3.27 of Chapter I), there exists a sequence $\{g_n\}$ in $L^2(\Gamma)$ such that

$$
\begin{cases}
\|g_n\|_{0,\Gamma} = 1, \\
\psi_n \overset{\text{def}}{=} (B - \mu)g_n \longrightarrow 0 \quad \text{in} \quad L^2(\Gamma).
\end{cases}
$$

If u_n is the solution of (4.15) corresponding to g_n then we have

$$
\text{(4.27)} \qquad \psi_n = -(1 + \mu)g_n + 2i\omega u_{n|\Gamma}.
$$

From (4.22), it follows that $\{u_n\}$ is bounded in $H^1(\Omega)$ and therefore we can extract a subsequence (still denoted by n) such that

$$
\text{(4.28)} \qquad
\begin{cases}
u_n \rightharpoonup u \quad \text{in} \quad H^1(\Omega) \text{ weakly,} \\
g_n \rightharpoonup g \quad \text{in} \quad L^2(\Gamma) \text{ weakly.}
\end{cases}
$$

Passing to the limit in (4.27), we get

$$
\text{(4.29)} \qquad g = \frac{2i\omega}{1 + \mu} u_{|\Gamma}
$$

and so

$$
\|u\|_{0,\Gamma} = \frac{|1 + \mu|}{2|\omega|} \neq 0
$$

since in fact $g_n \longrightarrow g$ in $L^2(\Gamma)$ strongly.

Letting $n \to \infty$ in the variational equation (4.21) with $g = g_n$ and taking into account (4.29), we obtain

$$
\int_\Omega \nabla u \cdot \nabla \bar{v}\, dx - \omega^2 \int_\Omega u\bar{v}\, dx = i\omega \frac{1 - \mu}{1 + \mu} \int_\Gamma u\bar{v}\, dx \quad \forall v \in H^1(\Omega),
$$

i.e.,

$$
\begin{cases}
\Delta u + \omega^2 u = 0 \quad \text{in} \quad \Omega, \\
\dfrac{\partial u}{\partial n} + bu = 0 \quad \text{on} \quad \Gamma, \quad b = i\omega \dfrac{\mu - 1}{\mu + 1}.
\end{cases}
$$

Thus u is an eigenfunction of (4.12) and hence μ is an eigenvalue of B. This finishes the first step.

Second Step: Let us show that $\sigma(B)\backslash\{-1\}$ is a countably infinite sequence which converges to -1. Moreover, each element of $\sigma(B)\backslash\{-1\}$ is an eigenvalue of finite multiplicity.

Assume first ω^2 is not a Neumann eigenvalue of $(-\Delta)$ in Ω. Let $G: L^2(\Gamma) \longrightarrow L^2(\Gamma)$ be defined by the following rule:

$$(4.30) \qquad\qquad Gg = u_{|\Gamma} \quad \forall g \in L^2(\Gamma),$$

where $u \in H^1(\Omega)$ is the unique weak solution of

$$(4.31) \qquad \begin{cases} \Delta u + \omega^2 u = 0 & \text{in} \quad \Omega, \\ \dfrac{\partial u}{\partial n} = g & \text{on} \quad \Gamma. \end{cases}$$

It is easily checked that G is self-adjoint and compact. By applying the results of the general theory of self-adjoint compact operators developed in Paragraphs 3.2 and 3.3 of Chapter I and the standard arguments using Holmgren Theorem, one can check that $\sigma(G)$ consists of an infinite sequence of non-zero eigenvalues $\{\tau_m\}$, each of finite multiplicity, which converges to zero. The corresponding eigenfunctions satisfy

$$u_m = \tau_m \frac{\partial u_m}{\partial n} \quad \text{on} \quad \Gamma, \text{ i.e.,}$$

$$(4.32) \qquad \frac{\partial u_m}{\partial n} + b_m u_m = 0 \quad \text{on} \quad \Gamma \quad (\text{with } b_m = -\frac{1}{\tau_m}).$$

To each τ_m corresponds the eigenvalue μ_m of B given by (4.19), i.e.,

$$(4.33) \qquad\qquad \mu_m = \frac{i\omega + b_m}{i\omega - b_m}.$$

This shows that $\mu_m \longrightarrow -1$. Conversely, any eigenvalue τ of G comes from an eigenvalue μ of B that is different from -1. In fact, if μ $(\neq -1)$ is an eigenvalue of B then τ is an eigenvalue of G where

$$(4.34) \qquad\qquad \tau = \frac{\mu + 1}{\mu - 1} \frac{i}{\omega}.$$

The last expression for τ makes sense since 1 cannot be an eigenvalue of B in this case. This proves what we set out to prove in the second step if ω^2 is not a Neumann eigenvalue of $(-\Delta)$ in Ω.

Let us now examine the case when ω^2 is one such eigenvalue. We consider then the associated finite dimensional eigenspace W and we denote by $\gamma_1 W$ the image of W under the trace map γ_1. From Fredholm's Alternative (see

Theorem 3.13 of Chapter I), we know that (4.31) has a unique solution in $H^1(\Omega)/W$ iff g satisfies

$$(4.35) \qquad \int_\Gamma g w d\gamma = 0 \quad \forall w \in W,$$

i.e., $g \in (\gamma_1 W)^\perp$. In particular the trace $\gamma_1 u = u_{|\Gamma}$ is then defined up to an additive element of $\gamma_1 W$. We can thus choose $u_{|\Gamma} \in (\gamma_1 W)^\perp$ in a unique fashion. The above procedure enables us to define the operator G this time as a bounded operator from $(\gamma_1 W)^\perp$ into itself. We can apply the previous arguments to G and deduce the same conclusions as before. The only difference is that 1 is now an eigenvalue of B; it is of finite multiplicity.

Third Step: If -1 is not an eigenvalue of B then $\sigma_c(B) = \{-1\}$. This follows from the first step and the fact that B is unitary and so $\sigma_r(B) = \emptyset$ (see Exercise 2.2 of Chapter I).

Fourth Step: We now show that there are finitely many eigenvalues of B with negative imaginary part if $\omega > 0$. To this end, let us begin by regarding (4.12) as a 1-parameter family of eigenvalue problems with parameter $b \geq 0$. We know from Chapter I (see Paragraph 2.2, Example 3.5 and Exercise 3.13) that for all $b \geq 0$, problem (4.12) possesses a sequence of eigenvalues $\{\lambda_n(b)\}$ such that

$$(4.36) \qquad 0 \leq \lambda_1(b) \leq \cdots \leq \lambda_n(b) \leq \cdots \longrightarrow \infty.$$

Moreover, for any fixed n, $\lambda_n(b)$ is a non-decreasing continuous function of b (see Exercise 4.3 of Chapter I):

$$(4.37) \qquad \lambda_n(0) \leq \lambda_n(b) \leq \lambda_n(\infty)$$

where $\{\lambda_n(\infty)\}$ are the Dirichlet eigenvalues of $(-\Delta)$ in Ω.

From the correspondence established in (4.16)–(4.18), an eigenvalue μ of B with negative imaginary part is given by (4.19) where b is a solution of

$$(4.38) \qquad \omega = \sqrt{\lambda_n(b)}, \quad \omega > 0 \quad \text{for some } n \geq 1.$$

We claim now that there exist finitely many b's which satisfy (4.38). First, we observe that for any fixed $n \geq 1$, there exists at most one b_n such that $\omega^2 = \lambda(b_n)$. Indeed if $\lambda_n(b) = \lambda_n(c)$ for some b and c such that $b < c$ then, by monotonicity of $\lambda_n(\cdot)$, we get $\omega^2 = \lambda_n(d)$, for all $d \in [b,c]$. This implies that

$$\mu = \frac{i\omega + d}{i\omega - d}$$

is an eigenvalue of B, for all $d \in [b,c]$ (see (4.19)). In particular, this shows the existence of uncountably many eigenvalues for B which is impossible by the second step. Next, since the eigenvalues in (4.36) are of finite multiplicity, it follows that b_n's cannot be the same for infinitely many n's. This proves our claim and therefore the fourth step.

On the other hand, observe that if $\omega > 0$, then we have infinitely many eigenvalues μ of B with $\Im m(\mu) > 0$. This is a trivial consequence of the results of the second and fourth steps. Analogous statements hold good if $\omega < 0$. ∎

We conclude this paragraph by presenting a variant of the method described above and its application to the case of the *acoustic resonator* introduced in Paragraph 2.4 of Chapter I. We recall that in this example the boundary Γ of the domain Ω is divided into two pieces Γ_0 and Γ_1. Γ_0 represents a small opening through which the acoustic resonator is excited from the exterior, whereas Γ_1 is a rigid wall. It is difficult in practice to find exactly the nature of the boundary condition to be imposed on Γ_0 because the interior and the exterior of the resonator Ω interact with each other through Γ_0. In Paragraph 2.4, we have seen two boundary conditions which model this interaction (see equations (2.24) and (2.27) of Chapter I). It is completely natural to consider a boundary condition of the following type on Γ_0:

$$(4.39) \qquad \frac{\partial u}{\partial n} + bu = h \quad \text{on} \quad \Gamma_0 \times \mathbb{R},$$

where $h = h(x,t)$ is a given function and $b > 0$ is a parameter. With this model, we are led to the following eigenvalue problem which describes the free vibrations: Find $\lambda \in \mathbb{C}$ for which there exists a non-zero v such that

$$(4.40) \qquad \begin{cases} \Delta v + \lambda v = 0 \quad \text{in} \quad \Omega, \\ \dfrac{\partial v}{\partial n} = 0 \quad \text{on} \quad \Gamma_1, \\ \dfrac{\partial v}{\partial n} + bv = 0 \quad \text{on} \quad \Gamma_0. \end{cases}$$

In order to generalize the unitary operator method, we modify the operator $B = B(\omega)$ and define it as an element of $\mathcal{L}(L^2(\Gamma_0))$. More precisely, given $g \in L^2(\Gamma_0)$, we define $Bg = g'$, $g' = -\partial u/\partial n + i\omega u$, where u is now the solution of

$$(4.41) \qquad \begin{cases} \Delta u + \omega^2 u = 0 \quad \text{in} \quad \Omega, \\ \dfrac{\partial u}{\partial n} = 0 \quad \text{on} \quad \Gamma_1, \\ \dfrac{\partial u}{\partial n} + i\omega u = g \quad \text{on} \quad \Gamma_0. \end{cases}$$

◊ **Exercise 4.1.** Adapt the proof of Theorem 4.3 to conclude that problem (4.41) has a unique weak solution $u \in H^1(\Omega)$ for any $g \in L^2(\Gamma_0)$ and any real $\omega \neq 0$. ∎

Once B is well-defined, one can follow the procedure developed for solving problem (4.12). For the acoustic resonator this method seems to be particularly suitable since the value of the parameter b is not exactly known. However, this is not a handicap for the following reasons: It is intuitively clear that the dependence of the eigenvalues on b is weak if the size of Γ_0 is small with respect to that of Ω; this is the case in practice. Indeed, it can be proved that the spectrum of problem (4.40) converges to the Neumann eigenvalues of $(-\Delta)$ in Ω as the side of Γ_0 goes to zero (see CONCA, PLANCHARD, THOMAS and VANNINATHAN [1994]).

4.4. Practical Use of the Unitary Operator Method

As detailed in the preceding paragraph, this method consists of constructing the family of operators $B(\omega), \omega \in \mathbb{R}\backslash\{0\}$ and investigating the variation of the spectrum of $B(\omega)$ as ω varies. The main advantage of this procedure is that it gives simultaneously the eigenvalues of problems (4.12) for several boundary conditions since the operator $B(\omega)$ are independent of the parameter b. This is of great interest when, for example, a precise physical boundary condition is not well-known and some uncertainties exist in the value of the parameter b. This is the case, for instance, in the acoustic resonator examined above.

For each fixed ω, the construction of B is done by means of orthodox finite element techniques (see, for example, CIARLET [1978], RAVIART and THOMAS [1983]). The discretization of problem (4.21) leads to a system of the following form:

$$(4.42) \qquad M_h u_h = D_h g_h,$$

where h refers to the mesh size of the triangulation, M_h is a square matrix which represents the operator $\Delta + \omega^2 I$ with the boundary condition of (4.15) and $D_h g_h$ represents an approximation of the right hand side of (4.21). The analogue of $Bg = g'$ is the vector g'_h defined by

$$(4.43) \qquad g'_h(M) = -g_h(M) + 2i\omega u_h(M),$$

where M is a boundary node. The approximation to the operator B is then given by the matrix B_h which maps g_h to g'_h. It is constructed column by column, each one corresponding to a basis vector. The matrix thus obtained is unitary and its eigenvalues can be computed by a standard method, for instance, QR algorithm (see, e.g., WILKINSON [1965]).

Once the eigenvalues $\mu(\omega)$ of $B(\omega)$ are known, we go back to the relation (4.19). If b is known then we compute the values of ω which when squared gives the required eigenvalues of problem (4.12). From a practical point of view, it is advantageous to represent the relation (4.19) in polar coordinates. To this end, observe that if we put $b + i\omega = \rho e^{i\theta}$, then a simple computation shows that

$$(4.44) \qquad\qquad\qquad \mu = -e^{2i\theta}$$

and (4.19) is reduced to

$$(4.45) \qquad\qquad\qquad \arg(\mu(\omega)) = -\pi + 2\arctan(\frac{\omega}{b}).$$

We are now in a position to illustrate the application of the above method in a numerical example. This concerns the Neumann eigenvalue problem in a square domain $\Omega \subset \mathbb{R}^2$ of side π, i.e., we take $b = 0$ in (4.12). An explicit computation yields the eigenvalues $\lambda_{m,n} = m^2 + n^2, m \geq 0, n \geq 0$.

In Figure 4.3, we traced the curves $\arg(\mu_m(\omega))$ as ω varies. Their intersections with the ω-axis give different eigenvalues λ for problem (4.12) with $b = 0$. Observe that some of these eigenvalues lie on two curves $\arg(\mu_m(\omega))$ and $\arg(\mu_{m'}(\omega))$ thus showing that the multiplicity of such eigenvalues is two. This provides a numerical confirmation of Proposition 4.5.

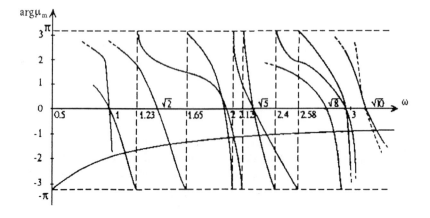

Figure 4.3: *The curves* $\arg(\mu_m(\omega))$.

4.5. Application to Helmholtz Model

The Helmholtz model was introduced in Paragraph 1.2 and the existence results were obtained in Paragraph 2.2 by following the Green's operator technique. In the present paragraph, we will show how to adapt the unitary operator method to compute the eigenvalues of the Helmholtz model. At first sight, one can think of applying this method to establish existence of eigenvalues. However, as we shall see below, this method avoids the frequency values NF defined in (2.1) and thus it is less general than the Green's operator method successfully carried out in Paragraph 2.2. Our opinion is that the unitary operator method has definite advantages in numerical computation of eigenvalues. Because of the scanning done in the frequency space, the fact that the values in NF are avoided does not play any important role.

To simplify the matters, we take $c = 1$ and consider the case of a single tube whose physical parameters are k and m. The corresponding unitary operator $A = A(\omega) \in \mathcal{L}(L^2(\Gamma_0))$ is introduced as follows: For real $\omega \neq 0$ and $\omega \neq \pm\sqrt{k/m}$, let u be the unique weak solution in $H^1(\Omega)$ of

$$(4.46) \qquad \begin{cases} \Delta u + \omega^2 u = 0 & \text{in} \quad \Omega, \\[2mm] \dfrac{\partial u}{\partial n} = \dfrac{\rho_0 \omega^2}{k - m\omega^2}\Big(\displaystyle\int_{\Gamma_1} u \, nd\gamma\Big) \cdot \mathbf{n} & \text{on} \quad \Gamma_1, \\[4mm] \dfrac{\partial u}{\partial n} + i\omega u = g & \text{on} \quad \Gamma_0, \end{cases}$$

where g is a given element in $L^2(\Gamma_0)$. Then define

$$(4.47) \qquad Ag = g' \overset{\text{def}}{=} -\frac{\partial u}{\partial n} + i\omega u.$$

A variational formulation of problem (4.46) is the following:

$$(4.48) \qquad \begin{cases} u \in H^1(\Omega), \\[2mm] a(u,v) - \omega^2 \displaystyle\int_{\Omega} u\bar{v}dx + i\omega \int_{\Gamma_0} u\bar{v}d\gamma - \frac{\rho_0\omega^2}{k-m\omega^2}b(u,v) = \int_{\Gamma_0} g\bar{v}d\gamma, \end{cases}$$

where the linear forms $a(\cdot,\cdot)$ and $b(\cdot,\cdot)$ are defined by

$$a(u,v) = \int_{\Omega} \nabla u \cdot \nabla \bar{v}dx, \quad b(u,v) = \int_{\Gamma_1} u \, nd\gamma \cdot \int_{\Gamma_1} \bar{v} \, nd\gamma.$$

Lemma 4.17. *If u is a weak solution of* (4.48) *with $g = 0$ then $u = 0$.*

Proof. Taking $v = u$ as a test function and separating the imaginary part of the resulting relation we obtain $u = 0$ on Γ_0. Thus u satisfies

$$\begin{cases} \Delta u + \omega^2 u = 0 & \text{in} \quad \Omega, \\ u = \dfrac{\partial u}{\partial n} = 0 & \text{on} \quad \Gamma_0. \end{cases}$$

We conclude $u \equiv 0$ by the Uniqueness Theorem of Holmgren. ∎

Theorem 4.18. *For any $g \in L^2(\Gamma_0)$ and for any real $\omega \notin \{0, \pm\sqrt{k/m}\}$, problem (4.48) has a unique solution $u \in H^1(\Omega)$ and we have the estimate $\|u\|_{1,\Omega} \le c\|g\|_{L^2(\Gamma_0)}$ for some constant $c > 0$.*

Proof. Uniqueness is proved in Lemma 4.17 above. For the existence, we use a Galerkin procedure as in Theorem 4.3. We work with a basis $\{v_n\}$ of $H^1(\Omega)$ consisting of Neumann eigenfunctions of $(-\Delta)$ in Ω. For each $n \ge 1$, let us consider the following approximation of problem (4.48): Find u_n in the linear span V_n of $\{v_1, ..., v_n\}$ such that for all v in V_n,

$$(4.49) \quad a(u_n, v) - \omega^2 \int_\Omega u_n \bar{v} dx + i w \int_{\Gamma_0} u_n \bar{v} d\gamma - \frac{\rho_0 \omega^2}{k - m\omega^2} b(u_n, v) = \int_{\Gamma_0} g\bar{v} d\gamma.$$

Exactly as in Theorem 4.3, we conclude that there exists a solution u_n of (4.49) by establishing its uniqueness and deduce that

$$(4.50) \qquad\qquad \{u_{n|\Gamma_0}\} \text{ is bounded in } L^2(\Gamma_0).$$

Our next step is to prove that

$$(4.51) \qquad\qquad \{u_n\} \text{ is bounded in } L^2(\Omega).$$

To this end, let us suppose the contrary and assume that $\|u_n\|_{0,\Omega} \longrightarrow \infty$. Define $\hat{u}_n = u_n/\|u_n\|_{0,\Omega}$. Taking $v = u_n$ in (4.49), separating real parts and dividing by $\|u_n\|_{0,\Omega}^2$, we get

$$\int_\Omega |\nabla \hat{u}_n|^2 dx - \omega^2 - \frac{\rho_0 \omega^2}{k - m\omega^2} \Big| \int_{\Gamma_1} \hat{u}_n \mathrm{n} d\gamma \Big|^2 = \frac{1}{\|u_n\|_{0,\Omega}} \Re\Big(\int_{\Gamma_0} g \hat{u}_n d\gamma \Big).$$

Letting $n \to \infty$ in this relation, we get

$$(4.52) \qquad \int_\Omega |\nabla \hat{u}_n|^2 dx - \frac{\rho_0 \omega^2}{k - m\omega^2} \Big| \int_{\Gamma_1} \hat{u}_n \mathrm{n} d\gamma \Big|^2 \longrightarrow \omega^2.$$

We now divide the proof of (4.51) into two cases.

• Case (i): $\omega^2 > (k/m)$.

In this case, it follows from (4.52) that $\{\hat{u}_n\}$ is bounded in $H^1(\Omega)$. Passing to a subsequence, we can suppose that

$$(4.53) \qquad \hat{u}_n \rightharpoonup \hat{u} \quad \text{in} \quad H^1(\Omega) \text{ weakly}.$$

Thus we get $\|\hat{u}\|_{0,\Omega} = 1$ and $\hat{u} = 0$ on Γ_0. On the other hand, dividing (4.49) by $\|u_n\|_{0,\Omega}$ and passing to the limit, we get

$$a(\hat{u}, v) - \omega^2 \int_\Omega \hat{u}\bar{v}dx - \frac{\rho_0\omega^2}{k - m\omega^2} b(\hat{u}, v) = 0 \quad \forall v \in H^1(\Omega).$$

Since $\hat{u} = 0$ on Γ_0, we see that $\hat{u}$ is a solution of (4.48) with $g = 0$. Appealing to Lemma 4.17, we conclude $\hat{u} = 0$ which is a contradiction. Thus (4.51) is proved in this case.

• Case (ii): $\omega^2 < (k/m)$.

In this case, we claim that there exists a constant M independent of n such that

$$(4.54) \qquad \left| \int_{\Gamma_1} \hat{u}_n \mathbf{n} d\gamma \right| \leq M.$$

This in conjunction with (4.52) implies that $\{\hat{u}_n\}$ is bounded in $H^1(\Omega)$ and so we can proceed as in case (i) above and conclude that (4.51) holds.

To prove (4.54), we argue by contradiction and suppose (along a subsequence) that

$$(4.55) \qquad \left| \int_{\Gamma_1} \hat{u}_n \mathbf{n} d\gamma \right| \longrightarrow \infty.$$

We introduce

$$(4.56) \qquad \tilde{u}_n = \hat{u}_n / \left| \int_{\Gamma_1} \hat{u}_n \mathbf{n} d\gamma \right|.$$

Dividing (4.52) by $\left| \int_{\Gamma_1} \hat{u}_n \mathbf{n} d\gamma \right|^2$ and passing to the limit, we obtain

$$\int_\Omega |\nabla \tilde{u}_n|^2 dx \longrightarrow \frac{\rho_0\omega^2}{k - m\omega^2}$$

and conclude that $\{\tilde{u}_n\}$ is bounded in $H^1(\Omega)$. Passing to a subsequence, if necessary, we can suppose that

$$(4.57) \qquad \tilde{u}_n \rightharpoonup \tilde{u} \quad \text{in} \quad H^1(\Omega) \text{ weakly}.$$

Thus we get

(4.58) $$\left| \int_{\Gamma_1} \tilde{u}\,\mathbf{n}\,d\gamma \right| = 1 \quad \text{and} \quad \tilde{u} = 0 \quad \text{on} \quad \Gamma_0.$$

On the other hand, dividing (4.49) by $\|u_n\|_{0,\Omega}\left| \int_{\Gamma_1} \hat{u}_n \mathbf{n}\,d\gamma \right|$ and passing to the limit, we get

$$a(\tilde{u}, v) - \omega^2 \int_{\Omega} \tilde{u}\bar{v}\,dx - \frac{\rho_0 \omega^2}{k - m\omega^2} b(\tilde{u}, v) = 0 \quad \forall v \in H^1(\Omega).$$

and therefore we conclude as before that $\tilde{u} = 0$ which is in contradiction with (4.50). This proves (4.54) and so case (ii) is finished.

Our next claim is that

(4.59) $$\{u_n\} \text{ is bounded in } H^1(\Omega).$$

As usual, the choice $v = u_n$ in (4.49) and the corresponding real part give

(4.60) $$\int_{\Omega} |\nabla u_n|^2\,dx - \omega^2 \int_{\Omega} |u_n|^2\,dx - \frac{\rho_0 \omega^2}{k - m\omega^2}\left| \int_{\Gamma_1} u_n \mathbf{n}\,d\gamma \right|^2 = \Re\left(\int_{\Gamma_0} g\bar{u}_n\,d\gamma \right).$$

Next, if $\omega^2 > (k/m)$ then (4.59) is a consequence of (4.50), (4.51) and (4.60). On the other hand, if $\omega^2 < (k/m)$ the idea consists of proving that

(4.61) $$\left\{ \left| \int_{\Gamma_1} u_n \mathbf{n}\,d\gamma \right| \right\} \text{ is bounded.}$$

For this purpose, we suppose the contrary and follow the renormalization technique used in deducing (4.51) in case (ii). Our claim (4.59) is now an easy consequence of (4.60) and (4.61). Once we get (4.59), the usual classical arguments enable us to pass to the limit in (4.49) and complete the proof of the existence part. ■

Thanks to Lemma 4.17 and Theorem 4.18 the operator A is well-defined as an element of $\mathcal{L}(L^2(\Gamma_0))$. We have

Theorem 4.19. *If $\omega \notin \{0, \pm\sqrt{(k/m)}\}$ then the operator $A = A(\omega)$ is unitary in $L^2(\Gamma_0)$. Moreover, we have $A^*(\omega) = A(-\omega)$.*

Proof. We follow the same method as in Theorem 4.4. The new property to be used this time is that

$$\int_{\Gamma_1} \left[\frac{\partial u}{\partial n} \bar{u} - \frac{\partial \bar{u}}{\partial n} u \right] d\gamma = 0,$$

which follows immediately from the boundary condition satisfied by u. ∎

The remaining properties of the unitary operator A are exactly the same as the one listed in Theorem 4.16 and the proof also follows the same steps. The important modifications needed have already been indicated in the previous two results. For this reason, we do not intend to restate Theorem 4.16 and neither go through its proof.

Finally, we remark that the above analysis can be readily generalized to the case of several tubes. From a numerical point of view, this method seems particularly well-adapted since it can be combined with suitable domain decomposition method to reduce the problem to the case of several single tubes and thereby reduce considerably the computation time. For the implementation of these ideas, see GRÉGOIRE [1973] and PLANCHARD [1982].

4.6. A Numerical Example Concerning Stokes Model

Several numerical experiments have been performed by DURÁN [1991] (see also CONCA and DURÁN [1994]). One of his examples will be seen below. This was carried out in a test geometry which consists of a square cavity with a single tube located at the center. Finite element method was used to discretize the Stokes system (for a general presentation of the finite element method in the resolution of Stokes and Navier-Stokes equations, the reader can consult GIRAULT and RAVIART [1986]). He used P_2 elements for the velocity field and P_1 elements for the pressure. The resulting discrete version of the quadratic eigenvalue problem (2.51) has been transformed into a generalized eigenvalue problem as indicated in Paragraph 2.3 (see also Paragraph 4.3 of Chapter I) and solved using a standard computer software package. The main features of the triangulation τ_h used by him are summarized below:

Number of triangles	128
Total numbers of vertices and middle points	288
Number of degrees of freedom for the velocity	450
Number of degrees of freedom for the pressure	80

Table 4.4 shows the computed eigenvalues of problem (2.51) for different values of (k/m). The viscosity of the fluid ν was taken to be unity. Figure 4.5

shows three of the eigenvectors represented by the velocity field and the isobaric lines in the case $(k/m) = 1$. It is noted that there are at most four non-real eigenvalues. This is not a surprise and it only confirms our theoretical result described by Theorem 3.18 (recall we have $K = 1$ here). In addition, the numerical results provide valuable information on the behaviour of the spectrum as the parameter k varies when m is fixed. In particular, it can be observed that for small values of the parameter k, the spectrum is only made up of real numbers. This is consistent with equation (2.51). As its value increases, the real eigenvalues collapse at a certain critical value of k and afterwards we see the appearance of non-real eigenvalues in conjugate pairs.

i	$k/m = 0.01$ ω_i	$k/m = 0.1$ ω_i	$k/m = 1.0$ ω_i	$k/m = 10.0$ ω_i	$k/m = 100.0$ ω_i
1	-1.868×10^{-4}	-1.869×10^{-3}	-1.876×10^{-2}	-0.196	$-2.067 + 1.752\,i$
2	-1.868×10^{-4}	-1.869×10^{-3}	-1.876×10^{-2}	-0.196	$-2.067 - 1.752\,i$
3	-2.625	-2.625	-2.625	-2.625	$-2.067 + 1.752\,i$
4	-3.863	-3.862	-3.847	-3.696	$-2.067 - 1.752\,i$
5	-3.863	-3.862	-3.847	-3.696	-2.625
6	-7.909	-7.909	-7.909	-7.909	-7.909
7	-7.934	-7.934	-7.934	-7.934	-7.934
8	-8.167	-8.167	-8.167	-8.164	-8.138
9	-8.167	-8.167	-8.167	-8.164	-8.138
10	-9.223	-9.223	-9.223	-9.223	-9.223
11	-9.554	-9.554	-9.554	-9.554	-9.554
12	-9.844	-9.844	-9.844	-9.841	-9.818
13	-9.844	-9.844	-9.844	-9.841	-9.818
14	-10.318	-10.318	-10.318	-10.318	-10.318
15	-11.574	-11.574	-11.573	-11.567	-11.504
16	-11.574	-11.574	-11.573	-11.567	-11.504
17	-11.924	-11.924	-11.924	-11.924	-11.924
18	-12.173	-12.173	-12.173	-12.173	-12.173
19	-13.724	-13.724	-13.724	-13.722	-13.701
20	-13.724	-13.724	-13.724	-13.722	-13.701
21	-14.096	-14.096	-14.096	-14.096	-14.096
22	-17.256	-17.256	-17.256	-17.256	-17.256
23	-17.758	-17.758	-17.758	-17.757	-17.748
24	-17.758	-17.758	-17.758	-17.757	-17.748
$\vdots$	$\vdots$	$\vdots$	$\vdots$	$\vdots$	$\vdots$
etc.	etc.	etc.	etc.	etc.	etc.
372	-431.518	-431.518	-431.518	-431.518	-431.518
373	-431.518	-431.518	-431.518	-431.518	-431.518

Table 4.4: *Eigenvalues of the discrete version of the Stokes model.*

Eigenfunction (velocity field and isobaric lines) for $\omega = \omega_3 = -2.625$

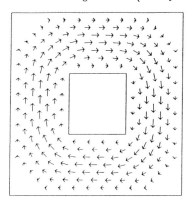 

Eigenfunction (velocity field and isobaric lines) for $\omega = \omega_4 = -3.847$

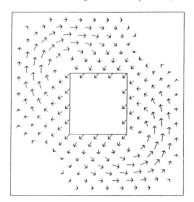 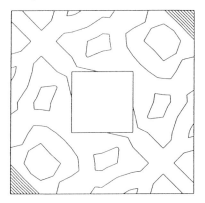

Eigenfunction (velocity field and isobaric lines) for $\omega = \omega_7 = -7.934$

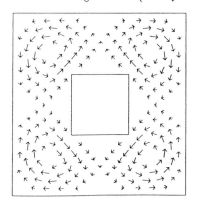 

Figure 4.5: *Graphical view of the eigenfunctions of the discrete version of the Stokes model.*

Chapter III

Asymptotic Methods in Fluid-Solid Structures

This chapter is divided into six sections. The first one defines the so-called Beppo-Levi spaces of functions which are slightly different from the usual Sobolev spaces. Section 1 also examines their local and global properties and these are used in the asymptotic analysis of the various models of fluid-solid structures. Section 2 is devoted to the description of the Bloch wave method in an example occurring in the quantum mechanics of an electron in a crystal. Based on this, we propose a *non-standard homogenization* procedure for spectral problems. Application of this new method to Laplace and Helmholtz models is the subject matter of Sections 3 and 4 wherein one also finds a discrete version of the method along with some numerical results. In Sections 5 and 6, we apply two techniques of the *standard homogenization method* to these models. The first is the so-called *two-scale convergence* technique and the second one is the *two-scale expansion* procedure.

§1. Beppo-Levi Spaces and their Properties

In our analysis of coupling of fluid-solid structures, we will have occasion to use other spaces of functions, called Beppo-Levi spaces, apart from the usual Sobolev spaces $H^m(\Omega)$ and the space $H(\Omega; \Delta)$. We remark that they appear in a natural way in our approach to study the asymptotic behaviour of the vibrations of such structures. In this paragraph, we will restrict our efforts to the definition of these spaces and some of their essential properties. In the literature, one often finds the treatment of these spaces on bounded and exterior domains. The domains of our interest are of different type; they are complements of unbounded domains. Therefore, even though the material is classical, we shall provide some proofs which will also cover the class of domains with which we will be concerned. For more details on the subject, the reader can consult the paper by DENY and LIONS [1953-54] and NIKODYM [1933] or the more recent books of DAUTRAY and LIONS [1988-90] Volume 4 and MAZ'JA [1985].

1.1. The Space $BL(\Omega)$

Let Ω be a connected open subset of $\mathbf{R}^N$ which is not necessarily bounded nor an exterior domain, i.e., the complement of a compact set in $\mathbf{R}^N$. The *Beppo-Levi space* (of order 1) on Ω is defined by

$$(1.1) \qquad BL(\Omega) = \left\{ u \in \mathcal{D}'(\Omega) \ \Big| \ \frac{\partial u}{\partial x_\ell} \in L^2(\Omega) \quad \forall \ell = 1, ..., N \right\}.$$

Since the map

$$(1.2) \qquad u \in BL(\Omega) \longmapsto |u|_{1,\Omega} \overset{\text{def}}{=} \|\nabla u\|_{0,\Omega}$$

defines a semi-norm on $BL(\Omega)$ which vanishes only on constants (because Ω is connected), it induces a norm on the quotient space $BL(\Omega) \overset{\text{def}}{=} BL(\Omega)/\mathbf{C}$. $BL(\Omega)$ endowed with the inner product

$$(1.3) \qquad (\dot{u}, \dot{v}) \overset{\text{def}}{=} \int_\Omega \nabla \dot{u} \cdot \nabla \bar{\dot{v}} dx \quad \forall \dot{u}, \dot{v} \in BL(\Omega)$$

is a Hilbert space. The first property of interest of the space $BL(\Omega)$ is the following local regularity of the functions in this space which shows, in particular, that the distributions in $BL(\Omega)$ are given by functions:

Theorem 1.1. *We have the inclusion*

$$BL(\Omega) \subset L^2_{loc}(\Omega).$$

Proof. Let ω be a bounded open subset of Ω such that $\omega \subset \bar{\omega} \subset \Omega$. We use the standard technique of localization by introducing two open sets Ω_1, Ω_2 of Ω and a cut-off function $\phi \in \mathcal{D}(\Omega)$ such that

$$\begin{cases} \bar{\omega} \subset \Omega_1 \subset \Omega_2 \subset \Omega, \\ \phi = 0 \quad \text{off} \quad \Omega_2, \\ \phi = 1 \quad \text{off} \quad \Omega_1. \end{cases}$$

Let $u \in BL(\Omega)$ and put $v = \phi u$ and consider this as an element of $\mathcal{D}'(\mathbf{R}^N)$ with compact support. On the other hand, let us choose $\gamma \in \mathcal{D}(\mathbf{R}^N)$ which is 1 in a neighbourhood of the origin. Denote by $E = E(x)$ the following fundamental solution of Δ in $\mathbf{R}^N$:

$$E(x) = \begin{cases} a_n |x|^{2-N} & \text{if} \quad N \geq 3, \\ \frac{1}{2\pi} \log |x| & \text{if} \quad N = 2, \end{cases}$$

where a_n is a constant which depends only on n. Then we have

$$\Delta(\gamma E) = \zeta + \delta,$$

where $\zeta \in \mathcal{D}(\mathbb{R}^N)$ and δ is the usual Dirac mass at the origin. Taking convolution of the above relation with v, we obtain

$$v + \zeta * v = \sum_{\ell=1}^{N} \frac{\partial}{\partial x_\ell}(\gamma E) * \frac{\partial v}{\partial x_\ell} \quad \text{in} \quad \mathbb{R}^N.$$

By our choice of ϕ, we see that $\frac{\partial \phi}{\partial x_\ell} u$ is zero on Ω_1 and so $[\frac{\partial}{\partial x_\ell}(\gamma E)] * [\frac{\partial \phi}{\partial x_\ell} u]$ will vanish on ω provided the support of γ is sufficiently small. Thus we arrive at the following relation on ω:

$$v + \zeta * v = \sum_{\ell=1}^{N} \left[\frac{\partial}{\partial x_\ell}(\gamma E)\right] * \left[\phi \frac{\partial u}{\partial x_\ell}\right] \quad \text{on} \quad \omega.$$

Since $\zeta * v \in C^\infty(\mathbb{R}^N)$ and $\frac{\partial}{\partial x_\ell}(\gamma E) \in L^1(\mathbb{R}^N)$, it follows that $v \in L^2(\omega)$, thanks to *Young's Inequality* for convolution product (see, e.g., BREZIS [1983] p.66). ∎

After the above local regularity result, we are now interested in a global property. First, we suppose that Ω is bounded. In this case, we would like to know whether $BL(\Omega) = H^1(\Omega)$. More precisely, we seek conditions on Ω which guarantee that, for all $u \in BL(\Omega)$, there exists a constant $c = c(u)$ such that $(u + c) \in H^1(\Omega)$. The following result is due to DENY and LIONS [1953-54]:

Theorem 1.2. *Assume that Ω is a bounded open subset of $\mathbb{R}^N$. Then $BL(\Omega) = H^1(\Omega)$ iff there exists a constant $M = M(\Omega)$ such that*

(1.4) $\inf_{c \in \mathbb{C}} \|u + c\|_{0,\Omega} \leq M|u|_{1,\Omega} \quad \forall u \in H^1(\Omega).$

Proof. *First Step:* We prove that (1.4) is necessary. Let $\dot{H}^1(\Omega)$ be the quotient space $H^1(\Omega)/\mathbb{C}$ which is a Hilbert space for the quotient norm

(1.5) $\|\dot{u}\| = \inf_{c \in \mathbb{C}} \|u + c\|_{0,\Omega} + |u|_{1,\Omega}.$

Since $BL(\Omega) = \dot{H}^1(\Omega)$, the canonical injection $\dot{u} \longmapsto \dot{u} = \{u + c\}$ from $\dot{H}^1(\Omega)$ is onto $BL(\Omega)$. Since it is also continuous and injective, the inverse is bounded by virtue of the Open Mapping Theorem. This proves (1.4).

Second Step: We show that (1.4) is sufficient. To this end, let us reconsider the above canonical injection which is always continuous. Now, (1.4) ensures that $\dot{H}^1(\Omega)$ is closed in $BL'(\Omega)$. To complete the proof, it suffices to show that $\dot{H}^1(\Omega)$ is dense in $BL'(\Omega)$. Let then $u \in BL(\Omega)$. Without loss of generality, we can assume that u is real-valued. We construct a sequence $\{u_k\}_k$ in $H^1(\Omega)$ by the method of truncation:

$$(1.6) \qquad u_k(x) = \begin{cases} -k & \text{if} \quad u(x) < -k, \\ u(x) & \text{if} \quad |u(x)| \leq k, \\ k & \text{if} \quad u(x) > k. \end{cases}$$

Since u_k can be expressed also as $u_k = \inf\{\sup(u, -k), k\}$, it is easy to check that $u_k \in H^1(\Omega)$ and its gradient is given by

$$\nabla u_k = \chi_k \nabla u,$$

where χ_k is the characteristic function of the set $\{x \in \Omega \mid |u(x)| \leq k\}$. By Dominated Convergence Theorem, we see that $\nabla u_k \longrightarrow \nabla u$ in $L^2(\Omega)^N$, i.e., $\dot{u}_k \longrightarrow \dot{u}$ in $BL'(\Omega)$. This completes the proof. ∎

◇ **Exercise 1.1.** (Bramble-Hilbert Lemma) If Ω satisfies the regularity condition (1.5) of Chapter I, using Rellich's Lemma, prove that the inequality (1.4) holds. (*Hint*: Argue by contradiction). ∎

Among the bounded open sets, we now have, thanks to Theorem 1.2 and Exercise 1.1, a huge collection of sets Ω for which $H^1(\Omega)$ and $BL(\Omega)$ coincide.

1.2. The Space $W^1(\Omega)$

Let us now consider the case of unbounded domains Ω. The study of the behaviour of elements of $BL(\Omega)$ in bounded parts of Ω has just been carried out. It is therefore interesting to know their behaviour as $|x| \to \infty$, $x \in \Omega$. As we shall see, this depends on the dimension N and to this end, we introduce the following *Sobolev spaces with weights*:

$$(1.7) \quad W^1(\Omega) \overset{\text{def}}{=} \left\{ u \in \mathcal{D}'(\Omega) \;\middle|\; \rho u \in L^2(\Omega), \; \frac{\partial u}{\partial x_\ell} \in L^2(\Omega) \quad \forall \ell = 1, ..., N \right\},$$

where the weight $\rho(x)$ depends on N and is defined by

$$(1.8) \qquad \rho(x) = \begin{cases} \dfrac{1}{(1 + |x|^2)^{1/2}} & \text{if} \quad N \geq 3, \\[3mm] \dfrac{1}{(1 + |x|^2)^{1/2}\left[1 + \log(1 + |x|^2)\right]} & \text{if} \quad N = 2. \end{cases}$$

The above space endowed with the norm

$$(1.9) \qquad \|u\|_{\rho,1,\Omega}^2 \overset{\text{def}}{=} \int_{\Omega} |\rho(x)u(x)|^2 dx + |u|_{1,\Omega}^2$$

is a Hilbert space. Clearly the following inclusions are continuous:

$$(1.10) \qquad H^1(\Omega) \hookrightarrow W^1(\Omega) \hookrightarrow BL(\Omega).$$

We use the notation $W_0^1(\Omega)$ to denote the closure of $\mathcal{D}(\Omega)$ in $W^1(\Omega)$. For every $R > 0$, we shall denote by $B(0,R)$ or simply by B_R the open ball centered at the origin and with radius R. B_R' will stand for the interior of the complement of B_R. Now, we have

Lemma 1.3. *There exists a constant $c = c(N)$ such that for all $R > 0$, we have*

$$(1.11) \qquad \|u\|_{\rho,1,B_R'} \leq c|u|_{1,B_R'} \quad \forall u \in W_0^1(B_R').$$

Proof. *First Step*: We suppose $N \geq 3$. By density, it is enough to prove that there exists a constant $c > 0$ such that

$$\int_{B_R'} |\rho(x)u(x)|^2 dx \leq c|u|_{1,B_R'}^2 \quad \forall u \in \mathcal{D}(B_R').$$

Using the following change of variables $x \longmapsto [r,\eta]$, where $r = |x|$ and $\eta = (x/|x|) \in S^{N-1}$, we obtain

$$\int_{B_R'} |\rho(x)u(x)|^2 dx \leq \int_{B_R'} \frac{|u(x)|^2}{|x|^2} dx = \int_R^{\infty} \int_{S^{N-1}} \frac{|u(r,\eta)|^2}{r^2} r^{N-1} dr d\eta =$$

$$= \int_R^{\infty} \int_{S^{N-1}} |u(r,\eta)|^2 r^{N-3} dr d\eta.$$

We now apply the following *generalized Hardy's Inequality*

$$(1.12) \qquad \int_R^{\infty} |f(r)|^2 r^{\beta} dr \leq \left(\frac{2}{|\beta+1|}\right)^2 \int_R^{\infty} |f'(r)|^2 r^{\beta+2} dr,$$

which holds for all $f \in \mathcal{D}(]R,\infty[)$, $\beta \neq -1$ and $R > 1$. We take $\beta = N - 3$. Since $N \geq 3$, we have $\beta \geq 0$ and hence choosing $f(r) = u(r,\eta)$ with η as a parameter, we get

$$\int_{B_R'} |\rho(x)u(x)|^2 \leq c \int_R^{\infty} \int_{S^{N-1}} |\frac{\partial u}{\partial r}(r,\eta)|^2 r^{N-1} dr d\eta = c \int_{B_R'} |\nabla u(x)|^2 dx,$$

where $c = (4/(N-2))^2$. To complete the proof of (1.11), it therefore remains to establish (1.12). In fact, the left side of (1.12) can be rewritten as

$$\int_R^\infty |f(r)|^2 \frac{1}{\beta+1} \frac{d}{dr} r^{\beta+1} dr$$

and by integration by parts, we see that it is equal to

$$-\frac{2}{\beta+1} \int_R^\infty f(r) f'(r) r^{\beta+1} dr.$$

Applying Cauchy-Schwarz's Inequality, we see that the left side of (1.12) is dominated by

$$\frac{2}{|\beta+1|} \left(\int_R^\infty |f(r)|^2 r^\beta dr \right)^{1/2} \left(\int_R^\infty |f'(r)|^2 r^{\beta+2} dr \right)^{1/2}$$

which yields the desired inequality (1.12).

Second Step: We suppose now $N = 2$. Following the arguments of the first step we see that to finish the proof it suffices to use the following *generalized Hardy's Inequality* (instead of (1.12)),

$$(1.13) \qquad \int_R^\infty |f(r)|^2 \frac{1}{r} |\log r|^\beta dr \le \left(\frac{2}{|\beta+1|} \right)^2 \int_R^\infty |f'(r)|^2 r |\log r|^{\beta+2} dr$$

which one can easily verify for all $f \in \mathcal{D}(]R, \infty[)$ and $\beta \ne -1$. ∎

Remark. An obvious consequence of Lemma 1.3 is the fact that if Ω is any open set such that either Ω or any translate of Ω is contained in B'_R for some R then

$$(1.14) \qquad \|u\|_{\rho,1,\Omega} \le c|u|_{1,\Omega} \quad \forall u \in W_0^1(\Omega).$$

Here c is the same constant that appears in (1.11). ∎

We turn now our attention to those domains Ω for which the inequality (1.11) remains valid for all $u \in W^1(\Omega)$. If Ω is bounded then $W^1(\Omega)$ coincides with $H^1(\Omega)$ and so this cannot be true. However, for the following class of unbounded domains, we have a positive answer.

Definition. Let Ω be a connected open subset of $\mathbb{R}^N$. Ω is said to satisfy *condition* (P) iff there exist $R > 0$, $x_0 \in \mathbb{R}^N$ and an extension operator $P \in \mathcal{L}(BL(\Omega), BL(B'(x_0, R))) \cap \mathcal{L}(W^1(\Omega), W^1(B'(x_0, R)))$ such that

$$(1.15) \quad \begin{cases} \text{(i)} & \Omega \subset B'(x_0, R), \\ \text{(ii)} & |Pu|_{1,B'(x_0,R)} \leq c|u|_{1,\Omega} \quad \forall u \in BL(\Omega), \\ \text{(iii)} & \|Pu\|_{\rho,1,B'(x_0,R)} \leq c\|u\|_{\rho,1,\Omega} \quad \forall u \in W^1(\Omega), \end{cases}$$

for some constant c depending only on Ω. ∎

◇ **Exercise 1.2.** If Ω is bounded then there cannot exist an extension operator P satisfying (1.15) (ii) and (iii). ∎

A simple example of an open set satisfying condition (P) is an exterior domain Ω of the form $\Omega = \mathbb{R}^N \backslash \bar{\mathcal{O}}$, where $\mathcal{O}$ is a bounded open set having the regularity specified by (1.5) of Chapter I. Another example which is suggested by future applications is the case of the domain Ω which is obtained from $\mathbb{R}^N$ by removing a bundle of infinite number of holes which are periodically distributed and disjoint. Observe that this example is not an exterior domain; however, it satisfies condition (P), a fact which we check later.

Theorem 1.4. *Let Ω satisfy condition (P). Then there is a constant $c > 0$ which depends on Ω, such that*

$$(1.16) \quad \|u\|_{\rho,1,\Omega} \leq c|u|_{1,\Omega} \quad \forall u \in W^1(\Omega).$$

Proof. We argue by contradiction and suppose that there is a sequence $\{u_n\}$ in $W^1(\Omega)$ such that

$$(1.17) \quad \begin{cases} |u_n|_{1,\Omega} \longrightarrow 0 \quad \text{as } n \to \infty, \\ \|u_n\|_{\rho,1,\Omega} = 1 \quad \forall n. \end{cases}$$

Let us put $v_n = Pu_n$ and using (1.15), we see that v_n satisfies conditions similar to (1.17). More precisely,

$$(1.18) \quad \begin{cases} |v_n|_{1,B'_R} \longrightarrow 0 \quad \text{as } n \to \infty, \\ 1 \leq \|v_n\|_{\rho,1,\Omega} \leq c \quad \forall n. \end{cases}$$

Without loss of generality, we can assume that $x_0 = 0$ and $R = 1$ in (1.15). Choose ϕ and ψ in $C^\infty(\mathbb{R}^N)$ such that

$$(1.19) \quad \begin{cases} \phi + \psi = 1 \quad \text{in } \mathbb{R}^N, \\ \phi = 0 \quad \text{for } |x| \geq 2, \\ \psi = 0 \quad \text{for } |x| \leq 1, \\ \phi \geq 0 \quad \text{and} \quad \psi \geq 0. \end{cases}$$

Taking $v \in W^1(B'_1)$ and differentiating ψv and ϕv, we get

$$(1.20) \qquad \begin{cases} |\psi v|_{1,B'_1} \le |v|_{1,B'_1} + c_1 \|v\|_{0,B_2 \cap B'_1}, \\ |\phi v|_{1,B'_1} \le |v|_{1,B_2 \cap B'_1} + c_1 \|v\|_{0,B_2 \cap B'_1}. \end{cases}$$

By Rellich's Lemma, we know that the canonical embedding $H^1(B_2 \cap B'_1)$ into $L^2(B_2 \cap B'_1)$ is compact and hence we can extract a subsequence from $\{v_n\}$, which we still denote $\{v_n\}$, such that

$$v_n \longrightarrow v \quad \text{in} \quad L^2(B_2 \cap B'_1).$$

It is now a consequence from (1.18) and (1.20) that the sequences $\{\phi v_n\}$ and $\{\psi v_n\}$ are Cauchy in $H^1(B_2 \cap B'_1)$ and $W^1_0(B'_1)$ respectively. Hence

$$\begin{cases} \phi v_n \longrightarrow w_1 \quad \text{in} \quad H^1(B_2 \cap B'_1), \\ \psi v_n \longrightarrow w_2 \quad \text{in} \quad W^1_0(B'_1). \end{cases}$$

Therefore, $v_n = (\phi + \psi)v_n \longrightarrow w_1 + w_2 = w, w \in W^1(B'_1)$. From (1.18), we deduce then that $\nabla w = 0$ and so $w = c$, constant. The asymptotic behaviour of functions in $W^1(B'_1)$ impose that $c = 0$ which contradicts (1.18). ∎

1.3. Behaviour at Infinity

We are now in a position to prove the following result which describes the asymptotic behaviour of the functions of $BL(\Omega)$ as $|x| \to \infty, x \in \Omega$:

Theorem 1.5. *Let $\Omega \subset \mathbb{R}^N$ satisfy condition (P) and $N \ge 2$. Then the canonical embedding $u \in W^1(\Omega) \longmapsto \dot{u} = \{u + c\} \in BL(\Omega)$ defines an isomorphism which is both topological and algebraic.* ∎

By means of this isomorphism, we will always identify $W^1(\Omega)$ with $BL(\Omega)$ provided Ω satisfies condition (P). It is an easy consequence of this identification that given any $u \in BL(\Omega)$ there is a constant $c = c(u)$ such that $(u + c) \in W^1(\Omega)$ and in particular $\rho(u + c) \in L^2(\Omega)$.

Proof of Theorem 1.5. It is evident that the canonical embedding of $W^1(\Omega)$ into $BL'(\Omega)$ is a linear continuous map. Theorem 1.4 ensures that its image is closed in $BL'(\Omega)$. Therefore, to conclude the proof, it is sufficient to show that $W^1(\Omega)$ is dense in $BL'(\Omega)$. Without loss of generality, assume that $x_0 = 0$ and $R = 1$ as before. Assume also that Theorem 1.5 is true for the exterior domain B_1'. Now, it is a simple matter to finish the proof because, for all $\dot u \in BL'(\Omega)$, we have $P\dot u \in BL'(B_1')$ and so there exists an element $\dot v \in W^1(B_1')$ such that $\dot v = P\dot u$. Obviously, $v_{|\Omega} \in W^1(\Omega)$ and $(v_{|\Omega})' = \dot u$ in Ω. This is the desired result.

Thus we are reduced to show Theorem 1.5 in the case of the exterior domain B_1'. Since $W^1(B_1')$ has already been identified with a closed subspace of $BL'(B_1')$ it is sufficient to prove that its orthogonal complement $W^1(B_1')^\perp$ in $BL'(B_1')$ is reduced to zero. To this end, let $u \in W^1(B_1')^\perp$ be arbitrary. This means that

$$\int_{B_1'} \nabla u \cdot \nabla \bar v \, dx = 0 \quad \forall v \in W^1(B_1').$$

Taking suitable test functions v in $W^1(B_1')$, we see that u satisfies

(1.21)
$$\begin{cases} \Delta u = 0 & \text{in} \quad B_1', \\ \dfrac{\partial u}{\partial n} = 0 & \text{on} \quad \partial B_1, \quad \displaystyle\int_{B_1'} |\nabla u|^2 \, dx < \infty. \end{cases}$$

We will solve this system explicitly and show that constants are the only solutions. To do this, we make a change of variable $x \in B_1' \longmapsto [r, \eta]$, where $r = |x| > 0$ and $\eta = (x/|x|) \in S^{N-1}$. Let $w = w(r, \eta)$ be the transformed solution of (1.21). We can express w in terms of spherical harmonics $\{Y_n\}_{n \geq 0}$ which are nothing but the eigenfunctions of the *Laplace-Beltrami operator* Δ_B on S^{N-1}:

(1.22)
$$w(r, \eta) = \sum_{n \geq 0} f_n(r) Y_n(\eta).$$

It is classically known that there is a countable number of eigenvalues $\{\lambda_n\}_{n \geq 0}$ with complete eigenvectors $\{Y_n\}_{n \geq 0}$ for Δ_B in the space $L^2(S^{N-1})$ (see, e.g., DAUTRAY and LIONS [1988-90] Volume 2 or MULLER [1966]). The eigenvalues form a decreasing sequence

$$\lambda_0 > \lambda_1 \geq \lambda_2 \geq \cdots \longrightarrow -\infty$$

where the first eigenvalue $\lambda_0 = 0$ is simple and the corresponding eigenvectors are constant functions. The other eigenvalues have multiplicities greater than 1†.

---

† If $N = 3$, it is known that $\lambda_n = -n(n+1)$, $n \geq 0$ and its multiplicity is exactly $(2n+1)$. If $N = 2$ then the eigenvalues λ_n are equal to $-n^2$; λ_0 is simple whereas for $n \geq 1$, λ_n has multiplicity 2.

In order to find the coefficients $f_n(r)$ in (1.22), we write Δ in polar coordinates:

$$(1.23) \qquad \Delta u = \frac{\partial^2 w}{\partial r^2} + \frac{N-1}{r}\frac{\partial w}{\partial r} + \frac{1}{r^2}\Delta_B w.$$

Substituting the expression (1.22) into (1.21) we get the following set of equations for f_n:

$$(1.24) \qquad \begin{cases} f_n'' + \dfrac{N-1}{r}f_n' - \dfrac{\lambda_n}{r^2}f_n = 0 \quad r > 1, \\[3mm] f_n'(1) = 0, \quad \displaystyle\int_1^\infty |f_n'(r)|^2 r^{N-1}\,dr < \infty. \end{cases}$$

We make one more change of variable $r > 1 \longmapsto t > 0$, where $r = e^t$. Let $g_n(t) = f_n(r)$. An easy computation shows that g_n's satisfy the following set of equations:

$$(1.25) \qquad \begin{cases} g_n'' + (N-2)g_n' - \lambda_n g_n = 0 \quad \text{for} \quad t > 0, \\[3mm] g_n'(0) = 0, \quad \displaystyle\int_0^\infty |g_n'(t)|^2 e^{(N-2)t}\,dt < \infty. \end{cases}$$

We can easily solve these ordinary differential equations and find that the solutions to be of the form:

$$(1.26) \qquad g_n(t) = A_n e^{m_1 t} + B_n e^{m_2 t}$$

where m_1, m_2 are the roots of the equation

$$(1.27) \qquad m^2 + (N-2)m - \lambda_n = 0$$

and A_n, B_n are constants restricted by the relation

$$(1.28) \qquad A_n m_1 + B_n m_2 = 0.$$

It is easily seen that the integrability condition together with the explicit form (1.26) for the solution and (1.28) show that

$$A_n = B_n = 0 \quad \text{if} \quad n \geq 1.$$

This means that the only solutions of (1.21) are constants and this completes the proof of Theorem 1.5. ∎

We remark that if $N = 1$ then the spaces $W^1(\Omega)$ and $BL'(\Omega)$ do not coincide even if $\Omega = B_1'$. Indeed, the function u defined by $u(x) = 1$ if $x > 1$ and $u(x) = -1$ if $x < -1$ belongs to $BL(\Omega)$ and it is not possible to choose a constant c such that $(u + c) \in W^1(\Omega)$.

The last result in which we are interested concerns the approximation of elements of $W^1(\Omega)$ or $BL'(\Omega)$ by smooth functions.

Theorem 1.6. *We suppose that either Ω is a bounded open set satisfying (1.5) of Chapter I or Ω is an unbounded open set satisfying condition (P). Then $\mathcal{D}(\bar{\Omega})$ is dense in $W^1(\Omega)$.*

Proof. If Ω is bounded then the space $W^1(\Omega)$ coincides with $H^1(\Omega)$ and the result is classical. In the case of unbounded domains, using the extension operator P, we are reduced to the case of the exterior domain B_1' as shown in the proof of Theorem 1.5. Afterwards, the proof proceeds in two stages: *truncation* followed by *regularization*.

For the first step, let $u \in W^1(B_1')$ be given and choose $\zeta \in \mathcal{D}(\mathbb{R}^N)$ such that $\zeta(x) = 1$ if $|x| \leq 1$ and $\zeta(x) = 0$ if $|x| \geq 2$. Next, we construct the sequence $u_n(x) = u(x)\zeta(x/n)$ which clearly belongs to $H^1(B_1')$ and has compact support. We will now show that this sequence converges to u in the semi-norm $|\cdot|_{1,B_1'}$. To simplify, we consider the case $N \geq 3$. We have

$$|u - u_n|^2_{1,B_1'} \leq \int_{B_1'} |\zeta(\frac{x}{n}) - 1|^2 |\nabla u|^2 dx + \int_{\{n \leq |x| \leq 2n\}} |u|^2 |\nabla \zeta|^2 \frac{1}{n^2} dx.$$

The first integral on the right side converges to zero by Dominated Convergence Theorem. The second one can be majorized by

$$c \int_{B_1'} \chi_{\{|x| \geq n\}} \frac{|u(x)|^2}{|x|^2} dx.$$

Since $(u(x)/|x|) \in L^2(B_1')$, this integral also converges to zero by Dominated Convergence Theorem. This concludes the truncation step.

The second step, namely regularization involves convolution and proceeds as in the case of $\mathbb{R}^N$. This is classical and so we omit it. ∎

§2. Bloch Wave Method in a Classical Example

In the existence theory developed in Chapter II, we worked with arbitrary distribution of solid structures in the fluid. In the qualitative theory that follows, periodic distribution of solids is supposed. This justifies the title of the book. In such a situation, it is natural to think of Bloch wave method which is concerned with the spectral theory of partial differential equations in domains with periodic structures. For example, the domain can have a periodic distribution of holes and/or the equations can have periodic coefficients. In the mathematics literature, the study of ordinary differential equations with periodic coefficients is known as *Floquet theory*. In the physics side, it goes under the name *Bloch wave method* because it was BLOCH [1928] who first introduced this technique in his study of motion of electrons in a crystalline solid. Since crystals have periodic structures this study amounts to propagation of waves in such media and more exactly, to the spectral analysis of Schrödinger equation with periodic potential. Afterwards, this method has been well-developed and applied in various situations. The existing large physical literature on these topics can be obtained from the books of BRILLOUIN [1953], CRACKNELL and WONG [1973], REED and SIMON [1978] Volume IV and ZIMAN [1972].

Needless to say that the corresponding mathematical theory is also very big and we sketch the classical approach in Paragraph 2.1 below. Here we do not provide many details of proofs because this approach can be found in several references including BENSOUSSAN, LIONS and PAPANICOLAOU [1978], EASTHAM [1973], KUCHMENT [1993], SÁNCHEZ-HUBERT and SÁNCHEZ-PALENCIA [1989]. Moreover the set-up used in this classical approach is not directly applicable to the class of problems treated in this book. On the contrary, we intend to present in §3 a somewhat different and original approach to the theory of Bloch waves in another framework which is well adapted to the applications we have in mind. Of course, we provide full details of the proofs in §3.

Paragraph 2.1 apart from presenting the classical approach in the case of *Schrödinger operator* also offers some remarks concerning the historical development of the theory. There the reader can find the existence of the so-called *Bloch waves*, the corresponding decomposition and their main properties. In Paragraph 2.2, we present the dual approach to Bloch wave decomposition which consists of simply translating the approach of the previous paragraph in terms of Fourier transformation. As a consequence of these results, we establish, in Paragraph 2.3, the continuity of the spectral family of the Schrödinger operator. Based on Bloch wave decomposition, we propose a non-standard homogenization procedure to eigenvalue problems which is illustrated in the case of Schrödinger operator.

2.1. Bloch Waves and their Properties

Let us begin by recalling that we have studied the propagation of waves in the free space (see Paragraph 2.5 of Chapter I) which is obviously a particular case of a domain with a periodic structure. The central problem was to obtain the spectral resolution of the self-adjoint operator $(-\Delta)$ in $\mathbb{R}^n$ and we did this by means of plane waves and Fourier analysis. More exactly, we proved that the spectrum of this operator consisted of the non-negative real axis and that the plane waves $e^{i\zeta \cdot x}$ with $|\zeta|^2 = \lambda$ can be considered as "approximate eigenvectors" with "eigenvalue" $\lambda \geq 0$. These functions are not elements of $L^2(\mathbb{R}^N)$ but they span all of $L^2(\mathbb{R}^N)$ in the sense of Fourier inversion (see (2.36) of Chapter I).

In the example just cited, the medium does not vary at all. By perturbation techniques, one can treat the case when the medium is slowly varying (see BENSOUSSAN, LIONS and PAPANICOLAOU [1978] Chapter IV). At the other extreme lie media which oscillate rapidly. For instance, the coefficients representing the medium can have small period. The spectral analysis of such operators have been thoroughly investigated following the methods presented in the first chapter of the above cited reference. The relevant works, in this context, are KESAVAN [1979] and VANNINATHAN [1981].

In contrast with the above cases, we are going to analyze, in this paragraph, an intermediate situation. More precisely, we consider the operator $A = (-\Delta + W)$ in $L^2(\mathbb{R}^N)$ where W is a Y-periodic function in the sense to be defined below. The above analysis is generalized and its spectral decomposition is obtained.

The functions which now play the role of approximate eigenvectors are known as (classical) Bloch waves. The first result in this subject is due to GELFAND [1950]. In his paper, a proof is outlined of the so-called *generalized Parseval's Identity* for functions in $L^2(\mathbb{R}^N)$. A more detailed discussion of Bloch waves was published by ODEH and KELLER [1964]. On the other hand, the regularity results of these waves can be found in WILCOX [1978].

From a physical point of view, $(-\Delta + W)$ in $L^2(\mathbb{R}^3)$ is the one-electron Hamiltonian and W is the multiplication operator by the function $W(x)$ which represents the crystal potential. W is therefore real-valued and periodic with the periodicity of the crystal lattice. In the general situation of $\mathbb{R}^N$, we express the periodicity condition as follows:

(2.1) $W(x + 2\pi p) = W(x)$ for a.e. x in $\mathbb{R}^N$, $\forall p \in \mathbb{Z}^N$.

A function satisfying the above condition is said to be Y-*periodic*, where

(2.2) $Y =]0, 2\pi[^N$.

Y is referred to as the *reference cell*. Further it is assumed that $W \in L^\infty(Y)$ and we consider A to be an unbounded operator in $L^2(\mathbb{R}^N)$ with domain $D(A) = H^2(\mathbb{R}^N)$. Then we have

Theorem 2.1. $A = (-\Delta + W)$ *acting on* $D(A) = H^2(\mathbb{R}^N)$ *is a self-adjoint operator in* $L^2(\mathbb{R}^N)$. *Moreover, A is bounded below, i.e., there exists* $m_- \in \mathbb{R}$ *such that* $(Au, u) \geq m_- \|u\|^2_{0,\Omega}$ *for all* $u \in D(A)$.

Proof. The self-adjointness of A is an easy consequence of the fact that $(-\Delta)$ is a self-adjoint operator proved in Example 3.7 of Chapter I. On the other hand, an easy computation shows that A is bounded below by $m_- I$ with $m_- = -\|W\|_{0,\infty,Y}$. ∎

In the sequel, we will assume, without loss of generality, that $W \geq a_0 > 0$; otherwise one can replace W by $(W + c)$, where c is any large constant.

As a consequence, we see that the constant m_- in Theorem 2.1 is positive. As mentioned above, our goal is to obtain the spectral decomposition of A. The above modification in W results only in a translation of the spectrum of A without affecting its eigenfunctions. We consider thus the eigenvalue equation of A:

$$(2.3) \qquad A\psi \stackrel{\text{def}}{=} -\Delta\psi + W\psi = \lambda\psi \quad \text{in} \quad \mathbf{R}^N.$$

Definition. The *Bloch waves* for the operator A are the solutions $\psi(x)$ of (2.3) that have the form

$$(2.4) \qquad \psi(x) = e^{i\zeta \cdot x}\phi(x) \quad \text{for a.e. } x \text{ in } \mathbf{R}^N,$$

where ϕ is a non-zero Y-periodic function and $\zeta \in \mathbf{R}^N$ is a parameter. Alternatively, we can look for solutions of (2.3) that have the following property:

$$(2.5) \qquad \psi(x + 2\pi p) = e^{2\pi i \zeta \cdot p}\psi(x) \quad \text{for a.e. } x \text{ in } \mathbf{R}^N, \forall p \in \mathbf{Z}^N.$$

Functions which satisfy (2.5) will be called (ζ, Y)-*periodic*. ∎

    If the medium were homogeneous, i.e., W were a constant, it is customary to solve (2.3) by Fourier analysis and look for plane wave solutions of the form $\psi(x) = e^{i\zeta \cdot x}$. These obviously satisfy (2.5). Thus in passing from the constant coefficient case to that of periodic coefficients, it seems natural to impose (2.5). Thus the Bloch waves can be thought of being born out of the interaction between the plane waves and the periodic medium. Condition (2.5) is a replacement of *Sommerfeld radiation condition* which is usually imposed in the presence of compact obstacles in the medium (see Remark following Example 3.18 in Chapter I). Another motivation for (2.5) will be provided later in Paragraph 3.1.

    The idea of finding (ζ, Y)-periodic solutions of (2.3) is independently due to G. Floquet and F. Bloch. It is clear from (2.5) that if ζ is replaced by $(\zeta + q)$ with $q \in \mathbf{Z}^N$, then property (2.5) is unaltered and we can therefore restrict ζ to the cell

$$(2.6) \qquad Y' = [0, 1[^N.$$

We refer to Y' as the *reciprocal cell* of Y. In the physics literature, Y' is known as the *first Brillouin zone*.

Remark. In the general case where the reference cell is $\prod_{\ell=1}^{N}]0, \tau_\ell[, \tau_\ell > 0$, then the corresponding reciprocal cell is given by $\prod_{\ell=1}^{N}]0, (2\pi/\tau_\ell)[$. ∎

Substituting (2.4) into (2.3) leads us to the following equation for the Y-periodic function ϕ:

$$(2.7) \qquad\qquad A(\zeta)\phi = \lambda\phi \quad \text{in} \quad \mathbb{R}^N,$$

where the operator $A(\zeta)$ is defined as

$$(2.8) \qquad\qquad A(\zeta) = -(\Delta + 2i\zeta \cdot \nabla - |\zeta|^2) + W.$$

This is known as the *shifted operator* in the literature because $A(\zeta)$ is formally obtained from A by replacing the derivatives $\frac{\partial}{\partial x_\ell}$ by $(\frac{\partial}{\partial x_\ell} + i\zeta_\ell)$. Thus we are reduced to studying the spectral properties of the family of operators $\{A(\zeta)\}_{\zeta \in Y'}$. To resolve (2.7), we follow the Green's operator technique studied extensively in Chapter I. With this in mind, we introduce the following Lebesgue and Sobolev spaces:

$$(2.9) \qquad\qquad L^2_\#(Y) \stackrel{\text{def}}{=} \big\{v \in L^2_{loc}(\mathbb{R}^N) \mid v \text{ satisfies (2.1)}\big\},$$

$$(2.10) \qquad H^1_\#(Y) \stackrel{\text{def}}{=} \Big\{v \in L^2_\#(Y) \mid \frac{\partial v}{\partial x_\ell} \in L^2_\#(Y) \quad \forall \ell = 1, ..., N\Big\}.$$

They are Hilbert spaces with respect to the inner products in $L^2(Y)$ and $H^1(Y)$ respectively. For each fixed ζ, we introduce the bilinear form corresponding to the operator $A(\zeta)$:

$$(2.11) \qquad a(\zeta; u, v) = \int_Y \big[\nabla u \cdot \nabla \bar{v} - 2i(\zeta \cdot \nabla u)\bar{v} + |\zeta|^2 u\bar{v} + Wu\bar{v}\big]dx,$$

for each $u, v \in H^1_\#(Y)$. To resolve (2.7), we introduce $G(\zeta): L^2_\#(Y) \to L^2_\#(Y)$ defined as $G(\zeta)f = u$, where $u \in H^1_\#(Y)$ is the unique solution of the following variational problem

$$(2.12) \qquad\qquad a(\zeta; u, v) = \int_Y f\bar{v}dx \quad \forall v \in H^1_\#(Y).$$

To establish the existence and uniqueness of u, we use Lax-Milgram Lemma and the following result:

Proposition 2.2. *For all $\zeta \in \mathbb{R}^N$, there exists a constant $\alpha > 0$ such that we have the following ellipticity inequality:*

$$\mathfrak{Re}\, a(\zeta; u, u) \geq \alpha\|u\|^2_{1,Y} \quad \forall u \in H^1_\#(Y).$$

Proof. A simple integration by parts shows that

$$(2.13) \qquad -2i \int_Y (\zeta \cdot \nabla u)\bar{v}\,dx = 2i \int_Y (\zeta \cdot \nabla \bar{v})u\,dx.$$

Therefore, we have

$$(2.14) \qquad \mathfrak{Re}\, a(\zeta; u, u) = \|\nabla u + iu\zeta\|_{0,Y}^2 + \int_Y W|u|^2\,dx.$$

This proves that the bilinear form $a(\zeta; \cdot, \cdot)$ satisfies the three conditions of Exercise 2.1 below. Hence the ellipticity follows. ∎

On one hand, by Rellich's Lemma, the injection $H_\#^1(Y) \hookrightarrow L_\#^2(Y)$ is compact and on the other hand, we check easily that the bilinear form $a(\zeta; \cdot, \cdot)$ is Hermitian (see (2.13)). As a consequence, $G(\zeta)$ is a compact self-adjoint operator on $L_\#^2(Y)$. Moreover, it is easily seen that $G(\zeta)$ is injective. Therefore by applying the spectral theory of compact self-adjoint operators (see Paragraph 3.3 of Chapter I), we conclude the following

Theorem 2.3. *Let $\zeta \in \mathbb{R}^N$ be fixed. Then there exists a sequence of eigenvalues for problem (2.7) each of which is of finite multiplicity. As usual, we arrange them in increasing order repeating each value as many times as its multiplicity:*

$$(2.15) \qquad 0 < \lambda_1(\zeta) \leq \cdots \leq \lambda_n(\zeta) \leq \cdots \longrightarrow \infty.$$

The corresponding eigenfunctions denoted by $\{\phi_n(\zeta, \cdot)\}$ form an orthonormal basis in $L_\#^2(Y)$. ∎

◇ **Exercise 2.1.** Let $(V, \|\cdot\|)$ and $(H, |\cdot|)$ be Hilbert spaces such that we have the inclusion $V \hookrightarrow H$ which is compact. Suppose that $a(\cdot, \cdot)$ defines a continuous sesquilinear form on V with the following properties:

$$\begin{cases} \text{(i)} & \mathfrak{Re}\, a(u, u) \geq 0 \quad \forall u \in V, \\ \text{(ii)} & \mathfrak{Re}\, a(u, u) \geq c_1\|u\|^2 - c_2|u|^2 \quad \forall u \in V, \\ \text{(iii)} & u \in V, \ \mathfrak{Re}\, a(u, u) = 0 \Rightarrow u = 0, \end{cases}$$

where $c_1 > 0$ and $c_2 \in \mathbb{R}$. Show then that there exists a constant $c_3 > 0$ such that

$$\mathfrak{Re}\, a(u, u) \geq c_3\|u\|^2 \quad \forall u \in V.$$ ∎

◇ **Exercise 2.2.** Show that the result of Proposition 2.2 is valid for $\zeta \in \mathbf{C}^N$ also. More precisely, extend the definition of $a(\zeta; \cdot, \cdot)$ to $\zeta \in \mathbf{C}^N$ and show that there exists a constant $\alpha > 0$ such that

$$\mathfrak{Re}\, a(\zeta; u, u) \geq \alpha \|u\|_{1,Y}^2 \quad \forall u \in H_{\#}^1(Y).$$

Conclude therefore that $A(\zeta)$ has compact inverse for $\zeta \in \mathbf{C}^N$. ∎

◇ **Exercise 2.3.** (i) Show that there exists a constant $c > 0$ independent of $\zeta \in Y'$ such that

$$|\zeta| \|u\|_{0,Y} \leq c \|\nabla u + iu\zeta\|_{0,Y}$$

for all $\zeta \in Y'$ and $u \in H_{\#}^1(Y)$. (*Hint:* The reader may carry out the following steps: (a) Define the space $H_{\#}^1(\zeta, Y) = \{\tilde{u} \in H_{loc}^1(\mathbf{R}^N) \mid \tilde{u} \text{ is } (\zeta, Y)\text{-periodic}\}$ and show that smooth elements of this space are dense. (b) Obtain the relation

$$\tilde{u}(x) = \frac{1}{e^{2\pi i \zeta_j} - 1} \int_0^1 \frac{\partial \tilde{u}}{\partial x_j}(x + te_j)dt \quad \forall j = 1, \cdots N$$

for smooth elements of $H_{\#}^1(\zeta, Y)$ and deduce that for all $\tilde{u} \in H_{\#}^1(\zeta, Y)$ the following estimate:

$$|\zeta| \|\tilde{u}\|_{0,Y} \leq c \|\nabla \tilde{u}\|_{0,Y}$$

holds with c independent of $\zeta \in Y'$).

(ii) As a consequence of (i), deduce that there exists a constant $c > 0$ independent of $\zeta \in Y'$ such that

$$\|\nabla u\|_{0,Y} + |\zeta| \|u\|_{0,Y} \leq c \|\nabla u + iu\zeta\|_{0,Y} \quad \forall u \in H_{\#}^1(Y), \ \zeta \in Y'. \ \blacksquare$$

According to our definition of Bloch waves, they are given by

(2.16) $$\psi_n(\zeta, x) = e^{i\zeta \cdot x} \phi_n(\zeta, x) \quad \forall n \geq 1.$$

The continuity of the functions $\lambda_n : \bar{Y}' \longrightarrow \mathbf{R}$ can be derived from perturbation theory and it is done, for example, in EASTHAM [1973]. (The arguments needed to prove this fact are more sophisticated than the ones advanced in §4 of Chapter I). The study of the analyticity properties of λ_n with respect to ζ requires special attention. A standard application of Rellich's Theorem (see KATO [1966] p.392) shows the existence of branches of eigenvalues of $A(\zeta)$ which are real-analytic with respect to each individual variable ζ_ℓ. However, perturbation theory is inadequate to study the real-analyticity of these branches with respect to all the variables ζ (see KATO [1966] p.177). This point has been analyzed by WILCOX [1978] who proved the following sharp result:

Theorem 2.4. *To each potential $W \in L^\infty(Y)$ that satisfies (2.1) and to each $\rho > 0$ there corresponds a function $D(\zeta, \lambda)$ complex-analytic for $|\mathfrak{Im}(\zeta)| \leq \rho$ and $\lambda \in \mathbb{C}$ which is Y'-periodic in the ζ variable and for each $\zeta \in \mathbb{R}^N$ the sequence $\{\lambda_n(\zeta)\}$ in (2.15) is precisely the set of zeros of $D(\zeta, \lambda)$, enumerated by magnitude and repeated according to their multiplicities.* ∎

As already pointed out in §4 of Chapter I, the enumeration of eigenvalues by magnitude destroys, in general, the regularity of the functions $\lambda_n(\zeta)$ due to the multiplicity. However, this regularity is not completely lost as shown by yet another result of WILCOX [1978].

Theorem 2.5. *For each $n \geq 1$, there exists a closed null set X_n of $\mathbb{R}^N$ such that the function $\zeta \longmapsto \lambda_n(\zeta)$ is real-analytic in $\bar{Y}'\backslash X_n$.* ∎

Let us now say a few words about the regularity of the eigenfunctions ϕ_n. To this end, let us observe that the eigenfunctions are not uniquely fixed even if the corresponding eigenvalue is simple. This is because the phase of such eigenfunction can be chosen to be an arbitrary function of ζ. In the case of multiple eigenvalues, the range of choices is much wider. WILCOX [1978] constructs explicitly a family of eigenfunctions ϕ_n having the regularity announced in the following result:

Theorem 2.6. *There exist sequences of functions $\phi_n\colon \bar{Y}' \times \bar{Y} \longrightarrow \mathbb{C}$ and closed null sets $Z_n \subset \bar{Y}'$ such that the map $\zeta \longmapsto \phi_n(\zeta, \cdot) \in C^0(\bar{Y})$ is real-analytic on $\bar{Y}'\backslash Z_n$ and $\{\phi_n(\zeta, \cdot)\}$ is a complete orthonormal sequence of eigenfunctions of $A(\zeta)$ for all $\zeta \in \bar{Y}'\backslash Z$ where $Z = \bigcup_{n=1}^\infty Z_n$.* ∎

Since ζ has several components, we cannot, in general, choose eigenvectors which are continuous functions of ζ even in the case of a complex analytic family of operators acting on a finite dimensional space, as may be shown by simple examples (see Exercise 2.4 below). In this sense, the above result is close to the best possible and it implies, in particular, that the selection of eigenvectors can be made in such a way that they are measurable with respect to the variables (ζ, x).

◇ **Exercise 2.4.** Let $N = 2$ and define

$$T(\zeta_1, \zeta_2) = \begin{pmatrix} \zeta_1 & \zeta_2 \\ \zeta_2 & -\zeta_1 \end{pmatrix} \qquad \forall \zeta = (\zeta_1, \zeta_2) \in \mathbb{R}^2.$$

Find the eigenvalues of this matrix and show that one cannot choose eigenvectors such that they define continuous functions of $\zeta \in \mathbb{R}^2$. ∎

We are now in a position to announce the classical result of GELFAND [1950] and ODEH and KELLER [1964] which is the main element for the construction of the spectral family of the operator A.

Theorem 2.7. (Continuous Bloch Wave Decomposition) *Let* $v \in L^2(\mathbf{R}^N)$ *be arbitrary. Then the following limits*

$$(2.17) \qquad \lim_{M \to \infty} \int_{\{|x| \leq M\}} v(x)\bar{\psi}_n(\zeta, x)dx$$

exist in $L^2(Y')$ *for* $n \geq 1$. *These limits are denoted as* $\hat{v}_n(\zeta)$ *and they are called Bloch coefficients of* v *and they satisfy the (generalized) Parseval's Identity:*

$$(2.18) \qquad \|v\|^2_{0,\mathbf{R}^N} = \sum_{n=1}^{\infty} \|\hat{v}_n\|^2_{0,Y'}.$$

Moreover, we have the following inversion formula:

$$(2.19) \qquad v(x) = \lim_{M \to \infty} \sum_{n=1}^{M} \int_{Y'} \hat{v}_n(\zeta)\psi_n(\zeta, x)d\zeta,$$

where the limit is understood in the sense of $L^2(\mathbf{R}^N)$. ∎

The relation (2.19) giving the expansion of f in terms of Bloch waves was first obtained by ODEH and KELLER [1964] wherein one can also find a proof of (2.18). Their paper also contains a discussion of the functions $\zeta \longmapsto \lambda_n(\zeta)$ and a proof of the fact that the spectrum of A has a *band structure*. More exactly, they derived the following relation:

Theorem 2.8.

$$\sigma(A) = \bigcup_{n=1}^{\infty} \lambda_n(\bar{Y}'). \quad \blacksquare$$

The construction of the spectral family of A is now immediate.

Theorem 2.9. *The spectral family* $E_\lambda(A)$ *of* A *is given by*

$$(2.20) \qquad (E_\lambda v)(x) = \begin{cases} 0 & \text{if} \quad \lambda < 0, \\ \int_{Y'} \sum_{n;\lambda_n \leq \lambda} \hat{v}_n(\zeta)\psi_n(\zeta, x)d\zeta & \text{if} \quad \lambda \geq 0; \end{cases}$$

the addition being taken over those indices $n = n(\zeta)$ *for which* $\lambda_n(\zeta) \leq \lambda$. ∎

Thus, we have achieved our goal of obtaining the spectral decomposition of A. From a practical point of view, the advantage is the reduction of a spectral problem in $\mathbb{R}^N$ to a family of eigenvalue problems in the reference cell Y. This family is parametrized by $\zeta \in Y'$ and so the cost of computation of the eigenvalues is considerably reduced.

In order to see more clearly how the Bloch wave method works in a general set-up, not necessarily involving Schrödinger operator, let us now give some guidelines which constitute the major steps of the proof of Theorems 2.7–2.9. The starting point is to associate to every $v \in L^2(\mathbb{R}^N)$ and to every $\zeta \in Y'$ the following function of two variables (ζ, x):

$$(2.21) \qquad v_{\#}(\zeta, x) = \sum_{p \in \mathbb{Z}^N} v(x + 2\pi p) e^{-i\zeta \cdot (x + 2\pi p)}.$$

It is easily checked that the right side series converges and defines an element of $L^2_{\#}(Y)$ as a function of x. The main point is that we recover v from $v_{\#}$ by the following formula:

$$(2.22) \qquad v(x) = \int_{Y'} e^{i\zeta \cdot x} v_{\#}(\zeta, x) d\zeta.$$

This gives a decomposition (integral) of v in terms of the functions $e^{i\zeta \cdot x} v_{\#}(\zeta, x)$ which are (ζ, Y)-periodic. So, it is natural to introduce, for all $\zeta \in Y'$, the space

$$(2.23) \qquad L^2_{\#}(\zeta, Y) \overset{\text{def}}{=} \{ \psi \in L^2_{loc}(\mathbb{R}^N) \mid \psi \text{ is } (\zeta, Y)\text{-periodic} \}$$

which is a Hilbert space for the inner product of $L^2(Y)$. Observe that

$$L^2_{\#}(\zeta, Y) \cap L^2_{\#}(\zeta', Y) = \{0\} \quad \text{if} \quad \zeta \neq \zeta'.$$

Therefore, (2.22) amounts to say that $L^2(\mathbb{R}^N)$ can be written as a direct integral decomposition:

$$(2.24) \qquad L^2(\mathbb{R}^N) = \int_Y^{\oplus} L^2_{\#}(\zeta, Y) d\zeta.$$

Observe that what we have done up to now is completely independent of the operator A. The advantage of the above decomposition is that it also provides a splitting of a given operator. Indeed, once A is given, then in a natural way one can associate to it two families of operators $\{A(\zeta)\}_{\zeta}$ and $\{B(\zeta)\}_{\zeta}$ one acting on Y-periodic functions and the other acting on (ζ, Y)-periodic functions and they are defined as follows:

$$(2.25) \qquad \begin{cases} A(\zeta)\phi(x) = e^{-i\zeta \cdot x} A(e^{i\zeta \cdot x}\phi(x)) \quad \forall \phi \in L^2_{\#}(Y), \\ B(\zeta)\psi(x) = A\psi(x) \quad \forall \psi \in L^2_{\#}(\zeta, Y). \end{cases}$$

These operators are obviously related by

(2.26) $$B(\zeta) = e^{i\zeta \cdot x} A(\zeta) e^{-i\zeta \cdot x}.$$

The operator $A(\zeta)$ coincides with the one introduced in (2.8). The main motivation to introduce these operators is that they provide the following decompositions of A which follow from (2.22):

(2.27)
$$\begin{cases} Av(x) = \displaystyle\int_{Y'} e^{i\zeta \cdot x} A(\zeta) v_{\#}(\zeta, x) d\zeta, \\[2em] Av(x) = \displaystyle\int_{Y'} B(\zeta)(e^{i\zeta \cdot x} v_{\#}(\zeta, x)) d\zeta. \end{cases}$$

Now, if we expand $v_{\#}(\zeta, \cdot)$ in terms of the eigenfunctions $\phi_n(\zeta, \cdot)$ and use (2.16), we obtain

(2.28) $$v_{\#}(\zeta, x) = \sum_{n=1}^{\infty} \hat{v}_n(\zeta) \phi_n(\zeta, x).$$

Thanks to the decomposition (2.27), the analysis of the operator A is reduced to that of $\{B(\zeta)\}_\zeta$. Fixing ζ, we deduce from (2.26) and (2.28) that

$$B(\zeta)(e^{i\zeta \cdot x} v_{\#}(\zeta, x)) = \sum_{n=1}^{\infty} \lambda_n(\zeta) \hat{v}_n(\zeta) e^{i\zeta \cdot x} \phi_n(\zeta, x)$$

which gives the spectral decomposition of $B(\zeta)$.

The decomposition (2.28) leads to another splitting of A. More precisely, we have

$$A = \bigoplus_{n=1}^{\infty} A_n$$

where A_n is defined by

$$A_n v(x) = \int_{Y'} \lambda_n(\zeta) \hat{v}_n(\zeta) e^{i\zeta \cdot x} \phi_n(\zeta, x) d\zeta.$$

All the conclusions of Theorems 2.7–2.9 follow relatively easily from the above considerations.

Theorem 2.7 says that the mapping $v \longmapsto \{\hat{v}_n(\zeta)\}_n$ is a unitary isomorphism of $L^2(\mathbb{R}^N)$ onto $L^2(Y', \ell^2)$. This transforms our operator A in $L^2(\mathbb{R}^N)$ into the operator $\tilde{A}$ in $L^2(Y', \ell^2)$ defined by

$$\tilde{A}\{\hat{v}_n(\zeta)\}_n = \{\lambda_n(\zeta) \hat{v}_n(\zeta)\}_n.$$

Further A_n goes into $\tilde{A}_n \in \mathcal{L}(L^2(Y'))$ which is given by the multiplication by $\lambda_n(\zeta)$. As a consequence, the spectral projection $E_\lambda(\tilde{A}_n)$ of $\tilde{A}_n$ is nothing but the operator given by the multiplication of the characteristic function of the set $\{y \in Y' \mid \lambda_n(\zeta) \leq \lambda\}$.

To illustrate the power of the method, let us consider the situation of Paragraph 2.5 and Example 3.18 of Chapter I.

◇ **Example 2.5.** (Laplace operator in $\mathbb{R}^N$) This corresponds to the zero-potential W. In order to obtain the associated Bloch waves, we need to solve the following family of eigenvalue problems parametrized by $\zeta \in Y'$:

(2.29)
$$\begin{cases} -\Delta\psi(\zeta,\cdot) = \lambda(\zeta)\psi(\zeta,\cdot) & \text{in } Y, \\ \psi(\zeta,\cdot) \text{ is } (\zeta,Y)\text{-periodic.} \end{cases}$$

One can explicitly solve this problem by separation of variables and obtain

(2.30)
$$\begin{cases} \lambda_q(\zeta) = |\zeta + q|^2, & q \in \mathbb{Z}^N \\ \psi_q(\zeta, x) = \dfrac{1}{(2\pi)^{N/2}} e^{i(\zeta+q)\cdot x}, & q \in \mathbb{Z}^N. \end{cases}$$

With the help of these expressions, we see that the Bloch coefficients $\hat{v}_q(\zeta)$ of a function $v \in L^2(\mathbb{R}^N)$ coincide with the translate Fourier transformation of v. More precisely, we have

(2.31)
$$\hat{v}_q(\zeta) = \frac{1}{(2\pi)^{N/2}} \int_{\mathbb{R}^N} v(x) e^{-i(\zeta+q)\cdot x} dx = \hat{v}(\zeta + q).$$

The resolution of identity given by (2.19) then coincides with the Fourier inversion formula for $L^2(\mathbb{R}^N)$ functions. This is just the same conclusion we reached in Example 3.18 of Chapter I (see also (2.36) of Chapter I). ∎

We conclude this paragraph by citing a complementary result due to EASTHAM [1973] on the behaviour of the eigenvalues given by Theorem 2.3.

Theorem 2.10. *Under the hypothesis of Theorem 2.3, there exists a constant $c > 0$ independent of ζ such that*

(2.32)
$$\lambda_n(\zeta) \sim cn^{\frac{2}{N}} \quad \text{as } n \to \infty. \blacksquare$$

Indeed, by the use of Courant-Fischer characterization of eigenvalues (see Theorem 3.18 of Chapter I) one obtains interlacing inequalities between $\lambda_n(\zeta)$, Dirichlet and Neumann eigenvalues. The behaviour (2.32) follows then from the classical *Weyl's asymptotic formula* (see COURANT and HILBERT [1953] p.442).

2.2. Momentum Space Picture of Bloch Wave Decomposition

In this paragraph, we reconsider the Schrödinger operator $A = (-\Delta + W)$ with a potential which is Y-periodic introduced in Paragraph 2.1. The Bloch wave decomposition of this operator obtained therein constitutes what is known as the *physical space picture*. We now present the so-called *momentum space picture* which, in fact, consists of just translating the method in terms of Fourier transformation.

We begin by recalling that since W is Y-periodic it admits Fourier expansion

$$(2.33) \qquad W(x) = \sum_{p \in \mathbf{Z}^N} W_p e^{ip \cdot x},$$

where the Fourier coefficients $\{W_p\}_p$ of W are defined by

$$(2.34) \qquad W_p = \frac{1}{(2\pi)^N} \int_Y W(x) e^{-ip \cdot x} dx \quad \text{for } p \in \mathbf{Z}^N.$$

Since $W \in L^2(Y)$, we have also the Parseval's Identity:

$$(2.35) \qquad \frac{1}{(2\pi)^N} \int_Y |W(x)|^2 dx = \sum_{p \in \mathbf{Z}^N} |W_p|^2.$$

This identity permits to identify the space $L^2(Y)$ with the space $\ell^2(\mathbf{Z}^N, \mathbf{C})$ of square summable sequences with values in $\mathbf{C}$. Moreover, the multiplication operator $u \longmapsto Wu$ goes over, under Fourier transformation, to the convolution operator. We have

$$(2.36) \qquad \widehat{Wu}(\xi) = \sum_{p \in \mathbf{Z}^N} W_p \hat{u}(\xi - p),$$

where $\xi \in \mathbf{R}^N$ denotes the variable dual to $x \in \mathbf{R}^N$ via Fourier transformation. Analogously, the operator A becomes

$$(2.37) \qquad \widehat{Au}(\xi) = |\xi|^2 \hat{u}(\xi) + \sum_{p \in \mathbf{Z}^N} W_p \hat{u}(\xi - p).$$

The right hand side of (2.37) can be considered as an unbounded operator in $L^2(\mathbf{R}^N)$ acting on $\hat{u}$ which we denote by $\hat{A}$. Evidently, its domain is given by

$$D(\hat{A}) = \left\{ \hat{u} \in L^2(\mathbf{R}^N) \mid |\xi|^2 \hat{u}(\xi) \in L^2(\mathbf{R}^N) \right\} = H^2(\mathbf{R}^N).$$

Let us now translate, in terms of Fourier transformation, the direct integral decomposition (2.24) of the space $L^2(\mathbf{R}^N)$ in terms of the spaces

$L^2_\#(\zeta, Y)$. Recalling that ψ belongs to $L^2_\#(\zeta, Y)$ iff $e^{-i\zeta \cdot x}\psi(x) \in L^2_\#(Y)$, we can thus express ψ as follows:

$$\psi(x) = \sum_{p \in \mathbf{Z}^N} (e^{-i\zeta \cdot x}\psi)_p e^{i(\zeta + p) \cdot x} = \sum_{\eta \in (\zeta + \mathbf{Z}^N)} (e^{-i\zeta \cdot x}\psi)_{\eta - \zeta} e^{i\eta \cdot x}.$$

Observe that since $(\eta - \zeta) \in \mathbf{Z}^N$, the coefficients appearing in the above series are well-defined. We interpret this as the Fourier series for ψ on the translated lattice $(\zeta + \mathbf{Z}^N)$, i.e.,

$$(2.39) \qquad \psi(x) = \sum_{\eta \in (\zeta + \mathbf{Z}^N)} \psi_\eta e^{i\eta \cdot x}, \quad \psi_\eta \overset{\text{def}}{=} (e^{-i\zeta \cdot x}\psi)_{\eta - \zeta}.$$

The corresponding Parseval's Identity is written as

$$(2.40) \qquad \frac{1}{(2\pi)^N} \int_Y |\psi(x)|^2 dx = \sum_{\eta \in (\zeta + \mathbf{Z}^N)} |\psi_\eta|^2.$$

Consequently, we can identify the space $L^2_\#(\zeta, Y)$ with the space $\ell^2(\zeta + \mathbf{Z}^N, \mathbb{C})$ of square summable sequences indexed by the translated lattice $(\zeta + \mathbf{Z}^N)$ and with values in $\mathbb{C}$. With the help of this identification, we see that formula (2.24) becomes

$$(2.41) \qquad L^2(\mathbb{R}^N) = \int_Y^\oplus \ell^2(\zeta + \mathbf{Z}^N, \mathbb{C}) d\zeta.$$

Corresponding to the operator $\hat{A}$ acting on $L^2(\mathbb{R}^N)$ we have, as before, two families of operators $\{\hat{A}(\zeta)\}_\zeta$ and $\{\hat{B}(\zeta)\}_\zeta$ one acting on $\ell^2(\mathbf{Z}^N, \mathbb{C})$ and the other one acting on $\ell^2(\zeta + \mathbf{Z}^N, \mathbb{C})$. They are defined by

$$(2.42) \qquad \begin{cases} \left[\hat{A}(\zeta)\{\phi_q\}\right]_q = |q|^2\phi_q + 2(\zeta \cdot q)\phi_q + |\zeta|^2\phi_q + \sum_{p \in \mathbf{Z}^N} W_{q-p}\phi_p \quad \forall q, \\ \left[\hat{B}(\zeta)\{\psi_\eta\}\right]_\eta = |\eta|^2\psi_\eta + \sum_{\eta' \in (\zeta + \mathbf{Z}^N)} W_{\eta - \eta'}\psi_{\eta'} \quad \forall \eta \in (\zeta + \mathbf{Z}^N), \end{cases}$$

where $\{\phi_q\}_{q \in \mathbf{Z}^N}$ and $\{\psi_\eta\}_{\eta \in (\zeta + \mathbf{Z}^N)}$ are in $\ell^2(\mathbf{Z}^N, \mathbb{C})$ and $\ell^2(\zeta + \mathbf{Z}^N, \mathbb{C})$ and belong to the domains of $\hat{A}(\zeta)$ and $\hat{B}(\zeta)$ respectively. To find the relation between $\hat{A}(\zeta)$ and $\hat{B}(\zeta)$ corresponding to (2.26), we introduce the unitary map

$$(2.43) \qquad \begin{cases} U(\zeta) \colon \ell^2(\zeta + \mathbf{Z}^N, \mathbb{C}) \longrightarrow \ell^2(\mathbf{Z}^N, \mathbb{C}), \\ U(\zeta)\{\psi_\eta\} = \{\psi_{\eta - \zeta}\}. \end{cases}$$

Then these operators are related by

(2.44) $$\hat{B}(\zeta) = U(\zeta)^{-1}\hat{A}(\zeta)U(\zeta).$$

By the very construction of $\hat{A}(\zeta)$, it is unitarily equivalent to $A(\zeta)$ via Parseval's Identity. Therefore its spectrum coincides with that of $A(\zeta)$ and hence it is described by Theorem 2.3. Of course, the corresponding eigenfunctions are given by the Fourier coefficients of the eigenvectors of $A(\zeta)$.

Thus, this description gives all necessary elements of the momentum space picture which allows us to obtain the spectral decomposition of A in an alternative way. To obtain this, it suffices to take Fourier transform of the relation (2.27) and use the definition of $\hat{B}(\zeta)$ given by (2.42). We do this in a formal way. Using (2.27), we can write

$$\hat{A} = \left(\int_{Y'} B(\zeta)d\zeta\right)^{\wedge} = \int_{Y'} \hat{B}(\zeta)d\zeta$$

and so, given any $v \in L^2(\mathbb{R}^N)$, we get

$$E_\lambda(\hat{A})\hat{v} = \int_{Y'} E_\lambda(\hat{B}(\zeta))\hat{v}d\zeta = \int_{Y'} \sum_{n;\lambda_n \leq \lambda} (\hat{v},\hat{\psi}_n)\hat{\psi}_n d\zeta;$$

the last equality being a consequence of the fact that $\{\lambda_n(\zeta)\}$ are all the eigenvalues of the operator $\hat{B}(\zeta)$ with corresponding eigenvectors $\{\hat{\psi}_n(\zeta;\cdot)\}$. We observe that the inner product $(\hat{v},\hat{\psi}_n)$ coincides with the Bloch coefficient $\hat{v}_n(\zeta)$ defined by (2.17). Thus we obtain the following result:

Theorem 2.11. *The spectral family $E_\lambda(\hat{A})$ of $\hat{A}$ is given by*

(2.45) $$E_\lambda(\hat{A})\hat{v}(\xi) = \begin{cases} 0 & if \quad \lambda < 0, \\ \int_{Y'} \sum_{n;\lambda_n \leq \lambda} \hat{v}_n(\zeta)\hat{\psi}_n(\zeta,\xi)d\zeta & if \quad \lambda \geq 0; \end{cases}$$

the addition being taken over those indices $n = n(\zeta)$ for which $\lambda_n(\zeta) \leq \lambda$. ∎

2.3. Absence of Eigenvalues for Schrödinger Operator

This paragraph is devoted to prove that the spectral family of Schrödinger operator A described in Paragraphs 2.2 and 2.3 is continuous. By Theorem 3.25 of Chapter I, the above continuity is equivalent to the absence of eigenvalues for the operator A. Recall that we have proved in §3 of Chapter I (see Examples 3.7 and 3.18) that the Laplace operator in $\mathbb{R}^N$ has no eigenvalues and in fact it has absolutely continuous spectrum (see Comments of Paragraph 3.5 of Chapter I). By a clever perturbation argument, THOMAS [1973] proves that the same properties continue to hold for the Schrödinger operator A. This constitutes a nice application of the Bloch wave decomposition and in particular of its momentum space picture. The proof requires several new concepts, arguments and results in the analytic perturbation theory of unbounded operators. Thus it lies somewhat outside the scope of the present book. Nevertheless, we intend to furnish the proof without going to much into the details of these auxiliary materials for which, of course, precise references will be given. We start by introducing the following

Definition. *A family* $\{T(z)\}_{z \in \mathbb{C}}$ *of unbounded closed operators with domains* $D(T(z))$ *in a complex Hilbert space* H *is said to be* complex-analytic *of type* (A) *iff*
(i) $D(T(z)) = D$ *is independent of* z.
(ii) $z \in \mathbb{C} \longmapsto T(z)u$ *is a complex analytic map with values in* H, *for all* $u \in D$. *(The analyticity here is understood in the sense of Paragraph 1.9 of Chapter I).* ∎

The above concept is precisely what is called a *holomorphic family of type* (A) in KATO [1966] p.375. In our case, let us consider the family of operators $A(\zeta)$ introduced in (2.8) which we rewrite as follows:

$$A(\zeta) = -\left[\Delta + 2i\zeta \cdot \nabla - (\zeta_1^2 + \cdots + \zeta_N^2)\right] + W.$$

For each $\zeta' = (\zeta_1, ..., \zeta_{N-1}) \in \mathbb{R}^{N-1}$, let us extend the definition of $A(\zeta)$ to all $\zeta_N = z \in \mathbb{C}$. The resulting operator will be denoted by $A(z)$. Its domain is, by definition,

$$D = H^2_{loc}(\mathbb{R}^N) \cap L^2_{\#}(Y)$$

and so independent of z and it is easily checked that it is closed. Finally, by the very definition of complex analyticity of a map, we see that condition (ii) above is also fulfilled by $A(z)$ because, for every $f \in L^2_{\#}(Y)$ and $\phi \in D$, the function

$$z \longmapsto \int_Y \bar{f} A(z)\phi \, dx$$

is a quadratic polynomial in $z \in \mathbb{C}$. We conclude therefore that $\{A(z)\}_{z \in \mathbb{C}}$ is complex-analytic of type (A) for each fixed $\zeta' \in \mathbb{R}^{N-1}$. Moreover, it is a *self-adjoint family* in the sense that $A(z)^* = A(\bar{z}) \; \forall z \in \mathbb{C}$.

We introduce now what is called the *Fermi surface* associated with the family of operators $A(\zeta)$. Let $\lambda \in \mathbb{R}$ be given. Define

(2.46) $\qquad F_\lambda = \{\zeta \in Y' \mid \lambda \text{ is an eigenvalue of } A(\zeta)\}.$

F_λ can also be written as

$$F_\lambda = \bigcup_{n \geq 1} \lambda_n^{-1}\{\lambda\}.$$

We remark that it is in fact a finite union as a consequence of Theorem 2.10. Since λ_n is a continuous function, it follows that F_λ is a closed subset of Y' (hence measurable). As a result of decomposition (2.24) we can prove

Theorem 2.12. λ *is an eigenvalue of* A *iff* meas $(F_\lambda) > 0$, *where* meas $(\cdot)$ *denotes the Lebesgue measure in* $\mathbb{R}^N$.

Proof. First, we observe that F_λ can be written as

(2.47) $F_\lambda = \{\zeta \in Y' \mid \lambda$ is an eigenvalue of $B(\zeta)\}.$

By definition, λ is an eigenvalue of A iff there exists $u \in D(A), u \neq 0$ such that $Au = \lambda u$. Let us decompose u as in (2.22):

$$u(\cdot) = \int_{Y'}^{\oplus} u(\zeta, \cdot) d\zeta,$$

where $u(\zeta, \cdot) \in L^2_\#(\zeta, Y)$. Since $u \neq 0$, there exists a measurable subset $E \subset Y'$ such that

$$\text{meas}(E) > 0, \quad u(\zeta, \cdot) \neq 0 \quad \text{for } \zeta \in E.$$

On the other hand, the equation $Au = \lambda u$ is equivalent to

$$B(\zeta)u(\zeta, \cdot) = \lambda u(\zeta, \cdot) \quad \text{for a.e. } \zeta \in Y'.$$

This implies that λ is an eigenvalue of $B(\zeta)$ for $\zeta \in E$ and so $E \subset F_\lambda$. Therefore meas $(F_\lambda) > 0$.

For the proof of the converse part, it suffices to invert the above arguments. ∎

In the sequel, we will show that F_λ is a null set for each λ and this prove that A has no eigenvalues. To achieve this, we will, in an essential way, use the analytic properties of branches of eigenvalues corresponding to a complex-analytic family of operators. Recall that if eigenvalues are ordered by magnitude then this property is lost. Existence of analytic branches representing the eigenvalues is guaranteed by the classical Rellich's Theorem (see KATO [1966] p.392). In our case, the operators $\{A(z)\}_{z \in \mathbb{C}}$ fulfill all the hypotheses of the above theorem; in particular, it is easily checked that $A(z)$ has a compact inverse for all $z \in \mathbb{C}$ (see Exercise 2.2). We can thus conclude that all eigenvalues of $A(z)$ can be represented by functions which are real-analytic on $\mathbb{R}$. More exactly, there exists a sequence $\{\mu_n(z)\}$ of scalar-valued functions, all real-analytic on $\mathbb{R}$, such that for $z \in \mathbb{R}$, $\{\mu_n(z)\}$ represent all the eigenvalues of $A(z)$ including multiplicities. Of course, the branches $\{\mu_n(z)\}$ are nothing but the eigenvalues $\{\lambda_n(\zeta)\}, \zeta = (\zeta_1, ..., \zeta_{N-1}, z)$, rearranged in a suitable way. The following property of these branches is crucial in the analysis of the spectrum of A:

Theorem 2.13. *Each eigenvalue $\mu_n(z)$ is not constant as a function of $z \in \mathbb{R}$.*

Proof. *First Step*: Let us begin by introducing the operator $T(z)$ by writing

$$A(z) = T(z) + W.$$

We then have, for $\mu \notin \sigma(T(z))$, that

$$
\begin{aligned}
A(z) - \mu I = T(z) - \mu + W &= \\
&= \left[I - W(T(z) - \mu I)^{-1} \right] (T(z) - \mu I) = \\
&= (I - C(z))(T(z) - \mu I),
\end{aligned}
$$

where $C(z)$ is the operator $W(T(z) - \mu I)^{-1}$. The above relation shows that $\mu \notin \sigma(A(z))$ provided that $\mu \notin \sigma(T(z))$ and $1 \notin \sigma(C(z))$.

Second Step: For each $\mu \in \mathbb{R}$, let us introduce the set

$$D_\mu = \{ z \in \mathbb{R} \mid \mu \notin \sigma(T(z)) \}.$$

Its complement in $\mathbb{R}$ can be described as follows:

(2.48)
$$
\begin{aligned}
\mathbb{R} \backslash D_\mu = \{ z \in \mathbb{R} \mid \mu \in \sigma(T(z)) \} &= \\
&= \{ z \in \mathbb{R} \mid \mu = |\zeta' + p'|^2 + (z + p_N)^2 \quad \text{for some } p \in \mathbb{Z}^N \}
\end{aligned}
$$

(see Example 2.5). Observe that for all $p \in \mathbb{Z}^N$, there exist at most two values of z in $\mathbb{R} \backslash D_\mu$ and so $\mathbb{R} \backslash D_\mu$ is countable. It follows that D_μ is uncountable and there exists $z_0 \in D_\mu$ which is not an integer, i.e.,

(2.49)
$$
\alpha \overset{\text{def}}{=} \inf_{p_N \in \mathbb{Z}} |z_0 + p_N| > 0.
$$

Third Step: To prove the theorem, we will argue by contradiction and suppose that $\mu_n(z) = \mu$ is a constant in $z \in \mathbb{R}$ for some $n \geq 1$. We consider the following subset of D_μ defined by

$$\tilde{D}_\mu = \{ z \in D_\mu \mid 1 \in \sigma(C(z)) \}.$$

We claim that $(\tilde{D}_\mu \cap I)$ is infinite for all bounded non-empty open interval I of $\mathbb{R}$. From the second step, it follows that $(D_\mu \cap I)$ is uncountable because $(\mathbb{R} \backslash D_\mu) \cap I$ is countable. If $(\tilde{D}_\mu \cap I)$ were finite then there exists $z_1 \in I$ such that $z_1 \in D_\mu$ but $z_1 \notin \tilde{D}_\mu$. In other words, $\mu \notin \sigma(T(z_1))$ and $1 \notin \sigma(C(z_1))$. By the first step, it then follows that $\mu \notin \sigma(A(z_1))$. This contradicts the fact that μ is an eigenvalue of $A(z)$ for all $z \in \mathbb{R}$. The conclusion of this step is therefore that $(\tilde{D}_\mu \cap I)$ is infinite for all I.

Fourth Step: The purpose here is to apply *Atkinson's Theorem* (see KATO [1966] p.370) to the family $\{C(z)\}_{z \in \mathcal{D}_\mu}$ where $\mathcal{D}_\mu$ is the superset of D_μ defined by

$$(2.50) \qquad\qquad \mathcal{D}_\mu = \{z \in \mathbb{C} \mid \mu \notin \sigma(T(z))\}.$$

Of course $D_\mu = \mathcal{D}_\mu \cap \mathbb{R}$. In order to apply the theorem, we verify its hypotheses. First of all, $\mathcal{D}_\mu$ is open in $\mathbb{C}$. Let $w \in \mathcal{D}_\mu$ be given. Then $\mu \in \rho(T(z))$, i.e., $(T(w) - \mu I)^{-1}$ exists and it is bounded. Since $z \longmapsto T(z)$ is complex-analytic in $\mathbb{C}$, $(T(z) - \mu I)^{-1}$ is also bounded and complex-analytic for all z such that $|z - w|$ is sufficiently small (for instance, one can also use Theorem 1.3 of KATO [1966] p.367). Above arguments show not only that $\mathcal{D}_\mu$ is open but also that $z \longmapsto C(z)$ is complex-analytic for $z \in \mathcal{D}_\mu$. Moreover, $C(z)$ is compact as $C(z)$ is a composition of a bounded operator W with a compact one. We are now in a position to apply Atkinson's Theorem and conclude that 1 is an eigenvalue of $C(z)$ for all $z \in \mathcal{D}_\mu$ or there are only a finite number of z in each compact subset of $\mathcal{D}_\mu$ for which 1 is an eigenvalue of $C(z)$. The second possibility is ruled out by our conclusion in the third step. Therefore 1 is an eigenvalue of $C(z)$ for all $z \in \mathcal{D}_\mu$. Since the spectrum of C is contained in $B(0, \|C\|)$ (see Proposition 3.2 of Chapter I), it follows that

$$(2.51) \qquad\qquad \|C(z)\| \geq 1 \quad \forall z \in \mathcal{D}_\mu.$$

Fifth Step: The aim here is to contradict (2.51) by showing the following: Let z_0 be a real number in D_μ having the property stated in (2.49). Its existence has been established in the second step. We show that $\{z_0 + ir \mid r \in \mathbb{R}\} \subset \mathcal{D}_\mu$ and

$$(2.52) \qquad\qquad \lim_{|r| \to \infty} \|C(z_0 + ir)\| = 0.$$

Since W is a bounded operator, it suffices to consider the operator $(T(z) - \mu I)^{-1}$ with $z = z_0 + ir$ to estimate the norm in (2.52). In order to show that the inverse $(T(z) - \mu I)^{-1}$ exists for $z = z_0 + ir, r \in \mathbb{R}$ and it is bounded, we take $f \in L^2_\#(Y)$ and consider the following boundary-value problem with periodic boundary conditions:

$$(2.53) \qquad \begin{cases} -\Delta\phi - 2i\zeta' \cdot \nabla_{x'}\phi - 2iz\dfrac{\partial\phi}{\partial x_N} + (|\zeta'|^2 + z^2 - \mu)\phi = f \quad \text{in} \quad \mathbb{R}^N, \\[2mm] \phi \in H^2_{loc}(\mathbb{R}^N) \cap L^2_\#(Y). \end{cases}$$

To solve this we pass to the momentum space picture and expand ϕ and f in terms of their Fourier series:

$$(2.54) \qquad \begin{cases} \phi(x) = \displaystyle\sum_{p \in \mathbb{Z}^N} \phi_p e^{ip \cdot x}, \\[3mm] f(x) = \displaystyle\sum_{p \in \mathbb{Z}^N} f_p e^{ip \cdot x}. \end{cases}$$

Substitution of these expressions into (2.53) yields the following relations between the Fourier coefficients of ϕ and f:

$$(2.55) \qquad a_p \phi_p = f_p \quad \forall p \in \mathbf{Z}^N,$$

where a_p is given by

$$(2.56) \qquad a_p = |p|^2 + 2(\zeta' \cdot p') + 2zp_N + |\zeta'|^2 + z^2 - \mu \quad \forall p \in \mathbf{Z}^N.$$

Let us observe first that

$$(2.57) \qquad |\Im m(a_p)| = |2r(z_0 + p_N)| \geq 2|r|\alpha$$

by our choice of z_0 (see (2.49)). In particular, it follows that $\{a_p^{-1}\}_{p \in \mathbf{Z}^N}$ belongs to $\ell^\infty(\mathbf{Z}^N, \mathbf{C})$ provided that $r \neq 0$ and so, we can solve (2.55) for ϕ_p in such a way that $\{\phi_p\}_{p \in \mathbf{Z}^N} \in \ell^2(\mathbf{Z}^N, \mathbf{C})$. By Parseval's Identity, this implies the existence of a unique solution $\phi \in L^2_\#(Y)$ for problem (2.53). In addition, we have the following estimate on the solution provided $r \neq 0$:

$$(2.58) \qquad \|\phi\|_{0,Y} \leq \frac{1}{2|r|\alpha} \|f\|_{0,Y}.$$

Next, we show that $\phi \in H^1_\#(Y)$. Once again, by Parseval's Identity, this is equivalent to proving that

$$(2.59) \qquad \{|p|\phi_p\}_{p \in \mathbf{Z}^N} \in \ell^2(\mathbf{Z}^N, \mathbf{C}).$$

To see this, let us multiply (2.55) by $\bar{\phi}_p$ and take the real part of the resulting relation. We obtain

$$\Re e(a_p)|\phi_p|^2 = f_p \bar{\phi}_p \quad \forall p \in \mathbf{Z}^N.$$

Our assertion (2.59) follows from this if we use the fact that there exists a constant $c > 0$ such that $\Re e(a_p) \geq c|p|^2$, for all $p \in \mathbf{Z}^N$.

Finally, the property that $\phi \in H^2_\#(Y)$ (which is equivalent to say that $\{|p|^2\phi_p\}_{p \in \mathbf{Z}^N} \in \ell^2(\mathbf{Z}^N, \mathbf{C})$) follows from (2.55) and the fact that $\phi \in H^1_\#(Y)$.

The above arguments show not only the existence of $(T(z) - \mu I)^{-1}$ for $z = z_0 + ir, r \in \mathbf{R}$, but also provides the following estimate (see (2.58)):

$$\|(T(z) - \mu I)^{-1}\| \leq \frac{1}{2|r|\alpha} \quad \text{if} \quad r \neq 0.$$

In particular, this implies (2.52) and leads to the required contradiction. This shows that our assumption that $\mu_n(z)$ is constant made in the third step is wrong and this completes the proof of the theorem. ∎

The fact that the point spectrum of A is empty is now an easy consequence of the above theorem.

Theorem 2.14. *The operator A has no eigenvalues.*

Proof. According to Theorem 2.12, it is enough to show that

$$(2.60) \qquad \qquad \text{meas}\,(F_\lambda) = 0 \quad \forall \lambda \in \mathbb{R},$$

where F_λ is the Fermi surface defined in (2.46). By Fubini's Theorem, we have

$$\text{meas}\,(F_\lambda) = \int_{Y'} \chi_{F_\lambda}(\zeta)\,d\zeta = \int_{[0,1]^{N-1}} d\zeta' \left(\int_0^1 \chi_{F_\lambda}(\zeta',\zeta_N)\,d\zeta_N \right)$$

$$= \int_{[0,1]^{N-1}} m(\zeta')\,d\zeta',$$

where $m(\zeta')$ is nothing but the one-dimensional Lebesgue measure of the set

$$S(\zeta') = \{\zeta_N \in \mathbb{R} \mid (\zeta',\zeta_N) \in F_\lambda\} =$$
$$= \{\zeta_N \in \mathbb{R} \mid \lambda \text{ is an eigenvalue of } A(\zeta',\zeta_N)\}.$$

Recalling our notation $\{\mu_n(z)\}$ for the eigenvalues of the operator $A(\zeta', z)$, we see immediately that $S(\zeta')$ can also be written as

$$(2.61) \qquad \qquad S(\zeta') = \bigcup_{n\geq 1} \mu_n^{-1}\{\lambda\}.$$

Since μ_n's are real-analytic functions which are not identically constants, it is classical that each set $\mu_n^{-1}\{\lambda\}$ is of measure zero and so $S(\zeta')$ is a null set, i.e., $m(\zeta') = 0$ for all ζ'. This proves (2.60) and hence the theorem. ∎

In the terminology introduced in Section 3 of Chapter I, the above result amounts to saying that $\sigma(A) = \sigma_c(A)$. One of the main ingredients that went into its proof is Rellich's Theorem which guarantees the existence of real-analytic branches of eigenvalues with respect to the single variable ζ_N when others are fixed. If we are ready to use the result concerning the real-analyticity with respect to all variables ζ (see Theorem 2.5), then we can improve our earlier result in the following way:

Theorem 2.15. *The spectrum of A is absolutely continuous.*

Proof. Since $A = \bigoplus_{n \geq 1} A_n$ and $\tilde{A} = \bigoplus_{n \geq 1} \tilde{A}_n$ and since A (respectively A_n) is similar to $\tilde{A}$ (respectively $\tilde{A}_n$) via the unitary isomorphism defined by Theorem 2.7, it is enough to prove that the spectrum of $\tilde{A}_n$ is absolutely continuous for each n.

It was already observed that the spectral family of $\tilde{A}_n$ is given by

$$E_\lambda(\tilde{A}_n)v(\zeta) = \chi_{\{\zeta | \lambda_n(\zeta) \leq \lambda\}}(\zeta)v(\zeta) \quad \forall v \in L^2(Y').$$

Thus we need to show that $\operatorname{meas}(\lambda_n^{-1}(F)) = 0$ whenever $\operatorname{meas}(F) = 0$.

According to Theorem 2.5, $\lambda_n(\zeta)$ is real-analytic (almost everywhere) and by Theorem 2.13, it is not identically constant. Thus $\nabla \lambda_n$ defines a real-analytic vector field which is not identically zero. It follows therefore that the set $\mathcal{C} = \{\zeta \in Y' | \nabla \lambda_n(\zeta) = 0\}$ is of measure zero (by virtue of Lemma 3.23). At each point ζ^0 of the complement of $\mathcal{C}$, the vector field $\nabla \lambda_n$ can be linearized locally by means of change of variables (see PALIS and DE MELO [1982] p.40). This means that $\nabla \lambda_n \simeq (1, 0, ..., 0)$, i.e., $\lambda_n \simeq \zeta_1$. If $\lambda_n(\zeta) = \zeta_1$, then $\lambda_n^{-1}(F) = \{\zeta | \zeta_1 \in F\}$ and so the desired conclusion follows from Fubini's Theorem. ∎

2.4. Non-Standard Homogenization of Spectral Problems

Even though we have studied the Schrödinger operator in $\mathbb{R}^N$ in the preceding paragraphs, we remark that in several practical situations the problem is really posed on bounded domains. Moreover, as it can be easily imagined that the period of the potential is small compared to the size of the domain, one is then naturally led to study the asymptotic behaviour of the spectrum when the ratio between these two physical parameters goes to zero. This is precisely the main goal of the theory of homogenization. One of the ways to do this is to fix the size of the domain and let the period go to zero. This is the usual procedure of homogenization followed, for example, by BENSOUSSAN, LIONS and PAPANICOLAOU [1978] and SÁNCHEZ-PALENCIA [1980]. Our goal in this paragraph is to propose a *new method* which consists of fixing the period and letting the size of the domain go to infinity. While the main techniques involved in the usual homogenization theory are, as we know, multiple scale, asymptotic expansions, energy method, etc., the *non-standard homogenization* procedure that is proposed here is based on Bloch wave decomposition, Theorem 4.7 of Chapter I (due to F. Rellich), Theorem 4.9 of Chapter I (due to E. Sánchez-Palencia), etc. We are going to illustrate both approaches. While the tools of standard homogenization theory can be seen in action in Sections 5 and 6, the purpose of this paragraph and the next two sections is to present the non-standard homogenization procedure to the examples studied in this book.

Although we present the method in the case of Schrödinger operator, we stress that the method extends to cover the case of operators with periodically oscillating coefficients as well as domains with periodic perforations. Indeed, we will see the application of this method in later sections to study the asymptotic behaviour of spectral problems in fluid-solid structures.

We feel that it is now appropriate to cite some earlier works on the usual homogenization method applied to eigenvalue problems. VANNINATHAN [1981] has studied homogenization of eigenvalue problems involving various standard boundary conditions in periodically perforated domains. KESAVAN [1979] has examined such problems for operators with periodically oscillating coefficients in a fixed domain. The homogenization analysis used in these articles is very different from the one adapted in this paragraph. This is essentially because the limit-problems in these articles are posed in a bounded domain; on the other hand, the limit-problem in our case is posed in an unbounded domain.

We begin by introducing the sequence of problems whose asymptotic behaviour is analyzed below. Recall that we define the reference cell as $Y =]0, 2\pi[^N$. For each $n \in \mathbb{N}$, we denote by Ω_n the rectangle in $\mathbb{R}^N$, defined by

$$(2.62) \qquad\qquad \Omega_n =] - 2\pi n, 2\pi n[^N.$$

Observe that Ω_n consists of $(2n)^N$ copies of Y. The boundary of Ω_n is denoted by Γ_n. We are interested in studying the asymptotic behaviour as $n \to \infty$ of the following eigenvalue problem: Find $\nu^{(n)} \in \mathbb{C}$ for which there exists $u^{(n)}$ which is not identically zero such that

$$(2.63) \qquad\qquad \begin{cases} -\Delta u^{(n)} + W u^{(n)} = \nu^{(n)} u^{(n)} & \text{in} \quad \Omega_n, \\ u^{(n)} = 0 \quad \text{on} \quad \Gamma_n. \end{cases}$$

Here W is the Y-periodic real potential which satisfies the conditions imposed in Paragraph 2.1, namely that

$$W \in L^\infty(Y), \quad W \geq a_0 > 0.$$

In order to solve (2.63) by Green's operator method, we introduce $G_n : L^2(\Omega_n) \longrightarrow L^2(\Omega_n)$, defined by the following rule:

$$(2.64) \qquad\qquad G_n f = v_n \quad \forall f \in L^2(\Omega_n),$$

where $v_n \in H_0^1(\Omega_n)$ is the unique weak solution of

$$(2.65) \qquad \int_{\Omega_n} [\nabla v_n \cdot \nabla \bar{w} + W v_n \bar{w}] \, dx = \int_{\Omega_n} f \bar{w} dx \quad \forall w \in H_0^1(\Omega_n).$$

As observed earlier in §2 of Chapter I, G_n is a well-defined self-adjoint compact operator. Moreover, we have

$$(2.66) \qquad\qquad \int_{\Omega_n} (G_n f) \bar{f} dx > 0 \quad \forall f \in L^2(\Omega_n), f \not\equiv 0.$$

Applying the spectral theory of compact self-adjoint operators developed in §3 of Chapter I, we conclude that the spectrum of problem (2.63) consists of a non-decreasing sequence of eigenvalues

$$(2.67) \qquad\qquad 0 < \nu_1^{(n)} \leq \cdots \leq \nu_\ell^{(n)} \leq \cdots \longrightarrow \infty.$$

The corresponding eigenvectors $\{u_\ell^{(n)}\}_\ell$ can be chosen to form an orthonormal basis for $L^2(\Omega_n)$.

Let us now describe the limiting spectral problem. Looking at (2.63), we see that the limit problem as $n \to \infty$ can formally be written as follows: Find $\nu \in \mathbb{C}$ for which there exists u not identically zero such that

$$-\Delta u + W u = \nu u \quad \text{in} \quad \mathbb{R}^N.$$

We shall identify the above problem as the spectral problem associated with a bounded operator G acting on $L^2(\mathbb{R}^N)$. It is defined by the following rule:

$$(2.68) \qquad\qquad Gf = v \quad \forall f \in L^2(\mathbb{R}^N),$$

where $v \in H^1(\mathbb{R}^N)$ is the unique weak solution of

$$(2.69) \qquad \int_{\mathbb{R}^N} [\nabla v \cdot \nabla \bar{w} + W v \bar{w}]\,dx = \int_{\mathbb{R}^N} f \bar{w}\,dx \quad \forall w \in H^1(\mathbb{R}^N).$$

Observe that G is nothing but the inverse of the Schrödinger operator A extensively studied in previous paragraphs. Either of these operators G or A is referred to as the limit operator associated to the sequence of problem (2.63). Using the vocabulary of the theory of homogenization, we call them also *homogenized operators*. As in the usual homogenization procedure, the study of the operators A and G is reduced to the reference cell as explained in the comments following Theorem 2.9.

In order to deduce that the spectral families of problems (2.63) converge as $n \to \infty$ towards that of the Schrödinger operator A, we need the strong convergence of the associated Green's operators. This is the object of the next theorem. We use the extension operator $g_n \longmapsto \tilde{g}_n$ from $L^2(\Omega_n)$ into $L^2(\mathbb{R}^N)$, where

$$(2.70) \qquad\qquad \tilde{g}_n(x) = \begin{cases} g_n(x) & \text{if} \quad x \in \Omega_n, \\ 0 & \text{if} \quad x \notin \Omega_n. \end{cases}$$

Theorem 2.16. *For every $f \in L^2(\mathbb{R}^N)$, let $f_n \in L^2(\Omega_n)$ denote the restriction of f to Ω_n. Then*

$$(G_n f_n)^{\sim} \longrightarrow Gf \quad in \quad H^1(\mathbb{R}^N).$$

Proof. Let us put $v_n = G_n f_n$ and take $w = v_n$ in (2.65). Using Cauchy-Schwarz's Inequality, we see that $\{\tilde{v}_n\}$ is bounded in $H^1(\mathbf{R}^N)$. Therefore, up to a subsequence we have $\tilde{v}_n \rightharpoonup v$ in $H^1(\mathbf{R}^N)$ weakly. Taking $w \in \mathcal{D}(\mathbf{R}^N)$ as a test function in (2.65), we can pass to the limit in (2.65) and conclude that v satisfies (2.69) for all $w \in \mathcal{D}(\mathbf{R}^N)$. Since $\mathcal{D}(\mathbf{R}^N)$ is dense in $H^1(\mathbf{R}^N)$, it follows that $v = Gf$ and the whole sequence $\{\tilde{v}_n\}$ converges to v in $H^1(\mathbf{R}^N)$ weakly.

On the other hand, if we put

$$(2.71) \qquad a(v, w) = \int_{\mathbf{R}^N} \left[\nabla v \cdot \nabla \bar{w} + W v \bar{w} \right] dx \quad \forall v, w \in H^1(\mathbf{R}^N),$$

then we have

$$(2.72) \ \ \|v - \tilde{v}_n\|_{1, \mathbf{R}^N}^2 \le ca(v - \tilde{v}_n, v - \tilde{v}_n) = c\left[a(v, v - \tilde{v}_n) - a(\tilde{v}_n, v) + a(\tilde{v}_n, \tilde{v}_n)\right].$$

Taking $w = v_n$ in (2.65), we see that

$$a(\tilde{v}_n, \tilde{v}_n) = \int_{\mathbf{R}^N} f \tilde{v}_n \, dx$$

and so $a(\tilde{v}_n, \tilde{v}_n) \longrightarrow a(v, v)$. We can now pass to the limit on the right side of (2.72) and conclude the strong convergence of $\tilde{v}_n$ in $H^1(\mathbf{R}^N)$. ∎

The convergence of the spectral families $\{E_\lambda(G_n)\}_n$ towards $E_\lambda(G)$ is now an easy consequence of Theorem 4.7 of Chapter I, since A has no eigenvalues. More precisely, we have

Theorem 2.17. *For all $f \in L^2(\mathbf{R}^N)$, the following convergence holds:*

$$(E_\lambda(G_n)f_n)^{\sim} \longrightarrow E_\lambda(G)f \quad in \quad L^2(\mathbf{R}^N), \ \forall \lambda \in \mathbf{R},$$

where, of course, $f_n = f_{|\Omega_n}$. ∎

The spectral family is described in terms of Bloch waves in Theorem 2.9 whereas $E_\lambda(G_n)$ can be expressed in terms of $\nu_\ell^{(n)}$ and $u_\ell^{(n)}$ as follows:

$$(2.73) \qquad E_\lambda(G_n)f_n = \begin{cases} 0 & \text{if} \quad \lambda < 0, \\ \displaystyle\sum_{\ell; \nu_\ell^{(n)} \le \lambda} (f_n, u_\ell^{(n)}) u_\ell^{(n)} & \text{if} \quad \lambda \ge 0. \end{cases}$$

Comments

1. In the above convergence analysis, we have exploited the Green's operators. In the examples to be seen in later sections, we will see that this is not indispensable and one can directly work with the unbounded operators A_n and A associated with problems (2.63) and (2.69) respectively. ∎

2. We have treated above the convergence of Dirichlet problems on Ω_n and the homogenized problem is posed in $\mathbb{R}^N$ and it does not contain boundary conditions explicitly. This suggests that the boundary condition on Γ_n does not play any essential role in the convergence analysis. Indeed, the result of Theorem 2.16 holds for other boundary conditions such as Neumann, periodic and others provided that uniformly bounded extension operators $P_n \in \mathcal{L}(H^1(\Omega_n), H^1(\mathbb{R}^N))$ exist. In the next section, we will illustrate this procedure with examples coming from fluid-solid structures. This phenomenon is close to the one in the usual homogenization where it is well-known that the homogenized operator is *independent* of the boundary conditions. ∎

3. One interesting question concerning the homogenization of problem (2.63) is the following: It is clear that a simple homothetic transformation allows us to transform (2.63) in Ω_n to a problem posed in the cube $\Omega =]-2\pi, 2\pi[^N$ where the period of the transformed potential is of order $(1/n)$. One might think of applying the usual homogenization procedure to this problem or to a suitable renormalized version and obtain a new homogenized operator A_c. The obvious question is to link A with A_c. ∎

§3. Bloch Wave Method in the Laplace Model

In various practical situations involved in nuclear industry, one is required to solve the mathematical models of Chapter II concerning the vibrations of tube bundles immersed in a fluid. These bundles exhibit periodic structure and contain a large number of tubes to the order of several hundreds or even thousands. As pointed out in the Preface, this is essentially to increase the heat exchanges in the reactor. The numerical computations of these models and the effective calculations of the eigenvalues are very expensive and difficult because the geometry involved is two complicated to be discretized. To overcome this difficulty, one is led, in a natural way, to study theoretically the asymptotic behaviour of the eigenvalues as the number of tubes increases to infinity. This is a typical situation of homogenization and as mentioned in the introduction of Paragraph 2.4, there are at least two ways of tackling this problem. In this section, we follow the non-standard homogenization method that we introduced in the previous paragraph. Our choice of the method is principally for two reasons: Experimental results of

WEAVER and KOROYANNAKIS [1982] show the evidence of continuous spectrum at the limit of large number of tubes. On the other hand, the numerical calculations done by IBNOU-ZAHIR and PLANCHARD [1983] confirm not only the presence of continuous spectrum but also point to its band structure. If we use the non-standard homogenization method based on Bloch wave decomposition, the band structure appears in a natural way. Moreover, it seems more likely to find continuous spectrum working in unbounded regions than in bounded domains. When continuous spectrum is expected at the limit, the individual values in the spectrum are not of much significance. What is important is then the distribution of eigenvalues and the associated spectral family serves precisely this purpose. We are then faced with the task of studying the asymptotic behaviour of the spectral family.

In order to carry out the tasks listed above, we plan to follow the program below: On one hand the practical problems are always posed on bounded domains and on the other hand we wish to give another motivation to the *Bloch wave method*. We, therefore, start with the Laplace model of fluid-solid structures in a rectangle with periodic boundary conditions instead of a Neumann condition. Even though the physical interpretation is somewhat lost with this change, the asymptotic behaviour as the number of tubes becomes large is preserved as was pointed out towards the end of the last section (see Comments of Paragraph 2.4). Using a discrete version of the Bloch wave method, we will be able to obtain an "explicit" solution of this problem in Paragraph 3.1. With the insight gained there, we attack the physical problem with a Neumann boundary condition and study its asymptotic behaviour. In Paragraphs 3.2 and 3.3 we describe the homogenized problem and the nature of its spectrum. We observe that this approach provides a motivation for the Bloch wave method which is different from the classical one presented in §2. Paragraph 3.4 is devoted to the convergence analysis and finally in Paragraph 3.5, we present some numerical results along with some comments on the qualitative nature of the spectrum.

Though desirable, it is not necessary to go through §2 to understand the contents of the present section. The necessary notions and the concepts are re-defined here for the sake of self-sufficiency.

3.1. Discrete Bloch Wave Decomposition of the Laplace Model

As mentioned in the introduction of this section, the object of this paragraph is to solve "explicitly" the Laplace model with periodic boundary conditions. For the description of the Laplace model, the reader is referred to Paragraphs 1.1 and 2.1 of Chapter II. We start, therefore, by describing the geometry of the domain in which this model will be considered. For ease of notations and writing, we consider this model in two dimensions but needless to repeat that the entire analysis goes through in higher dimensions.

Let us recall the definition of the basic reference cell $Y =]0, 2\pi[^2$ which represents the periodic structure of the domain. In each period of the domain we assume, without loss of generality, the presence of only one tube whose cross-section is represented by T. T is then a connected open subset of Y ·such that $\bar{T} \subset Y$ and with a smooth boundary. The region occupied by the fluid is denoted as Y^*, i.e.,

$$Y^* = Y \backslash \bar{T}.$$

If X is any subset of $\bar{Y}$ and $j = (j_1, j_2) \in \mathbb{Z}^2$ is a multi-integer, we denote by X_j the translated image of X by the vector $(2\pi j_1, 2\pi j_2)$, i.e.,

$$X_j \overset{\text{def}}{=} (2\pi j_1, 2\pi j_2) + X.$$

With this notation, for any two integers $n_1, n_2 \in \mathbb{N}$, we introduce the domain $\Omega = \Omega(n_1, n_2)$ of interest

$$\Omega = \mathcal{R} \backslash \bigcup_{j_1=0}^{n_1-1} \bigcup_{j_2=0}^{n_2-1} \bar{T}_j,$$

where $\mathcal{R}$ is the rectangle $]0, 2\pi n_1[\times]0, 2\pi n_2[$. Observe that Ω consists of $n_1 n_2$ copies of the perforated cell Y^*. The number of tubes considered will be frequently referred to throughout and we shall denote it by K, i.e., $K = n_1 n_2$. Apart from Ω, we also consider the unbounded domain

$$\mathcal{O} = \mathbb{R}^2 \backslash \bigcup_{j \in \mathbb{Z}^2} \bar{T}_j.$$

3.1.1. Description of the Spectrum

Since the tubes are identical, the physical parameters, namely the mass and the stiffness constant of the tube T_j are independent of j. We denote them by m and k respectively. In such circumstances, we know that $\sqrt{k/m}$ cannot be an admissible frequency of the fluid-solid structure (see Exercise 2.1 of Chapter II). Therefore, the Laplace model can be formulated by eliminating the displacement vector s_j of the tubes T_j. More precisely, by introducing the eigenvalue parameter

$$\lambda = \frac{\rho_0 \omega^2}{k - m\omega^2}$$

(ρ_0 being the density of the fluid and ω being the frequency of the coupled system) the Laplace model with periodic boundary conditions can be written as follows (see (1.9) of Chapter I): Find $\lambda \in \mathbb{C}$ for which there exists a non-identically constant function ϕ such that

$$(3.1) \quad \begin{cases} \Delta\phi = 0 \quad \text{in} \quad \mathcal{O}, \\ \dfrac{\partial\phi}{\partial n} = \lambda \left(\displaystyle\int_{\Gamma_j} \phi n\, d\gamma \right) \cdot \mathbf{n} \quad \text{on} \quad \Gamma_j, \; j \in \mathbb{Z}^2, \\ \phi \quad \Omega\text{-periodic.} \end{cases}$$

The last property in (3.1) means that

$$(3.2) \quad \phi(x_1 + 2\pi j_1, x_2 + 2\pi j_2) = \phi(x_1, x_2) \quad \text{for a.e. } (x_1, x_2) \in \mathcal{O}, \; \forall j_1, j_2 \in \mathbb{Z}.$$

In order to give the variational formulation of problem (3.1), we introduce the spaces

$$(3.3) \quad \begin{cases} L^2_{\#}(\Omega) \overset{\text{def}}{=} \{\psi \in L^2_{loc}(\mathcal{O}) \mid \psi \text{ is } \Omega\text{-periodic}\}, \\ H^1_{\#}(\Omega) \overset{\text{def}}{=} \left\{\psi \in L^2_{\#}(\Omega) \mid \dfrac{\partial\psi}{\partial x_\ell} \in L^2_{\#}(\Omega) \quad \forall \ell = 1, 2\right\} \end{cases}$$

which are Hilbert spaces with respect to the inner products in $L^2(\Omega)$ and $H^1(\Omega)$ respectively. In $H^1_{\#}(\Omega)$, the constant functions will be identified with $\mathbb{C}$ and we shall denote by $\dot{H}^1_{\#}(\Omega)$ the quotient space $H^1_{\#}(\Omega)/\mathbb{C}$. If $\psi \in H^1_{\#}(\Omega)$, its equivalence class in $\dot{H}^1_{\#}(\Omega)$ is denoted by $\dot\psi$. Multiplying (3.1) by a function $\psi \in H^1_{\#}(\Omega)$ and integrating by parts, we arrive at the following variational formulation: Find $\lambda \in \mathbb{C}$ and a non-zero $\dot\phi \in \dot{H}^1_{\#}(\Omega)$ such that

$$(3.4) \quad \int_\Omega \nabla\dot\phi \cdot \nabla\overline{\dot\psi}\, dx = \lambda \sum_{j_1=0}^{n_1-1} \sum_{j_2=0}^{n_2-1} \left(\int_{\Gamma_j} \dot\phi n\, d\gamma \right) \cdot \left(\overline{\int_{\Gamma_j} \dot\psi n\, d\gamma} \right) \quad \forall \dot\psi \in \dot{H}^1_{\#}(\Omega).$$

Since the map

$$\dot\psi \longmapsto |\dot\psi|_{1,Y^*} \overset{\text{def}}{=} \left(\int_\Omega |\nabla\dot\psi|^2\, dx \right)^{1/2}$$

defines a norm on $\dot{H}^1_{\#}(\Omega)$ equivalent to the quotient norm, we can follow the Green's operator technique (as in Paragraph 2.1 of Chapter II; Theorem 2.3) and prove the following result:

Theorem 3.1. *There exist $2K$ (strictly) positive numbers $\lambda_1, ..., \lambda_{2K}$ (which are not necessarily distinct) and $2K$ equivalence classes of functions $\dot{\phi}_1, ..., \dot{\phi}_{2K}$ in $\dot{H}^1_{\#}(\Omega)$ with the following properties:*

(i) *For each $\ell = 1, ..., 2K$, the pair $(\lambda_\ell, \dot{\phi}_\ell)$ is a solution of (3.4).*
(ii) *The set $\{\dot{\phi}_\ell\}_{\ell=1}^{2K}$ is orthonormal in the sense that*

$$\int_\Omega \nabla \dot{\phi}_\ell \cdot \nabla \bar{\dot{\phi}}_{\ell'} dx = \delta_{\ell\ell'} \quad \forall \ell, \ell' = 1, ..., 2K.$$

(iii) *The pairs $\{(\lambda_\ell, \dot{\phi}_\ell)\}_{\ell=1}^{2K}$ are all the solutions of problem (3.4), i.e., if $(\lambda, \dot{\phi}) \in \mathbb{C} \times \dot{H}^1_{\#}(\Omega)$ satisfies (3.4) then there exists $\ell \geq 1$ such that $\lambda = \lambda_\ell$ and $\dot{\phi}$ can be written as a linear combination of all $\dot{\phi}_{\ell'}$'s for which $\lambda_{\ell'}$ is equal to λ.* ∎

As in the Neumann case, we see that there are two eigenvalues, including multiplicities, for each tube. The above theorem describes completely the solutions of (3.4).

3.1.2. The Orthogonal Decomposition

As was mentioned earlier, in applications, the number of tubes K will be very large and therefore the computation of eigenvalues is expensive. Our aim in this subparagraph is to characterize the eigenvalues and eigenvectors provided by Theorem 3.1 in another way which will simplify their computations. This is achieved by splitting problem (3.4) into a family of K subproblems. Each of these is an eigenvalue problem in the perforated cell Y^* and provides two of the solutions of (3.4). To obtain these subproblems, we decompose $H^1_{\#}(\Omega)$ into K orthogonal subspaces.

With this in mind, we introduce the following notion:

Definition. Let S^1 be the unit circle in $\mathbb{C}$ and $w = (w_1, w_2) \in S^1 \times S^1$ be given. A function ψ defined in $\mathcal{O}$ is said to be (w, Y^*)-*periodic* iff

(3.5)
$$\begin{cases} \psi(x_1 + 2\pi, x_2) = w_1 \psi(x_1, x_2), \\ \psi(x_1, x_2 + 2\pi) = w_2 \psi(x_1, x_2) \end{cases}$$

for almost all $(x_1, x_2) \in \mathcal{O}$. ∎

Let us remark that if $w = (1, 1)$ then the above definition coincides with the standard definition of Y^*-periodicity. The crucial observation is that if w_1 and w_2 are such that

(3.6)
$$w_1^{n_1} = w_2^{n_2} = 1$$

then (w, Y^*)-periodic functions are Ω-periodic in the sense of (3.2). Thus we foresee the role played by the n_1^{th} and n_2^{th} roots of unity in the decomposition of problem (3.4). It is also worth to note that the concept of (w, Y^*)-periodicity is the same as (ζ, Y^*)-periodicity provided $w_\ell = e^{2\pi i \zeta_\ell}, \ell = 1, 2$. Recall that the later concept was introduced and systematically used in §2.

Using the concept of (w, Y^*)-periodicity, let us now introduce the spaces in which the analysis of our eigenvalue problem will be carried out. For each $w = (w_1, w_2) \in S^1 \times S^1$, we consider the following spaces:

$$
(3.7) \quad
\begin{cases}
L^2_\#(w, Y^*) \overset{\text{def}}{=} \{\psi \in L^2_{loc}(\mathcal{O}) \mid \psi \text{ is } (w, Y^*)\text{-periodic}\}, \\
H^1_\#(w, Y^*) \overset{\text{def}}{=} \left\{\psi \in L^2_\#(w, Y^*) \mid \dfrac{\partial \psi}{\partial x_\ell} \in L^2_\#(w, Y^*) \quad \forall \ell = 1, 2\right\}
\end{cases}
$$

and we equip them with the inner products in $L^2(Y^*)$ and $H^1(Y^*)$ respectively. They are obviously Hilbert spaces. Further, there is an unitary isomorphism between these spaces and the ones defined by (3.3); more precisely, we have

$$
(3.8) \quad
\begin{cases}
\psi \in L^2_\#(Y^*) \quad \Longleftrightarrow \quad \psi(x) w_1^{\frac{x_1}{2\pi}} w_2^{\frac{x_2}{2\pi}} \in L^2_\#(w, Y^*), \\
\psi \in H^1_\#(Y^*) \quad \Longleftrightarrow \quad \psi(x) w_1^{\frac{x_1}{2\pi}} w_2^{\frac{x_2}{2\pi}} \in H^1_\#(w, Y^*).
\end{cases}
$$

Proposition 3.2. *If $w \neq (1, 1)$ then the map $\psi \longmapsto |\psi|_{1, Y^*}$ is a norm on $H^1_\#(w, Y^*)$ which is equivalent to the usual $H^1(Y^*)$ norm.*

Proof. It suffices to apply Exercise 2.1 with the following data:

$$
V = H^1_\#(w, Y^*), \quad H = L^2_\#(w, Y^*) \quad \text{and}
$$

$$
a(\phi, \psi) = \int_{Y^*} \nabla \phi \cdot \nabla \bar{\psi} dx \quad \forall \phi, \psi \in H^1_\#(w, Y^*).
$$

It is easily checked that $a(\cdot, \cdot)$ satisfies conditions (i) and (ii). Let us now verify (iii). Indeed, if ψ in $H^1_\#(w, Y^*)$ is such that $a(\psi, \psi) = 0$ then ψ is a constant function and therefore it follows from (3.5) that $\psi \equiv 0$ since $w \neq (1, 1)$. ∎

$\diamond$ **Exercise 3.1.** Refine the statement of Proposition 3.2 by proving that there exists a constant $c > 0$ independent of $w \in S^1 \times S^1 \setminus \{(1, 1)\}$ such that

$$
|\zeta| \|\psi\|_{0, Y^*} \leq c |\psi|_{1, Y^*} \quad \forall \psi \in H^1_\#(w, Y^*).
$$

Here $\zeta \in Y'$ is the point which corresponds to $w \in S^1 \times S^1$ via the change of variable: $w = (e^{2\pi i \zeta_1}, e^{2\pi i \zeta_2})$. (*Hint:* The reader may use Exercise 2.3 and an extension operator P in $\mathcal{L}(H^1(Y^*), H^1(Y))$ with the property that

$$
|P\psi|_{1, Y} \leq c |\psi|_{1, Y^*} \quad \forall \psi \in H^1(Y^*).
$$

Existence of such operators is proven in CIORANESCU and SAINT JEAN PAULIN [1979]). ∎

The above exercise already shows that the parameter value $(1, 1)$ is something special. Our attention will focus at this value in Paragraph 3.7 while we will be preoccupied with the other values in the earlier paragraphs.

We now come to the last point of preparation for the decomposition of $H^1_{\#}(\Omega)$. Let $w_0 = 1, w_1, ..., w_{n_1-1}$ be the n_1^{th} roots of unity. Similarly the n_2^{th} roots of unity are denoted as $w'_0 = 1, w'_1, ..., w'_{n_2-1}$. We then form K vectors in $S^1 \times S^1$ whose first components are one of the n_1^{th} roots of unity and its second components are one of the n_2^{th} roots of unity. More exactly, let

$$(3.9) \qquad w_{\ell k} = (w_\ell, w'_k) \quad \forall \ell = 0, ..., n_1 - 1, \ \forall k = 0, ..., n_2 - 1.$$

As observed earlier (see (3.6)), we have the inclusion

$$H^1_{\#}(w_{\ell k}, Y^*) \subset H^1_{\#}(\Omega) \quad \forall \ell, k.$$

In fact, more is true as shown by the following result:

Theorem 3.3. (Discrete Bloch Wave Decomposition) *The spaces of periodic functions $L^2_{\#}(w_{\ell k}, Y^*)$ (respectively $H^1_{\#}(w_{\ell k}, Y^*)$) for $\ell = 0, ..., n_1 - 1$ and $k = 0, ..., n_2 - 1$ where $w_{\ell k}$ is defined by (3.9) form a family of orthogonal subspaces of $L^2_{\#}(\Omega)$ (respectively $H^1_{\#}(\Omega)$) whose direct sum is equal to $L^2_{\#}(\Omega)$ (respectively $H^1_{\#}(\Omega)$):*

$$(3.10) \qquad \begin{cases} L^2_{\#}(\Omega) = \displaystyle\bigoplus_{\ell=0}^{n_1-1} \bigoplus_{k=0}^{n_2-1} L^2_{\#}(w_{\ell k}, Y^*), \\[2em] H^1_{\#}(\Omega) = \displaystyle\bigoplus_{\ell=0}^{n_1-1} \bigoplus_{k=0}^{n_2-1} H^1_{\#}(w_{\ell k}, Y^*). \end{cases}$$

The orthogonality conditions can be expressed as follows: If $\phi \in H^1_{\#}(w_{\ell k}, Y^*)$ and $\psi \in H^1_{\#}(w_{\ell' k'}, Y^*)$ with $(\ell, k) \neq (\ell', k')$ then

$$(3.11) \qquad \int_\Omega \phi \bar\psi \, dx = 0,$$

$$(3.12) \qquad \begin{cases} \displaystyle\int_\Omega \nabla\phi \cdot \nabla\bar\psi \, dx = 0, \\[1em] \displaystyle\sum_{j_1=0}^{n_1-1} \sum_{j_2=0}^{n_2-1} \left(\int_{\Gamma_j} \phi \mathbf{n} \, d\gamma\right) \cdot \left(\int_{\Gamma_j} \bar\psi \mathbf{n} \, d\gamma\right) = 0. \end{cases}$$

If $\phi \in L^2_{\#}(w_{\ell k}, Y^*)$ and $\psi \in L^2_{\#}(w_{\ell' k'}, Y^*)$ with $(\ell, k) \neq (\ell', k')$ then (3.11) holds.

Proof. *First Step*: We show here that given ψ which is Ω-periodic there exists $\psi_{\ell k}$ which is $(w_{\ell k}, Y^*)$-periodic for $\ell = 0, ..., n_1 - 1$ and $k = 0, ..., n_2 - 1$ such that

$$(3.13) \qquad \psi = \sum_{\ell=0}^{n_1-1} \sum_{k=0}^{n_2-1} \psi_{\ell k}.$$

Indeed, we define $\psi_{\ell k}$ by

$$(3.14) \quad \psi_{\ell k}(x_1, x_2) = \frac{1}{K} \sum_{j_1=0}^{n_1-1} \sum_{j_2=0}^{n_2-1} (w_\ell)^{-j_1} (w_k')^{-j_2} \psi(x_1 + 2\pi j_1, x_2 + 2\pi j_2).$$

First, let us verify that $\psi_{\ell k}$ is $(w_{\ell k}, Y^*)$-periodic. In fact, by replacing x_1 by $(x_1 + 2\pi)$ and making the change of variables $j_1' = (j_1 + 1)$ in the summation, we obtain

$$(3.15) \quad \psi_{\ell k}(x_1 + 2\pi, x_2) = \frac{w_\ell}{K} \sum_{j_1'=1}^{n_1} \sum_{j_2=0}^{n_2-1} w_\ell^{-j_1'} (w_k')^{-j_2} \psi(x_1 + 2\pi j_1', x_2 + 2\pi j_2).$$

Observe that the term of this summation corresponding to the value $j_1' = n_1$ is equal to $(w_k')^{-j_2} \psi(x_1, x_2 + 2\pi j_2)$ since w_ℓ is a n_1^{th} root of unity and ψ is Ω-periodic. Thus, comparing (3.15) with (3.14) we see that

$$\psi_{\ell k}(x_1 + 2\pi, x_2) = w_\ell \psi_{\ell k}(x_1, x_2).$$

Similarly, we prove $\psi_{\ell k}(x_1, x_2 + 2\pi) = w_k' \psi_{\ell k}(x_1, x_2)$. This means that $\psi_{\ell k}$ is $(w_{\ell k}, Y^*)$-periodic.

We now show that (3.13) holds. Substituting the expression (3.14) into the sum in (3.13) and interchanging the order of the summation, we obtain

$$(3.16) \qquad \sum_{\ell=0}^{n_1-1} \sum_{k=0}^{n_2-1} \psi_{\ell k} = \frac{1}{K} \sum_{j_1=0}^{n_1-1} \sum_{j_2=0}^{n_2-1} \left(\sum_{\ell=0}^{n_1-1} w_\ell^{-j_1} \right) \times$$
$$\left(\sum_{k=0}^{n_2-1} (w_k')^{-j_2} \right) \psi(x_1 + 2\pi j_1, x_2 + 2\pi j_2).$$

Let us now use the following identities satisfied by the n_1^{th} and n_2^{th} roots of unity:

$$(3.17) \qquad \sum_{\ell=0}^{n_1-1} (w_\ell)^{-j_1} = \begin{cases} n_1 & \text{if} \quad j_1 = 0, \\ 0 & \text{if} \quad j_1 \neq 0, \end{cases}$$

$$(3.18) \qquad \sum_{k=0}^{n_2-1} (w_k')^{-j_2} = \begin{cases} n_2 & \text{if} \quad j_2 = 0, \\ 0 & \text{if} \quad j_2 \neq 0. \end{cases}$$

Then relation (3.13) is an immediate consequence.

Concerning the regularity of $\psi_{\ell k}$, it is clear that $\psi_{\ell k} \in L_\#^2(w_{\ell k}, Y^*)$ (respectively $H_\#^1(w_{\ell k}, Y^*)$) if $\psi \in L_\#^2(\Omega)$ (respectively $H_\#^1(\Omega)$).

Second Step: We verify here the orthogonality properties (3.11) and (3.12). If $\phi \in L^2_\#(w_{\ell k}, Y^*)$ and $\psi \in L^2_\#(w_{\ell' k'}, Y^*)$, we can express

$$\int_\Omega \phi\bar{\psi}dx = \sum_{j_1=0}^{n_1-1} \sum_{j_2=0}^{n_2-1} \int_{Y^*_j} \phi\bar{\psi}dx$$

which is equal to, by periodicity,

$$\sum_{j_1=0}^{n_1-1} \sum_{j_2=0}^{n_2-1} (w_\ell)^{-j_1}(w'_k)^{-j_2}(\bar{w}_{\ell'})^{-j_1}(\bar{w}'_{k'})^{-j_2} \int_{Y^*(0,0)} \phi\bar{\psi}dx.$$

We use, this time, the following properties of the roots of unity:

$$(3.19) \qquad \sum_{j_1=0}^{n_1-1} (w_\ell \bar{w}_{\ell'})^{-j_1} = \begin{cases} n_1 & \text{if} \quad \ell = \ell', \\ 0 & \text{otherwise,} \end{cases}$$

$$(3.20) \qquad \sum_{j_2=0}^{n_2-1} (w'_k \bar{w}'_{k'})^{-j_2} = \begin{cases} n_2 & \text{if} \quad k = k', \\ 0 & \text{otherwise.} \end{cases}$$

The orthogonality property (3.11) now easily follows because $(\ell, k) \neq (\ell', k')$. The proof of (3.12) is similar and therefore omitted. ∎

3.1.3. Splitting into Subproblems in Y^*

The decomposition (3.10) of the space $H^1_\#(\Omega)$ established above leads, in a natural way, to a splitting of problem (3.4) into K subproblems. These are formulated in the spaces $H^1_\#(w_{\ell k}, Y^*)$ for $\ell = 0, ..., n_1 - 1, k = 0, ..., n_2 - 1$ and each one of them provides two eigenvalues of (3.4). Their variational formulations are slightly different depending on whether $(\ell, k) = (0, 0)$ or $(\ell, k) \neq (0, 0)$. For $(\ell, k) \neq (0, 0)$, the corresponding subproblem can be written as follows: Find $\lambda_{\ell k} \in \mathbb{C}$ for which there exists $\phi_{\ell k} \in H^1_\#(w_{\ell k}, Y^*)$, $\phi_{\ell k} \not\equiv 0$ such that

$$(3.21) \quad \int_{Y^*} \nabla\phi_{\ell k} \cdot \nabla\bar{\psi}_{\ell k}dx = \lambda_{\ell k}\left(\int_\Gamma \phi_{\ell k}\mathbf{n}d\gamma\right) \cdot \left(\int_\Gamma \bar{\psi}_{\ell k}\mathbf{n}d\gamma\right) \; \forall\psi_{\ell k} \in H^1_\#(w_{\ell k}, Y^*).$$

If $(\ell, k) = (0, 0)$ then the corresponding subproblem is: Find $\lambda_{00} \in \mathbb{C}$ and $\dot{\phi}_{00} \in \dot{H}^1_\#(Y^*)$, $\dot{\phi}_{00} \neq \dot{0}$ such that

$$(3.22) \quad \int_{Y^*} \nabla\dot{\phi}_{00} \cdot \nabla\bar{\dot{\psi}}_{00}dx = \lambda_{00}\left(\int_\Gamma \dot{\phi}_{00}\mathbf{n}d\gamma\right) \cdot \left(\int_\Gamma \bar{\dot{\psi}}_{00}\mathbf{n}d\gamma\right) \; \forall\dot{\psi}_{00} \in \dot{H}^1_\#(Y^*).$$

The subproblems (3.21), (3.22) can be, as usual, interpreted as:

(3.23)
$$\begin{cases} \Delta\phi_{\ell k} = 0 \quad \text{in} \quad \mathcal{O}, \\ \dfrac{\partial\phi_{\ell k}}{\partial n} = \lambda_{\ell k}(\displaystyle\int_{\Gamma_j} \phi_{\ell k}\, n\, d\gamma)\cdot \mathbf{n} \quad \text{on} \quad \Gamma_j, \ j \in \mathbf{Z}^2, \\ \phi_{\ell k} \quad (w_{\ell k}, Y^*)\text{-periodic.} \end{cases}$$

Observe that if $(\ell, k) = (0,0)$, then problem (3.22) is just the same type as the original problem (3.4) but posed on Y^* instead of Ω. Hence the existence of its solution follows from Theorem 3.1. Since Y^* has only one tube, this problem has two eigenvalues λ_{00}^1 and λ_{00}^2 with orthogonal eigenfunctions ϕ_{00}^1 and ϕ_{00}^2 respectively. Since Y^*-periodicity implies Ω-periodicity, these are also eigenfunctions of (3.4) with same eigenvalues. Thus (3.22) provides two of the $2K$ solutions of problem (3.4).

Since $\psi \longmapsto |\psi|_{1,Y^*}$ defines a norm on $H^1_\#(w_{\ell k}, Y^*)$ if $(\ell, k) \neq (0,0)$, we can follow the Green's operator technique, as in Paragraph 2.1 of Chapter II, and conclude the existence of two eigenvalues $\lambda_{\ell k}^1$ and $\lambda_{\ell k}^2$ with the corresponding orthogonal eigenvectors $\phi_{\ell k}^1$ and $\phi_{\ell k}^2$ in $H^1_\#(w_{\ell k}, Y^*)$. Since $(w_{\ell k}, Y^*)$-periodicity implies Y^*-periodicity, these are also eigenfunctions of (3.4) with same eigenvalues. Thus each problem (3.21) provides two of the $2K$ solutions of (3.4).

With these notations, we introduce the set

(3.24)
$$\sigma = \left\{\lambda_{00}^1, \lambda_{00}^2, ..., \lambda_{n_1-1,n_2-1}^1, \lambda_{n_1-1,n_2-1}^2\right\}$$

consisting of the eigenvalues of subproblems. Using the decomposition (3.10) for $H^1_\#(\Omega)$ and using the eigenfunctions of the subproblems, we construct the following orthogonal subset of $\dot{H}^1_\#(\Omega)$:

(3.25)
$$\mathcal{M} = \left\{\dot{\phi}_{00}^1, \dot{\phi}_{00}^2, ..., \dot{\phi}_{n_1-1,n_2-1}^1, \dot{\phi}_{n_1-1,n_2-1}^2\right\}.$$

Obviously, we can normalize these vectors as follows:

(3.26)
$$\int_\Omega \nabla\dot{\phi}_{\ell k}^s \cdot \nabla\overline{\dot{\phi}}_{\ell' k'}^t \, dx = \delta_{st}\delta_{\ell\ell'}\delta_{kk'}$$

for all $s, t = 1, 2$, and for all $\ell, \ell' = 0, ..., n_1 - 1$, $k, k' = 0, ..., n_2 - 1$. Thus, we see how the various subproblems provide solutions of the initial problem (3.4). Indeed, there is complete equivalence between (3.4) and the subproblems (3.23). To see this, let $(\lambda, \phi) \in \mathbb{C} \times H^1_\#(\Omega)$ be a solution of (3.4). By virtue of Theorem 3.3, we can decompose ϕ as follows:

(3.27)
$$\phi = \sum_{\ell=0}^{n_1-1} \sum_{k=0}^{n_2-1} \phi_{\ell k}$$

with $\phi_{\ell k} \in H^1_\#(w_{\ell k}, Y^*)$. It is easily checked that $(\lambda, \phi_{\ell k})$ is a solution of (3.21), (3.22). Since $\dot{\phi}$ is non-zero, there exist at least one (ℓ, k) such that $\phi_{\ell k} \not\equiv 0$. Therefore λ is an eigenvalue of one of the subproblems and hence it is in σ.

This completes all the ingredients to prove the following result which is the central theme of this paragraph. It shows that the sets σ and $\mathcal{M}$ formed out of subproblems provide all the solutions of our original problem (3.4):

Theorem 3.4. σ *constitutes the entire spectrum of eigenvalues of problem* (3.4) *and* $\mathcal{M}$ *consists of the corresponding eigenvectors. That is, for each* $\ell = 0, ..., n_1 - 1$, $k = 0, ..., n_2 - 1$ *and* $s = 1, 2$, $\dot{\phi}^s_{\ell k}$ *is an eigenfunction of* (3.4) *with eigenvalue* $\lambda^s_{\ell k}$. *These are all the solutions of problem* (3.4) *in the sense of Theorem* 3.1 (iii). ∎

It is clear that the foregoing result simplify considerably the computational efforts to solve problem (3.4). Indeed, (3.4) is reduced to a finite family of K eigenvalue problems in Y^*. It therefore suffices to discretize Y^* only once to solve (3.4). In the next paragraph, we will see how the discrete Bloch wave decomposition described in Theorem 3.3 leads to the more classical Bloch wave decomposition.

Comments

The contents of this paragraph are taken from the paper of AGUIRRE and CONCA [1988]. This work which introduces a *discrete* Bloch wave analysis of vibrations of fluid-solid structures was carried out independent of the works of ODEH and KELLER [1964] and EASTHAM [1973] on the (continuous) Bloch analysis of Schrödinger operator with periodic potential. ∎

3.2. Continuous Bloch Wave Decomposition for the Laplace Model

With our experience in discrete Bloch wave decomposition, we now begin to study the asymptotic behaviour of the (physical) Laplace model with Neumann boundary condition when the number of tubes goes to infinity. For this purpose, we are going to follow the method of *non-standard homogenization* proposed in Paragraph 2.4. Recall that this method consists essentially of three steps:
(i) Description of the limit problem.
(ii) Description of its spectral family in terms of the Bloch wave decomposition.
(iii) Convergence analysis.

In this paragraph and the next we shall concentrate our efforts to the points (i) and (ii) above whereas the convergence analysis is done in Paragraph 3.4. We will use the Beppo-Levi spaces $BL(\mathcal{O})$ extensively studied in §1 to describe the limit-problem.

To define the limit problem, we will increase the number of tubes to infinity by considering the following sequence of domains $\{\Omega_n\}$ defined as follows:

$$(3.28) \qquad \Omega_n = \mathcal{R}_n \backslash \bigcup_{j_1=-n}^{n-1} \bigcup_{j_2=-n}^{n-1} \bar{T}_j,$$

where $\mathcal{R}_n$ is the rectangle $]-2\pi n, 2\pi n[^2$. We recall that j stands for the multi-integer (j_1, j_2). Observe that Ω_n consists of $K_n = 4n^2$ copies of the perforated cell Y^*. Obviously, as $n \to \infty$, the limit of domains Ω_n is

$$(3.29) \qquad \mathcal{O} = \mathbb{R}^2 \backslash \bigcup_{j \in \mathbb{Z}^2} \bar{T}_j.$$

Since the tubes are identical, the Laplace model in Ω_n can be written as follows (see Subparagraph 3.1.1): Find $\lambda^{(n)} \in \mathbb{C}$ for which there is a non-constant $\phi^{(n)}$ such that

$$(3.30) \qquad \begin{cases} \Delta \phi^{(n)} = 0 \quad \text{in} \quad \Omega_n, \\ \dfrac{\partial \phi^{(n)}}{\partial n} = \lambda^{(n)} \left(\displaystyle\int_{\Gamma_j} \phi^{(n)} \mathbf{n} d\gamma \right) \cdot \mathbf{n} \quad \text{on} \quad \Gamma_j, \; j \in Q_n, \\ \dfrac{\partial \phi^{(n)}}{\partial n} = 0 \quad \text{on} \quad \Gamma_n, \end{cases}$$

where Γ_n is the boundary of $\mathcal{R}_n$ and Q_n is the following set of multi-integers:

$$(3.31) \qquad Q_n \overset{\text{def}}{=} \left\{ j = (j_1, j_2) \in \mathbb{Z}^2 \;\middle|\; -n \le j_\ell \le n-1, \; \ell = 1,2 \right\}.$$

Note that Q_n is the set of multi-integers labeling the holes in $\mathcal{R}_n$. We are interested in the behaviour of problems (3.30) as $n \to \infty$. As mentioned above, our first task is to guess the limit problem. Looking at (3.30), this can formally be written as follows: Find $\lambda \in \mathbb{C}$ for which there exists ϕ which is not identically constant such that

$$\begin{cases} \Delta \phi = 0 \quad \text{in} \quad \mathcal{O}, \\ \dfrac{\partial \phi}{\partial n} = \lambda \left(\displaystyle\int_{\Gamma_j} \phi \mathbf{n} d\gamma \right) \cdot \mathbf{n} \quad \text{on} \quad \Gamma_j, \; j \in \mathbb{Z}^2. \end{cases}$$

We shall identify the above problem as the spectral problem associated with a bounded operator S acting on $\ell^2(\mathbb{C}^2) \overset{\text{def}}{=} \ell^2(\mathbb{Z}^2, \mathbb{C}^2)$ (space of square summable sequences with values in $\mathbb{C}^2$ defined on $\mathbb{Z}^2$):

$$(3.32) \qquad \ell^2(\mathbb{C}^2) = \left\{ \mathbf{a} = \{\mathbf{a}_j\}_{j \in \mathbb{Z}^2} \;\middle|\; \mathbf{a}_j \in \mathbb{C}^2, \; \sum_{j \in \mathbb{Z}^2} |\mathbf{a}_j|^2 < \infty \right\}.$$

The usual inner product and the corresponding norm in this space are denoted by $(\cdot, \cdot)$ and $\|\cdot\|$ respectively. With this notation, the operator S is defined by the following rule:

(3.33)
$$\begin{cases} S: \ell^2(\mathbb{C}^2) \longrightarrow \ell^2(\mathbb{C}^2), \\ S\mathbf{a} = \left\{ \displaystyle\int_{\Gamma_j} \phi \mathbf{n} d\gamma \right\}_{j \in \mathbb{Z}^2} \quad \forall \mathbf{a} \in \ell^2(\mathbb{C}^2), \end{cases}$$

where ϕ is the unique solution of

$$\begin{cases} \Delta\phi = 0 \quad \text{in} \quad \mathcal{O}, \\ \dfrac{\partial\phi}{\partial n} = \mathbf{a}_j \cdot \mathbf{n} \quad \text{on} \quad \Gamma_j, \; j \in \mathbb{Z}^2, \\ \nabla\phi \in L^2(\mathcal{O})^N, \end{cases}$$

whose variational formulation is

(3.34)
$$\begin{cases} \phi \in BL(\mathcal{O}), \\ \displaystyle\int_{\Omega} \nabla\phi \cdot \nabla\bar{\psi} dx = \sum_{j \in \mathbb{Z}^2} \mathbf{a}_j \cdot \int_{\Gamma_j} \bar{\psi} \mathbf{n} d\gamma \quad \forall \psi \in BL(\mathcal{O}). \end{cases}$$

To see that S is well-defined, we apply Lax-Milgram Lemma and the proposition which follows:

Proposition 3.5. *The map*

$$\psi \longmapsto \left\{ \int_{\Gamma_j} \psi \mathbf{n} d\gamma \right\}_{j \in \mathbb{Z}^2}$$

is a well-defined continuous linear map from $BL(\mathcal{O})$ into $\ell^2(\mathbb{C}^2)$.

Proof. First by using Cauchy-Schwarz's Inequality and trace results cited in Chapter I, we see that there exists a constant $c > 0$ such that

$$\left| \int_{\Gamma} \psi \mathbf{n} d\gamma \right| \leq c \|\mathbf{n}\|_{0,\Gamma} \|\psi\|_{1,Y^*} \quad \forall \psi \in H^1(Y^*).$$

Since the left side of the above inequality is invariant if we replace ψ by $(\psi + d)$ where d is any complex constant, it follows that

$$\left| \int_{\Gamma} \psi \mathbf{n} d\gamma \right| \leq c \|\mathbf{n}\|_{0,\Gamma} \inf_{d \in \mathbb{C}} \|\psi + d\|_{1,Y^*} \quad \forall \psi \in H^1(Y^*).$$

A simple application of Bramble-Hilbert Lemma (see Exercise 1.1) in Y^* gives a constant $c > 0$ such that

(3.35)
$$|\int_\Gamma \psi \, n d\gamma| \le c|\psi|_{1,Y^*} \quad \forall \psi \in \dot{H}^1(Y^*).$$

By translation, the above inequality implies that

$$|\int_{\Gamma_j} \psi \, n d\gamma| \le c|\psi|_{1,Y_j^*} \quad \forall \psi \in H^1(Y_j^*), \; j \in \mathbb{Z}^2$$

with the same constant c which is therefore independent of $j \in \mathbb{Z}^2$. Adding up all these inequalities over $j \in \mathbb{Z}^2$, we get

$$\sum_{j \in \mathbb{Z}^2} |\int_{\Gamma_j} \psi \, n d\gamma|^2 \le c^2 |\psi|_{1,\mathcal{O}}^2 \quad \forall \psi \in BL'(\mathcal{O})$$

which proves the proposition. ∎

The above result shows, indeed, that the right side of (3.34) defines a continuous linear form on $BL'(\mathcal{O})$. The classical Lax-Milgram Lemma then guarantees the existence of a unique solution ϕ to problem (3.34) which satisfies

(3.36)
$$|\phi|_{1,\mathcal{O}} \le c\|\mathbf{a}\| \quad \forall \mathbf{a} \in \ell^2(\mathbb{C}^2)$$

with the constant c which is the same as the one appearing in (3.35). Thus S is well-defined and further we have

Proposition 3.6. *The operator S defined by (3.33) is a linear continuous operator which is self-adjoint and uniformly positive definite.*

Proof. It is easy to check that S is self-adjoint and so it remains to prove the existence of a constant $\alpha > 0$ such that

(3.37)
$$(S\mathbf{a}, \mathbf{a}) \ge \alpha \|\mathbf{a}\|^2 \quad \forall \mathbf{a} \in \ell^2(\mathbb{C}^2).$$

From the variational formulation (3.34) by taking $\psi = \phi$, it follows that

(3.38)
$$(S\mathbf{a}, \mathbf{a}) = |\phi|_{1,\mathcal{O}}^2.$$

The best constant in (3.37) is then obviously given by

(3.39)
$$\alpha = \min_{\mathbf{a} \in \ell^2(\mathbb{C}^2)} \frac{|\phi|_{1,\mathcal{O}}^2}{\|\mathbf{a}\|^2}.$$

Let $\mathbf{a} \in \ell^2(\mathbf{C}^2)$. It is clear that for all $j \in \mathbf{Z}^2$, there exists $\psi_j \in H^1(Y_j^*)$ such that

$$(3.40) \qquad \int_{\Gamma_j} \psi_j \mathbf{n} d\gamma = \mathbf{a}_j, \quad \int_{Y_j^*} |\nabla \psi_j|^2 dx \le c_0 |\mathbf{a}_j|^2, \quad \psi_j = 0 \quad \text{on} \quad \partial Y_j,$$

where $c_0 > 0$ is a constant independent of $\mathbf{a}$ and j. In fact, to construct ψ_j, it suffices to work in the reference perforated cell Y^* and use the fact that the following map is surjective from the space of $H^1(Y^*)$ functions that vanish on ∂Y into $\mathbf{C}^2$:

$$\psi \longmapsto \int_{\Gamma} \psi \mathbf{n} d\gamma.$$

Once ψ_j's are constructed as above, we can now define an element $\psi \in BL'(\mathcal{O})$ by setting $\psi = \psi_j$ on Y_j^*. Using this ψ as a test function in the variational formulation (3.34), we obtain

$$\|\mathbf{a}\|^2 = \int_{\mathcal{O}} \nabla \phi \cdot \nabla \bar{\psi} dx \le |\phi|_{1,\mathcal{O}} c_0^{1/2} \|\mathbf{a}\|$$

which implies

$$(3.41) \qquad \|\mathbf{a}\| \le c_0^{1/2} |\phi|_{1,\mathcal{O}}.$$

This shows that the constant α is not less than $c_0^{-1/2}$ and so we are done. ∎

After these main properties of S, we now turn our attention to describe its spectrum. It is here that the fact that the limit problem of (3.30) does not depend very much on the boundary condition on Γ_n will be used in an essential way (see Comment 2 in Paragraph 2.4). Thus it is natural to expect that S corresponds also to the limit problem of (3.30) with periodic boundary condition on Γ_n instead of Neumann. But the purpose of the preceding paragraph was precisely to solve "explicitly" such a problem for fixed n. Recall that we obtained the spectral decomposition of this problem via a family of subproblems posed on the space of $(w_{\ell k}, Y^*)$-periodic functions where $w_{\ell k} = (w_\ell, w_k)$, $\ell, k = 0, ..., 2n-1$ and $\{w_\ell\}_{\ell=0}^{2n-1}$ are the $2n^{th}$ roots of unity. To obtain a description of the spectrum of S, we merely have to let $n \to \infty$ in the above procedure. Since n^{th} roots of unity are dense in S^1, it is then natural to introduce a continuous family of subproblems indexed by $w = (w_1, w_2) \in S^1 \times S^1$ as follows: Find $\beta = \beta(w) \in \mathbf{C}$ for which there exists $\psi = \psi(w)$ which is not identically zero such that

$$(3.42) \qquad \begin{cases} \Delta \psi = 0 \quad \text{in} \quad \mathcal{O}, \\ \dfrac{\partial \psi}{\partial n} = \beta(\int_{\Gamma_j} \psi \mathbf{n} d\gamma) \cdot \mathbf{n} \quad \text{on} \quad \Gamma_j, \, j \in \mathbf{Z}^2, \\ \psi \quad (w, Y^*)\text{-periodic}. \end{cases}$$

As usual, a weak formulation of the above problem is easily obtained by multiplying (3.42) by functions $\chi \in H^1_\#(w, Y^*)$ and integrating by parts. More precisely, we see that β and ψ satisfy

$$(3.43) \quad \begin{cases} \psi \in H^1_\#(w, Y^*), \; \psi \neq 0, \\ \displaystyle\int_{Y^*} \nabla\psi \cdot \nabla\bar\chi \, dx = \beta\left(\int_\Gamma \psi \mathbf{n} d\gamma\right) \cdot \left(\int_\Gamma \bar\chi \mathbf{n} d\gamma\right) \quad \forall \chi \in H^1_\#(w, Y^*) \end{cases}$$

if $w \neq (1,1)$. As observed in Subparagraph 3.1.3, if $w = (1,1)$, the variational formulation of (3.42) is

$$(3.44) \quad \begin{cases} \dot\psi \in \dot H^1_\#(Y^*), \; \dot\psi \neq \dot 0, \\ \displaystyle\int_{Y^*} \nabla\dot\psi \cdot \nabla\bar{\dot\chi} \, dx = \beta\left(\int_\Gamma \dot\psi \mathbf{n} d\gamma\right) \cdot \left(\int_\Gamma \bar{\dot\chi} \mathbf{n} d\gamma\right) \quad \forall \dot\chi \in \dot H^1_\#(Y^*). \end{cases}$$

To solve these problems, we follow the same method as for the Neumann condition (see Paragraph 2.1 of Chapter II) and reduce the problem to an eigenvalue problem in a finite dimensional space. Thanks to the periodicity condition, this space has dimension two and more precisely this space is nothing but the space of w-periodic sequences defined by

$$(3.45) \quad \ell_\#(w, \mathbb{C}^2) = \left\{\mathbf{a} = \{\mathbf{a}_j\}_{j \in \mathbb{Z}^2} \mid \mathbf{a}_j \in \mathbb{C}^2, \quad \mathbf{a}_j = w_1^{j_1} w_2^{j_2} \mathbf{a}_0 \quad \forall j \in \mathbb{Z}^2\right\}.$$

This space can, of course, be identified with $\mathbb{C}^2$. We use the notation

$$\mathbf{a}(w) = \{\mathbf{a}_j(w)\}_{j \in \mathbb{Z}^2}$$

to denote a generic element of this space if there is a need to specify the period w. The inner product in this space is defined as

$$(3.46) \quad (\mathbf{a}(w), \mathbf{b}(w))_\# = \mathbf{a}_j(w) \cdot \bar{\mathbf{b}}_j(w)$$

and the corresponding norm is denoted as $\|\cdot\|_\#$. Note that the quantity on the right side of (3.46) is independent of $j \in \mathbb{Z}^2$.

Once the space is defined, we can follow the Green's operator technique and introduce the operator $S(w)$ acting on the space $\ell_\#(w, \mathbb{C}^2)$ into itself by the following rule:

$$(3.47) \quad S(w)\mathbf{a} = \left\{\int_{\Gamma_j} \vartheta \mathbf{n} d\gamma\right\}_{j \in \mathbb{Z}^2} \quad \forall \mathbf{a} \in \ell_\#(w, \mathbb{C}^2),$$

where $\vartheta = \vartheta(w)$ is the unique solution of the following variational problem:

$$(3.48) \quad \begin{cases} \vartheta \in H^1_\#(w, Y^*), \\ \displaystyle\int_{Y^*} \nabla\vartheta \cdot \nabla\bar\chi \, dx = \mathbf{a}_0 \cdot \int_\Gamma \bar\chi \mathbf{n} d\gamma \quad \forall \chi \in H^1_\#(w, Y^*) \end{cases}$$

provided that $w \neq (1,1)$. In case $w = (1,1)$, we make the usual modification of replacing the space $H^1_\#(w,Y^*)$ by $\dot{H}^1_\#(Y^*)$. Since the map $\chi \longmapsto |\chi|_{1,Y^*}$ defines a norm on $H^1_\#(w,Y^*)$ if $w \neq (1,1)$ and on $\dot{H}^1_\#(Y^*)$ if $w = (1,1)$ (see Proposition 3.2), we see that the operator $S(w)$ is well-defined and its main properties are summarized in

◇ **Exercise 3.2.** Adapt the proof of Proposition 3.6 and show that $S(w)$ is a linear continuous operator which is self-adjoint and that there exists $\tilde{\alpha}_1 > 0$ independent of $w \in S^1 \times S^1$ such that

$$(3.49) \qquad \big(S(w)\mathbf{a}, \mathbf{a}\big)_\# \geq \tilde{\alpha}_1 \|\mathbf{a}\|^2_\# \quad \forall \mathbf{a} \in \ell_\#(w, \mathbb{C}^2).$$

Moreover, establish the bound $\|S(w)\| \leq \tilde{\alpha}_2$ for some $\tilde{\alpha}_2 > 0$ independent of $w \in S^1 \times S^1$. (*Hint:* After constructing the test function in Y^* (see (3.40)), extend it to an element of $H^1_\#(w,Y^*)$ instead of $H^1_\#(Y^*)$). ∎

As a consequence, we deduce

Theorem 3.7. (i) *Let $w \in S^1 \times S^1, w \neq (1,1)$ be given. Then there exist two positive numbers $\{\beta_\ell(w)\}_{\ell=1,2}$ and two functions $\{\psi_\ell(w)\}_{\ell=1,2}$ in $H^1_\#(w,Y^*)$ with the following properties:*

(a) $$\beta_1(w) \leq \beta_2(w) \quad \forall w.$$

(b) *The pairs (β_1, ψ_1) and (β_2, ψ_2) are solutions of problem (3.43).*

(c) $$\int_{Y^*} \nabla \psi_\ell \cdot \nabla \bar{\psi}_{\ell'} dx = \beta_\ell \delta_{\ell\ell'} \quad \forall \ell, \ell' = 1, 2.$$

(d) *The set of vectors $\{\mathbf{b}^1(w), \mathbf{b}^2(w)\}$ defined by*

$$(3.50) \qquad \mathbf{b}^\ell(w) = \left\{ \int_{\Gamma_j} \psi_\ell(w) \mathbf{n} d\gamma \right\}_{j \in \mathbb{Z}^2} \quad \forall \ell = 1, 2$$

belong to the space $\ell_\#(w, \mathbb{C}^2)$ and they form an orthonormal basis for this space.

(e) *If $(\beta, \psi) \in \mathbb{C} \times H^1_\#(w,Y^*)$ is any solution of problem (3.43) then $\beta = \beta_\ell$ for some $\ell = 1, 2$ and ψ is a linear combination of ψ_1 and ψ_2.*

(ii) *If $w = (1,1)$, a statement analogous to (i) holds for problem (3.44) replacing the space $H^1_\#(w,Y^*)$ by $\dot{H}^1_\#(Y^*)$.*

(iii) *The functions β_1 and β_2 are bounded on $S^1 \times S^1$. In fact, there exist positive numbers $\alpha^{(1)}$ and $\alpha^{(2)}$ independent of $w \in S^1 \times S^1$ such that*

$$(3.51) \qquad 0 < \alpha^{(1)} \leq \beta_1(w) \leq \beta_2(w) \leq \alpha^{(2)} \quad \forall w \in S^1 \times S^1. \quad \blacksquare$$

Proof. Let $w \in S^1 \times S^1$ be fixed. The operator $S(w)$ being self-adjoint and positive definite, admits positive eigenvalues whose inverses define the functions $\{\beta_\ell(w)\}_{\ell=1,2}$ according to the requirement $\beta_1(w) \leq \beta_2(w)$. The corresponding eigenvectors of $S(w)$ give rise to $\{\psi_\ell(w)\}_{\ell=1,2}$ via (3.48). These functions can obviously be normalized according to the prescription (c) above. If we then define $\mathbf{b}^\ell(w)$ by (3.50), one can easily check that this is an eigenvector of $S(w)$ corresponding to the eigenvalue $\beta_\ell(w)^{-1}$, $\ell = 1, 2$. All other points of Theorem 3.7 follow from these observations and Exercise 3.2. This completes the proof. $\blacksquare$

Our next goal is to use the functions $\{\beta_\ell(w)\}_{\ell=1,2}$ and $\{\mathbf{b}^\ell(w)\}_{\ell=1,2}$ to obtain the spectral decomposition of the operator S. This will complete the second step of the *non-standard homogenization procedure* described in the introduction of this section. To do this, we will require some minimal regularity properties, such as measurability, of these functions with respect to $w \in S^1 \times S^1$. It is true that the matrix $S(w)$ depends analytically on $w \in S^1 \times S^1 \backslash \{(1,1)\}$ (see Paragraph 3.6). However, we cannot apply classical results in perturbation theory to deduce regularity properties of the eigenvalues and eigenvectors of $S(w)$. This is due to the presence of two parameters (w_1, w_2) and it is well-known that several pathologies can then arise. Examples showing them are given in KATO [1966].

We now examine the continuity of the eigenvalues $\beta_\ell(w)$ and the possibility of a *measurable selection* of the corresponding eigenvectors. It will be seen that the continuity of the eigenvalues follows straightforward arguments. On the contrary, it is classically known that even a measurable selection of eigenvectors is not obvious at all. This is one of the difficult points of the theory and it is usually avoided by the authors (see however the paper of WILCOX [1973]). The above mentioned difficulty is illustrated simply by the arbitrariness involved in the choice of the eigenvectors, namely they can be multiplied by a phase factor of the form $e^{i\theta(w)}$, where $\theta(w)$ is an arbitrary function of w without affecting any of the properties stated in Theorem 3.7.

In classical examples, such as Schrödinger equation seen in the previous section, the existence of a measurable selection of eigenvectors follows from other deep results which are based on sophisticated arguments in the classical theory of *Fredholm minors*. This work is due to WILCOX [1973]. Without going into this method, we exploit the structure of the Laplace model to construct explicitly a measurable selection of eigenvectors. We add that this selection procedure is very simple but may not be very general.

The desired properties of the eigenvalues and eigenvectors will be a consequence of the following result on the spaces $H^1_\#(w, Y^*)$ which shows that these vary continuously with w:

Lemma 3.8. *Let $w_0 \in S^1 \times S^1$ be arbitrary and $\{w_n\}$ be a sequence in $S^1 \times S^1$ which converges to w_0 as $n \to \infty$. Then for all $\chi \in H^1_\#(w_0, Y^*)$ there exists a sequence $\{\chi_n\}$ such that*

$$\chi_n \in H^1_\#(w_n, Y^*) \quad \forall n \geq 1, \tag{3.52}$$

$$\|\chi_n - \chi\|_{1, Y_j^*} \longrightarrow 0 \quad as \quad n \to \infty, \; \forall j \in \mathbb{Z}^2. \tag{3.53}$$

Proof. Let $\chi \in H^1_\#(w_0, Y^*)$ be given. For each $n \geq 1$, we define

$$e_n = \left(\frac{w_{1n}}{w_{10}}, \frac{w_{2n}}{w_{20}}\right)$$

where (w_{1n}, w_{2n}) and (w_{10}, w_{20}) are the components of w_n and w_0 respectively. Define χ_n by $\chi_n = \phi_n \chi$ where ϕ_n is the following smooth function:

$$\phi_n(x_1, x_2) = (e_{1n})^{\frac{x_1}{2\pi}} (e_{2n})^{\frac{x_2}{2\pi}}.$$

Let us verify that the sequence $\{\chi_n\}$ thus defined satisfy the condition of the lemma. Indeed, χ_n is (w_n, Y^*)-periodic, for

$$\chi_n(x_1 + 2\pi, x_2) = \phi_n(x_1 + 2\pi, x_2)\chi(x_1 + 2\pi, x_2)$$
$$= \phi_n(x_1, x_2)e_{1n}w_{01}\chi(x_1, x_2) = w_{1n}\chi_n(x_1, x_2).$$

A similar relation holds with respect to x_2. This establishes (3.52). Next, (3.53) will be an easy consequence if we show that

$$\phi_n \longrightarrow 1 \quad \text{in} \quad W^{1,\infty}_{loc}(\mathbb{R}^2) \text{ as } n \to \infty. \tag{3.54}$$

However, the latter property is easily established if we note that

$$\frac{\partial \phi_n}{\partial x_\ell} = \frac{1}{2\pi} \phi_n \log e_{\ell n}, \quad \ell = 1, 2 \tag{3.55}$$

and that $e_n \longrightarrow (1, 1)$ as $n \to \infty$. ∎

Using the above result, let us now prove the following property of the eigenvalues:

Theorem 3.9. *For each $w \in S^1 \times S^1$, let $\beta_1(w), \beta_2(w)$ be the eigenvalues of $S(w)$ arranged as follows: $\beta_1(w) \leq \beta_2(w)$. Then the functions $\beta_1(\cdot)$ and $\beta_2(\cdot)$ are continuous on $S^1 \times S^1 \setminus \{(1,1)\}$ and they are essentially bounded functions on $S^1 \times S^1$.*

Proof. First, let $w_0 \in S^1 \times S^1 \backslash \{(1,1)\}$ be given and $\{w_n\}$ be a sequence in $S^1 \times S^1 \backslash \{(1,1)\}$ such that $w_n \to w_0$. The estimates (3.51) show that the sequences $\{\beta_\ell(w_n)\}_n$, $\ell = 1,2$ are bounded. Hence we can extract a subsequence (which we denote again by n) such that

$$(3.56) \qquad \beta_\ell(w_n) \longrightarrow \beta_\ell^* \quad \text{as} \quad n \to \infty, \ \ell = 1,2,$$

where β_1^*, β_2^* satisfy $\beta_1^* \leq \beta_2^*$. Let $\psi_{\ell n} = \psi_\ell(w_n)$, $\ell = 1,2$ be the corresponding eigenfunctions defined via (3.43). According to our normalization condition in Theorem 3.7 (c), these sequences remain bounded in $H^1(Y^*)$ (see Exercise 3.1). Therefore, extracting a further subsequence, if necessary, and still denoting it by n, we get

$$(3.57) \qquad \psi_{\ell n} \rightharpoonup \psi_\ell^* \quad \text{in} \quad H^1(Y^*) \text{ weakly}, \ \ell = 1,2.$$

Since $\psi_{\ell n}$ is (w_n, Y^*)-periodic, it is natural to extend ψ_ℓ^* as an element of $H^1_\#(w_0, Y^*)$. To these sequences, we associate the sequences $\{\mathbf{r}_{\ell n}\}$ in $\mathbb{C}^2$ defined by

$$(3.58) \qquad \mathbf{r}_{\ell n} = \int_\Gamma \psi_{\ell n} \mathbf{n} \, d\gamma \quad \forall n \geq 1, \ \forall \ell = 1,2.$$

Thanks to (3.57) and the continuity of the trace map, we conclude that

$$(3.59) \qquad \mathbf{r}_{\ell n} \longrightarrow \mathbf{r}_\ell^* = \int_\Gamma \psi_\ell^* \mathbf{n} \, d\gamma \quad \forall \ell = 1,2.$$

The next step of the proof consists in showing that $(\beta_\ell^*, \psi_\ell^*)$, $\ell = 1,2$ are solutions of (3.43) with $w = w_0$. To this end, let us use χ_n of Lemma 3.8 as a test function in problem (3.43) defining $\psi_{\ell n}$. We obtain

$$\int_{Y^*} \nabla \psi_{\ell n} \cdot \nabla \bar{\chi}_n \, dx = \beta_\ell(w_n) \Big(\int_\Gamma \psi_{\ell n} \mathbf{n} \, d\gamma \Big) \cdot \Big(\int_\Gamma \bar{\chi}_n \mathbf{n} \, d\gamma \Big).$$

Using the convergence results (3.56)–(3.59), we can obviously pass to the limit in the above relation and get

$$(3.60) \qquad \int_{Y^*} \nabla \psi_\ell^* \cdot \nabla \bar{\chi} \, dx = \beta_\ell^* \Big(\int_\Gamma \psi_\ell^* \mathbf{n} \, d\gamma \Big) \cdot \Big(\int_\Gamma \bar{\chi} \mathbf{n} \, d\gamma \Big).$$

Since this relation is valid for all $\chi \in H^1_\#(w_0, Y^*)$ and $\psi_\ell^* \in H^1_\#(w_0, Y^*)$, we are done provided that $\psi_\ell^* \neq 0$. However this property follows from the following normalization condition implied by Theorem 3.7 (c):

$$(3.61) \qquad \Big| \int_\Gamma \psi_{\ell n} \mathbf{n} \, d\gamma \Big|^2 = 1 \quad \forall n \geq 1, \ \forall \ell = 1,2.$$

Thanks to (3.59), we can pass to the limit in the above condition and obtain

$$(3.62) \qquad |\int_\Gamma \psi_\ell^* \mathbf{n} d\gamma|^2 = 1 \quad \forall \ell = 1, 2.$$

In particular, this implies that ψ_ℓ^* is not a constant.

The above arguments show that $\beta_\ell^* = \beta_\ell(w_0)$ and $\psi_\ell^* = \psi_\ell(w_0)$ is an associated eigenvector for $\ell = 1, 2$. Since the limit of $\beta_\ell(w_n)$ is independent of the subsequence extracted, we conclude that the convergence (3.56) takes place for the whole sequence n and this establishes the continuity of the functions $\beta_1(\cdot)$ and $\beta_2(\cdot)$ on $S^1 \times S^1 \setminus \{(1,1)\}$. They are essentially bounded on $S^1 \times S^1$ thanks to Theorem 3.7 (iii). ∎

◇ **Exercise 3.3.** (Bounds on the eigenvalues $\beta_\ell(w)$) (i) Show that we have the following characterization for the eigenvalues $\beta_1(w), \beta_2(w)$:

$$(3.63) \qquad \beta_1^{-1}(w) = \max_{\mathbf{a} \in \mathbb{C}^2} \frac{1}{\|\mathbf{a}\|^2} \int_{Y_*} |\nabla \vartheta|^2 dx,$$

$$(3.64) \qquad \beta_2^{-1}(w) = \min_{\mathbf{a} \in \mathbb{C}^2} \frac{1}{\|\mathbf{a}\|^2} \int_{Y_*} |\nabla \vartheta|^2 dx,$$

where $\vartheta = \vartheta(w)$ is the solution of problem (3.48) with $\mathbf{a}_0 = \mathbf{a}$.

(ii) Using the characterizations (3.63) and (3.64), prove the inequalities

$$(3.66) \qquad \beta_1(w) \geq \beta_1^D > 0, \quad \beta_2(w) \geq \beta_2^D > 0 \quad \forall w \in S^1 \times S^1,$$

where β_1^D, β_2^D are constants independent of w and they depend only on Y^*. More precisely, they are the eigenvalues of the following problem in Y^*: Find $\beta^D \in \mathbb{C}$ for which there exists a ϕ^D not identically zero such that

$$(3.67) \qquad \begin{cases} \Delta \phi^D = 0 \quad \text{in} \quad Y^*, \\[2mm] \dfrac{\partial \phi^D}{\partial n} = \beta^D (\int_\Gamma \phi^D \mathbf{n} d\gamma) \cdot \mathbf{n} \quad \text{on} \quad \Gamma, \\[2mm] \phi^D = 0 \quad \text{on} \quad \partial Y. \end{cases}$$

The superscript D stands for the fact that we are having Dirichlet boundary condition on ∂Y in the above problem.

(iii) In the same manner, obtain the following upper bounds:

$$\beta_1(w) \le \beta_1^N, \quad \beta_2(w) \le \beta_2^N \quad \forall w \in S^1 \times S^1,$$

where β_1^N, β_2^N are the eigenvalues of the following problem in Y^*: Find β^N for which there exists a ϕ^N not identically constant such that

(3.68)
$$\begin{cases} \Delta \phi^N = 0 \quad \text{in} \quad Y^*, \\[2mm] \dfrac{\partial \phi^N}{\partial n} = \beta^N \left(\displaystyle\int_\Gamma \phi^N n d\gamma \right) \cdot n \quad \text{on} \quad \Gamma, \\[2mm] \dfrac{\partial \phi^N}{\partial n} = 0 \quad \text{on} \quad \partial Y. \end{cases}$$

The superscript N refers to Neumann condition on ∂Y in (3.68). ∎

◇ **Exercise 3.4.** (Locally Lipschitz continuity of the functions $\beta_\ell(w)$) Show that there exists a constant $c > 0$ such that, for all $w, w' \in S^1 \times S^1$, the following estimate holds:

$$|\beta_1(w) - \beta_1(w')| \le c \frac{|\zeta - \zeta'|}{|\zeta|^2 |\zeta'|^2}$$

and a similar one for $\beta_2(\cdot)$. Here ζ (respectively ζ') in Y' is related to w (respectively to w') by the same formula than in Exercise 3.1. (*Hint:* Prove first a similar estimate for the difference $\|S(w) - S(w')\|$.) ∎

So far we have been analyzing the continuity properties of the eigenvalues which are needed for our immediate use. Paragraphs 3.6 and 3.7 are reserved for the presentation of their deeper properties.

Let us now take up the issue of finding a *measurable selection* of eigenvectors. As mentioned already, this will be constructed explicitly. For this purpose, let us introduce the following subset of $S^1 \times S^1 \setminus \{(1,1)\}$:

(3.69) $$\mathcal{U} = \left\{ w \in S^1 \times S^1 \setminus \{(1,1)\} \mid \beta_1(w) < \beta_2(w) \right\}.$$

Since β_1 and β_2 are continuous functions, $\mathcal{U}$ is an open subset of $S^1 \times S^1 \setminus \{(1,1)\}$ and in particular measurable. Let $\mathcal{U}^c$ be its complement in $S^1 \times S^1$. Observe that for $w \in \mathcal{U}^c$, the eigenspace corresponding to the eigenvalue $\beta_1^{-1}(w) = \beta_2^{-1}(w)$ of $S(w)$ is the whole space $\mathbb{C}^2$ and so the choice of eigenvectors is easily done. Denoting them $\mathbf{b}^1(w), \mathbf{b}^2(w)$, we put

(3.70)
$$\begin{cases} \mathbf{b}^1(w) = (1,0), \\ \mathbf{b}^2(w) = (0,1) \quad \forall w \in \mathcal{U}^c. \end{cases}$$

(Here and in the sequel, we identify $\ell_\#(w, \mathbb{C}^2)$ with $\mathbb{C}^2$).

Our next step is to specify the choice in $\mathcal{U}$. To this end, we decompose $\mathcal{U}$ into two measurable disjoint subsets: $\mathcal{U} = \mathcal{G} \cup \mathcal{G}^c$, where

$$(3.71) \qquad \mathcal{G} \overset{\text{def}}{=} \left\{ w \in \mathcal{U} \,\middle|\, \begin{array}{l} \text{if } \mathbf{v}_1 = (v_{11}, v_{12}) \text{ is an eigenvector of } S(w) \\ \text{with eigenvalue } \beta_1^{-1}(w) \text{ then } v_{11} \neq 0 \end{array} \right\},$$

$$(3.72) \qquad \mathcal{G}^c \overset{\text{def}}{=} \mathcal{U} \setminus \mathcal{G}.$$

Observe that if $w \in \mathcal{G}^c$, then the eigenspace of the eigenvalue $\beta_1^{-1}(w)$ is nothing but $\{0\} \times \mathbb{C}$; thus $\mathcal{G}^c$ can also be written as

$$\mathcal{G}^c = \left\{ w \in \mathcal{U} \,\middle|\, \mathcal{N}\big(S(w) - \beta_1^{-1}(w)I\big) = \{0\} \times \mathbb{C} \right\}.$$

Note that $\mathcal{G}^c$ is a closed subset of $\mathcal{U}$. Indeed, if $\{w_n\}$ is a sequence in $\mathcal{G}^c$ converging to $w_0 \in \mathcal{U}$, then

$$S(w_n)\mathbf{e}_2 = \beta_1^{-1}(w_n)\mathbf{e}_2 \quad \forall n \geq 1,$$

where $\mathbf{e}_2 = (0, 1)$. Thanks to Theorem 3.9, we see that the limit of the right side is $\beta_1^{-1}(w_0)\mathbf{e}_2$.

On the other hand, using the variational formulation (3.48), we can easily pass to the limit on the left side and conclude that $S(w_0)\mathbf{e}_2 = \beta_1^{-1}(w_0)\mathbf{e}_2$. Therefore $w_0 \in \mathcal{G}^c$.

The choice of eigenvectors on $\mathcal{G}^c$ suggests itself:

$$(3.73) \qquad \begin{cases} \mathbf{b}^1(w) = (0, 1), \\ \mathbf{b}^2(w) = (1, 0) \quad \forall w \in \mathcal{G}^c. \end{cases}$$

To complete our selection of eigenvectors, it remains to specify the choice on $\mathcal{G}$. To this end, fix $w \in \mathcal{G}$ and let $\{\mathbf{v}_1, \mathbf{v}_2\}$ be any orthonormal basis of $\mathbb{C}^2$ formed by eigenvectors of $S(w)$; $\mathbf{v}_\ell$ is associated to $\beta_\ell^{-1}(w)$, $\ell = 1, 2$. Starting from this, we shall form another orthonormal basis $\{\mathbf{b}^1, \mathbf{b}^2\}$ $(\mathbf{b}^\ell = (b_{\ell 1}, b_{\ell 2}))$ of $\mathbb{C}^2$ consisting of eigenvectors of $S(w)$ such that

$$(3.74) \qquad b_{11} > 0, \quad b_{22} > 0$$

Indeed, since $v_{11} \neq 0$, there exists a unique phase factor $e^{i\theta_1}$ such that $e^{i\theta_1} v_{11} > 0$. We then define

$$(3.75) \qquad \mathbf{b}^1(w) = e^{i\theta_1} \mathbf{v}_1.$$

On the other hand, to define $\mathbf{b}^2(w)$, we first observe that $v_{22} \neq 0$ if $w \in \mathcal{G}$. In fact, if $v_{22} = 0$ then $\mathbf{v}_2$ is proportional to $\mathbf{e}_1 = (1, 0)$ and so it is easily seen that $w \in \mathcal{G}^c$. Thus $v_{22} \neq 0$ and therefore there exists a unique phase $e^{i\theta_2}$ such that $e^{i\theta_2} v_{22} > 0$. We then pose

$$(3.76) \qquad \mathbf{b}^2(w) = e^{i\theta_2} \mathbf{v}_2.$$

We observe that $\{\mathbf{b}^1(w), \mathbf{b}^2(w)\}$ defined above is an orthonormal basis of $\mathbb{C}^2$ and they are eigenvectors of $S(w)$ corresponding to $\beta_1^{-1}(w), \beta_2^{-1}(w)$ respectively. Moreover, condition (3.74) uniquely determines these eigenvectors for $w \in \mathcal{G}$. Indeed, if $\{\mathbf{c}_1, \mathbf{c}_2\}$ is any orthonormal basis of $\mathbb{C}^2$ consisting of eigenvectors of $S(w)$ associated to the eigenvalues $\beta_1^{-1}(w), \beta_2^{-1}(w)$ and satisfying (3.74) then because of the simplicity of the eigenvalues on $\mathcal{U}$, we see that there exists $\alpha_\ell \in \mathbb{C}, |\alpha_\ell| = 1, \ell = 1, 2$ such that

$$(3.77) \qquad \mathbf{c}_\ell = \alpha_\ell \mathbf{b}^\ell, \quad \ell = 1, 2.$$

This, in particular, implies that $0 < c_{\ell\ell} = \alpha_\ell b_{\ell\ell}, \ell = 1, 2$. Therefore, necessarily we must have $\alpha_\ell = 1$ for all ℓ, implying the desired uniqueness.

We are now in a position to announce the main result on the measurable selection of eigenvectors.

Theorem 3.10. *There exist measurable maps* $\psi_\ell \colon S^1 \times S^1 \longrightarrow H^1(Y^*)$, $\ell = 1, 2$ *such that for almost all* w, (β_ℓ, ψ_ℓ) *satisfy* (3.43) *and* (β_ℓ, ψ_ℓ) *has all the properties* (a)–(e) *stipulated in Theorem 3.7 (i).*

Proof. For all $\ell = 1, 2$ and $w \in S^1 \times S^1$, we define $\psi_\ell = \psi_\ell(w)$ to be the unique solution of (3.48) with $\mathbf{a}_0 = \beta_\ell(w) \mathbf{b}^\ell(w)$, where $\mathbf{b}^\ell$ is defined by (3.70) and (3.73)–(3.76). The proof of Theorem 3.10 would be complete if we show that $\mathbf{b}^\ell$ is a measurable map. Since $\mathcal{U}^c$ and $\mathcal{G}^c$ are measurable subsets of $S^1 \times S^1$ and $\mathbf{b}^\ell$ is constant therein, it suffices to establish the measurability of $\mathbf{b}^\ell$ on $\mathcal{G}$. In fact, b_ℓ is continuous on $\mathcal{G}$ as shown below. If $\{w_n\}$ is a sequence in $\mathcal{G}$ converging to $w_0 \in \mathcal{G}$ then there exist vectors $\mathbf{c}_1, \mathbf{c}_2$ in $\mathbb{C}^2$ such that for a subsequence (still denoted by n) we have

$$(3.78) \qquad \mathbf{b}^\ell(w_n) \longrightarrow \mathbf{c}_\ell \quad \forall \ell = 1, 2.$$

Moreover, $\{\mathbf{c}_1, \mathbf{c}_2\}$ forms an orthonormal basis of $\mathbb{C}^2$. Further, from the proof of Theorem 3.9, it follows that $\mathbf{c}_\ell$ is an eigenvector of $S(w_0)$ corresponding to the eigenvalue $\beta_\ell^{-1}(w_0), \ell = 1, 2$. Since $\mathbf{b}^\ell(w_n)$ satisfy (3.74) for all $n \geq 1$, it is true that

$$c_{\ell\ell} \geq 0 \quad \forall \ell = 1, 2.$$

Repeating the arguments of the uniqueness of a basis of eigenvectors of $S(w_0)$ satisfying (3.74), we see that (3.77) holds for some $\alpha_\ell \in \mathbb{C}$ with $|\alpha_\ell| = 1$, $\ell = 1, 2$. In particular,

$$0 \leq c_{\ell\ell} = \alpha_\ell b_{\ell\ell}(w_0) \quad \forall \ell = 1, 2.$$

Since $b_{\ell\ell}(w_0) > 0$ (because $w_0 \in \mathcal{G}$) it follows that $\alpha_\ell = 1$. Therefore $\mathbf{c}_\ell = \mathbf{b}^\ell(w_0), \ell = 1, 2$. This shows that the convergence (3.78) takes place for the entire sequence and completes the proof of the continuity of $\mathbf{b}^\ell$ on $\mathcal{G}$. ∎

Definition. Any measurable selection of orthonormal eigenvectors $\mathbf{b}^1(w)$, $\mathbf{b}^2(w)$ of $S(w), w \in S^1 \times S^1$ is called *Bloch waves* associated with S. The corresponding functions $\{\psi_1(w), \psi_2(w)\}$ (unique solutions of (3.48) with $\mathbf{a}_0 = \beta_\ell(w)\mathbf{b}^\ell(w))$ are also sometimes called *Bloch waves* associated with the Laplace model. ∎

In the sequel, we shall assume that we have made such a measurable selection of eigenvectors of $S(w)$ and deal with it throughout.

Remark. In the foregoing analysis, we have worked with the operator $S(w)$ acting on (w, Y^*)-periodic functions. Instead of working with these functions, one can as well work with functions which are just Y^*-periodic. These two methods are equivalent because of (3.8). We can transform the operator $S(w)$ into another operator $T(w)$ acting on $\ell_\#(\mathbb{C}^2)$ which is identified with $\mathbb{C}^2$. More precisely, $T(w)$ is defined by

$$(3.79) \qquad T(w) = V^{-1}(w)S(w)V(w),$$

where $V(w) \colon \mathbb{C}^2 \longrightarrow \ell_\#(w, \mathbb{C}^2)$ is a unitary operator defined as follows:

$$(3.80) \qquad V(w)\mathbf{a} = \left\{w_1^{j_1} w_2^{j_2} \mathbf{a}\right\}_{j \in \mathbb{Z}^2} \quad \forall \mathbf{a} \in \mathbb{C}^2. \ \blacksquare$$

Once the measurability of the eigenvectors and the continuity of the eigenvalues have been achieved, we are in a position to obtain the spectral decomposition of S in terms of its Bloch waves. This will complete the second step of our non-standard homogenization procedure.

Theorem 3.11. (Bloch Wave Decomposition of $\ell^2(\mathbb{C}^2)$). *Let $\mathbf{a} = \{\mathbf{a}_j\}_{j \in \mathbb{Z}^2}$ be given in $\ell^2(\mathbb{C}^2)$. Then for almost every $w \in S^1 \times S^1$, the element $\mathbf{a}_\#$ with components $\{(\mathbf{a}_\#)_j(w)\}_{j \in \mathbb{Z}^2}$ defined by*

$$(3.81) \qquad (\mathbf{a}_\#)_j(w) = \frac{1}{(2\pi)^2} \sum_{k \in \mathbb{Z}^2} \mathbf{a}_{j+k} w_1^{-k_1} w_2^{-k_2} \quad \forall j \in \mathbb{Z}^2$$

is an element of $\ell_\#(w, \mathbb{C}^2)$ and we have the following decomposition for $\mathbf{a}$:

$$(3.82) \qquad \mathbf{a}_j = \int_{S^1}\int_{S^1} (\mathbf{a}_\#)_j(w)dw \quad \forall j \in \mathbb{Z}^2.$$

Furthermore, the following Parseval's Identity holds:

$$(3.83) \qquad \frac{1}{(2\pi)^2} \|\mathbf{a}\|^2 = \int_{S^1}\int_{S^1} \sum_{\ell=1}^{2} |\alpha_\ell(w)|^2 dw_1 \, dw_2,$$

where $\alpha_1(w)$ and $\alpha_2(w)$ are complex constants given by

(3.84) $$\alpha_\ell(w) = \big(\mathbf{a}_\#(w), \mathbf{b}^\ell(w)\big)_\# \quad \forall \ell = 1, 2.$$

Finally, we have the following decomposition of the identity:

(3.85) $$\mathbf{a} = \int_{S^1} \int_{S^1} \sum_{\ell=1}^{2} \alpha_\ell(w) \mathbf{b}^\ell(w)\, dw.$$

Proof. Let $\mathbf{a} \in \ell^2(\mathbb{C}^2)$ be a given sequence with compact support. We remark that the series on the right side of (3.81) is absolutely convergent and so $\mathbf{a}_\#(w)$ is well-defined. Also it is easily seen that $\mathbf{a}_\#(w)$ is w-periodic and so $\mathbf{a}_\#(w) \in \ell_\#(w, \mathbb{C}^2)$. Furthermore, integrating (3.81) with respect to $w \in S^1 \times S^1$, we obtain

$$\int_{S^1} \int_{S^1} (\mathbf{a}_\#)_j(w)\, dw = \frac{1}{(2\pi)^2} \sum_{k \in \mathbb{Z}^2} a_{j+k} \int_{S^1} w_1^{-k_1}\, dw_1 \int_{S^1} w_2^{-k_2}\, dw_2.$$

We now use the formulae:

(3.86) $$\int_{S^1} w_\ell^{-k_\ell}\, dw_\ell = \begin{cases} 0 & \text{if } k_\ell \neq 0, \\ 2\pi & \text{if } k_\ell = 0, \end{cases}$$

which immediately leads us to the inversion formula (3.82).

Let us now prove (3.83). Indeed, since $\mathbf{a}_\#(w)$ belongs to $\ell_\#(w, \mathbb{C}^2)$ and $\{\mathbf{b}^1(w), \mathbf{b}^2(w)\}$ forms an orthonormal basis for $\ell_\#(w, \mathbb{C}^2)$, we can express

(3.87) $$\mathbf{a}_\#(w) = \sum_{\ell=1,2} \alpha_\ell(w) \mathbf{b}^\ell(w),$$

where $\alpha_1(w), \alpha_2(w)$ are defined by (3.84). Moreover, we have

(3.88) $$\|\mathbf{a}_\#(w)\|^2 = \sum_{\ell=1,2} |\alpha_\ell(w)|^2$$

which, on integration, implies that

$$\int_{S^1} \int_{S^1} \sum_{\ell=1}^{2} |\alpha_\ell(w)|^2\, dw = \frac{1}{(2\pi)^4} \sum_{k,k' \in \mathbb{Z}^2} a_{j+k} \cdot \bar{a}_{j+k'} \int_{S^1} w_1^{-k_1+k_1'}\, dw_1 \int_{S^1} w_2^{-k_1+k_2'}\, dw_2.$$

Using once again formulae (3.86), we conclude that (3.83) holds.

Substituting (3.87) into (3.82), we get easily (3.85) and this completes the proof of the theorem for all $\mathbf{a}$ with compact support. A standard density argument allows us to extend the above conclusions to all $\mathbf{a} \in \ell^2(\mathbb{C}^2)$ and for a.e. $w \in S^1 \times S^1$. ∎

Definition. The complex constants $\alpha_1(w), \alpha_2(w)$ defined by (3.84) are called the *Bloch coefficients* of $\mathbf{a} \in \ell^2(\mathbf{C}^2)$ with respect to the Bloch waves $\mathbf{b}^1(w), \mathbf{b}^2(w)$. ∎

$\Diamond$ **Exercise 3.5.** Show that the Parseval's Identity (3.83) can be generalized to the following *Plancherel's Identity*:

$$(3.89) \qquad \frac{1}{(2\pi)^2}(\mathbf{a}, \mathbf{a}') = \int_{S^1}\int_{S^1} \sum_{\ell=1,2} \alpha_\ell(w)\bar{\alpha}'_\ell(w)dw \quad \forall \mathbf{a}, \mathbf{a}' \in \ell^2(\mathbf{C}^2)$$

where $\alpha_\ell(w)$ and $\alpha'_\ell(w)$ are the Bloch coefficients of $\mathbf{a}$ and $\mathbf{a}'$ respectively. ∎

We now proceed to present a result which gives a decomposition for $BL'(\mathcal{O})$ functions. This corresponds to decomposition (3.82). Of course, one cannot expect a counter-part of (3.85) since the Bloch waves provide a basis for $\mathbf{C}^2$ and not for $H^1_\#(w, Y^*)$.

Theorem 3.12. *Let $\phi \in BL'(\mathcal{O})$ be given. Then for a.e. $w \in S^1 \times S^1$ the function $\phi_\# = \phi_\#(w, x)$, whose gradient has the following expression*

$$(3.90) \quad \nabla\phi_\#(w, x) = \frac{1}{(2\pi)^2} \sum_{j \in \mathbb{Z}^2} \nabla\phi(x + 2\pi j)w_1^{-j_1}w_2^{-j_2} \quad \text{for a.e. } x \in \mathcal{O}$$

is a well-defined element of $H^1_\#(w, Y^*)$. Moreover, the following inversion formula and the corresponding Parseval's Identity hold:

$$(3.91) \qquad \begin{cases} \nabla\phi(x) = \displaystyle\int_{S^1}\int_{S^1} \nabla\phi_\#(w, x)dw \quad \text{for a.e. } x \in \mathcal{O}. \\[2mm] \dfrac{1}{(2\pi)^2}\displaystyle\int_{\mathcal{O}} |\nabla\phi(x)|^2 dx = \int_{S^1}\int_{S^1}\int_{Y^*} |\nabla\phi_\#(w, x)|^2 dx dw. \end{cases} ∎$$

We are now in a position to describe the spectral family associated with S. This is done in terms of the Bloch decomposition obtained in Theorems 3.11 and 3.12.

Theorem 3.13. *(Spectral Decomposition of S) Let S be the operator defined by (3.33) and $\{\mathbf{b}^1(w), \mathbf{b}^2(w)\}$ be an orthonormal basis of $\ell_\#(w, \mathbf{C}^2)$ consisting of eigenvectors of $S(w)$ with the corresponding eigenvalues $\beta_1^{-1}(w)$ and $\beta_2^{-1}(w)$ respectively. Then the spectral resolution of S can be written as follows:*

$$(3.92) \qquad S\mathbf{a} = \int_{S^1}\int_{S^1} \sum_{\ell=1,2} \alpha_\ell(w)\beta_\ell^{-1}(w)\mathbf{b}^\ell(w)dw \quad \forall \mathbf{a} \in \ell^2(\mathbf{C}^2),$$

where $\alpha_1(w), \alpha_2(w)$ are the Bloch coefficients of $\mathbf{a}$ defined in (3.48).

Proof. Let $a \in \ell^2(\mathbb{C}^2)$ be given. Let us now consider the element $\phi_\#(w, \cdot)$ in $H^1_\#(w, Y^*)$ of Theorem 3.12 associated to ϕ, where $\phi \in BL'(\mathcal{O})$ is the solution of (3.34). Thanks to (3.91), we can write

$$(3.93) \qquad Sa = \left\{ \int\limits_{S^1} \int\limits_{S^1} \int\limits_{\Gamma_j} \phi_\#(w, x) \mathrm{n} d\gamma dw \right\}_{j \in \mathbb{Z}^2}.$$

The remaining part of the proof consists in decomposing $\int_{\Gamma_j} \phi_\#(w, x) \mathrm{n} d\gamma$ in terms of the eigenvectors $\mathbf{b}^1(w), \mathbf{b}^2(w)$. In order to this, let us deduce the equation satisfied by $\phi_\#$ from the one satisfied by ϕ. In fact, from the very definition (3.90), we get

$$(3.94) \qquad \begin{cases} \Delta \phi_\#(w, \cdot) = 0 \quad \text{in} \quad \mathcal{O}, \\ \dfrac{\partial \phi_\#}{\partial n}(w, \cdot) = (\mathbf{a}_\#)_j(w) \cdot \mathbf{n} \quad \text{on} \quad \Gamma_j, \ j \in \mathbb{Z}^2, \\ \phi_\#(w, \cdot) \in H^1_\#(w, Y^*) \end{cases}$$

where $\mathbf{a}_\#(w)$ is defined by (3.81). Comparing this system with the one satisfied by $\{\psi_1(w), \psi_2(w)\}$ and using the decomposition (3.87), we obtain by linearity that

$$\phi_\#(w, \cdot) = \sum_{\ell=1,2} \alpha_\ell(w) \beta_\ell^{-1}(w) \psi_\ell(w, \cdot).$$

Upon integrating this relation, we get

$$\int\limits_{\Gamma_j} \phi_\#(w, x) \mathrm{n} d\gamma = \sum_{\ell=1,2} \alpha_\ell(w) \beta_\ell^{-1}(w) \mathbf{b}^\ell(w).$$

Combining this with (3.93) we see that the proof is finished. ∎

It is worth to note that the decomposition (3.92) obtained in the earlier result can be formally written as

$$(3.95) \qquad S = \int\limits_{S^1} \int\limits_{S^1} S(w) dw.$$

This relation is evidently the counterpart of (2.27) and it is a natural consequence of the direct integral decomposition of $\ell^2(\mathbb{C}^2)$ provided by Theorem 3.11:

$$(3.96) \qquad \ell^2(\mathbb{C}^2) = \int\limits_{S^1} \int\limits_{S^1} \ell_\#(w, \mathbb{C}^2) dw.$$

This is to be compared with (2.24).

3.3. Consequences of the Bloch Wave Decomposition of S

The aim of this paragraph is to extract information from the spectral decomposition of S obtained in Theorem 3.13. We will be able to describe the spectral family of S, obtain the band structure of $\sigma(S)$ and the nature of the spectrum. All this will be done in terms of the subproblems (3.43).

Corollary 3.14. *Let $E_\tau(S)$ be the spectral family of the operator S. Then, for all $\mathbf{a} \in \ell^2(\mathbb{C}^2)$ and $j \in \mathbb{Z}^2$, we have*

$$(3.97) \qquad (E_\tau(S)\mathbf{a})_j = \begin{cases} 0 & \text{if } \tau \leq 0, \\ \displaystyle\int_{S^1}\int_{S^1} \sum_{\ell;\beta_\ell^{-1}(w)\leq\tau} \alpha_\ell(w)\mathbf{b}_j^\ell(w)dw & \text{if } \tau > 0. \end{cases} \blacksquare$$

The next result shows that the spectrum of S is nothing but the union of two intervals which are images of the continuous functions β_ℓ^{-1}, $\ell = 1,2$. It reveals the *band structure* of $\sigma(S)$. Further, it shows that there is no isolated eigenvalue of S which is of finite multiplicity. We introduce the notation

$$(3.98) \qquad \begin{cases} \alpha_{\min}^{(\ell)} \overset{\text{def}}{=} \displaystyle\inf_{w\in S^1\times S^1\setminus\{(1,1)\}} \beta_\ell^{-1}(w), \\ \alpha_{\max}^{(\ell)} \overset{\text{def}}{=} \displaystyle\sup_{w\in S^1\times S^1\setminus\{(1,1)\}} \beta_\ell^{-1}(w) \quad \forall\ell = 1,2. \end{cases}$$

Since $\beta_1(w) \leq \beta_2(w)$, we have the following easy inequalities:

$$(3.99) \qquad \alpha_{\min}^{(2)} \leq \alpha_{\min}^{(1)}, \quad \alpha_{\max}^{(2)} \leq \alpha_{\max}^{(1)}.$$

Theorem 3.15. *We have*

$$\sigma(S) = \sigma_{ess}(S) = [\alpha_{\min}^{(1)}, \alpha_{\max}^{(1)}] \cup [\alpha_{\min}^{(2)}, \alpha_{\max}^{(2)}].$$

Proof. *First Step*: We prove the inclusion

$$[\alpha_{\min}^{(1)}, \alpha_{\max}^{(1)}] \cup [\alpha_{\min}^{(2)}, \alpha_{\max}^{(2)}] \subset \sigma_{ess}(S).$$

Let $\alpha \in]\alpha_{\min}^{(\ell)}, \alpha_{\max}^{(\ell)}[$ for some $\ell = 1,2$ and $w_0 \in S^1 \times S^1\setminus\{(1,1)\}$ be such that $\beta_\ell^{-1}(w_0) = \alpha$. Because of the continuity of β_ℓ^{-1}, for all $\varepsilon > 0$ there exists $\delta > 0$ such that

$$(3.100) \qquad |\beta_\ell^{-1}(w) - \alpha| < \varepsilon \quad \forall w \in B(w_0, \delta) \cap (S^1 \times S^1).$$

Our goal is to show that $\alpha \in \sigma_{ess}(S)$. To this end, we will use *Weyl's criterion* for the essential spectrum of S (see Theorem 3.27 of Chapter I).

Let us consider a sequence $\{\varepsilon_n\}$ of positive numbers which converges to zero. We construct a sequence $\{\delta_n\}$ inductively as follows: Choose $\delta_0 > 0$ such that $(\varepsilon_0, \delta_0)$ satisfies (3.100). Then, for each $n \geq 1$, choose $\tilde{\delta}_n > 0$ such that $(\varepsilon_n, \tilde{\delta}_n)$ satisfy (3.100) and $\tilde{\delta}_n \leq \varepsilon_n$. Next, we set

$$\delta_n \stackrel{\text{def}}{=} \min\{\tilde{\delta}_n, \frac{\delta_{n-1}}{2}\} > 0.$$

To each δ_n, we associate the following subset of $S^1 \times S^1$:

$$D_n = \left\{ w \in S^1 \times S^1 \mid \frac{\delta_n}{2} < |w - w_0| < \delta_n \right\}.$$

By the very construction, D_n's are disjoint:

$$D_{n'} \cap D_n = \phi \quad \forall n' \neq n.$$

For each $n \geq 1$, we define now the vector $\mathbf{a}_n \in \ell^2(\mathbb{C}^2)$ by

$$\mathbf{a}_n = \frac{1}{2\pi |D_n|^{1/2}} \int\limits_{D_n} \mathbf{b}^\ell(w) dw = \frac{1}{2\pi |D_n|^{1/2}} \int\limits_{S^1} \int\limits_{S^1} \chi_{D_n}(w) \mathbf{b}^\ell(w) dw,$$

where $\mathbf{b}^\ell(w)$ is one of the Bloch waves associated with S and $|D_n|$ denotes the surface measure of D_n. By Parseval's Identity (3.83), we obtain

$$\|\mathbf{a}_n\|^2 = \frac{1}{|D_n|} \int\limits_{D_n} dw = 1 \quad \forall n.$$

On the other hand, by using Plancherel's Identity (3.89), we easily verify that $\{\mathbf{a}_n\}$ is an orthogonal sequence in $\ell^2(\mathbb{C}^2)$.

Now let us use the spectral decomposition provided by Theorems 3.11 and 3.13 to get

$$(S - \alpha I)\mathbf{a}_n = \frac{1}{2\pi |D_n|^{1/2}} \int\limits_{D_n} (\beta_\ell^{-1}(w) - \alpha) \mathbf{b}^\ell(w) dw$$

and therefore by Parseval's Identity (3.83), we get

$$\|(S - \alpha I)\mathbf{a}_n\|^2 = \frac{1}{|D_n|} \int\limits_{D_n} |\beta_\ell^{-1}(w) - \alpha|^2 dw.$$

which is clearly less than or equal to ε_n by (3.100) and the definition of D_n. In particular, this shows that $(S - \alpha I)\mathbf{a}_n$ converges to zero in $\ell^2(\mathbb{C}^2)$. By Weyl's criterion, we conclude therefore that $\alpha \in \sigma_{ess}(S)$. Since σ_{ess} is closed, this finishes the first step.

Second Step: We show here the inclusion

$$\sigma(S) \subset [\alpha_{\min}^{(1)}, \alpha_{\max}^{(1)}] \cup [\alpha_{\min}^{(2)}, \alpha_{\max}^{(2)}].$$

Let therefore $\alpha \notin [\alpha_{\min}^{(1)}, \alpha_{\max}^{(1)}] \cup [\alpha_{\min}^{(2)}, \alpha_{\max}^{(2)}]$ be given. We will show that α not belongs to $\sigma(S)$. Indeed, by applying Theorems 3.11 and 3.13 we can write the resolvent equation in terms of Bloch coefficients as follows: In fact, $(S - \alpha I)\mathbf{a} = \mathbf{a}'$ is equivalent to

$$(3.101) \qquad (\beta_\ell^{-1}(w) - \alpha)\alpha_\ell(w) = \alpha_\ell'(w) \quad \forall \ell = 1, 2, \ \forall w \in S^1 \times S^1.$$

Since α does not belong to the union of the image of β_ℓ^{-1}, it follows that

$$\frac{1}{\beta_\ell^{-1}(w) - \alpha} \in L^\infty(S^1 \times S^1) \quad \forall \ell = 1, 2.$$

Hence condition (3.101) defines the Bloch coefficients $\alpha_1(w), \alpha_2(w)$ of $\mathbf{a}$ in a unique way and this element $\mathbf{a}$ is in $\ell^2(\mathbb{C}^2)$ by Parseval's Identity (3.83). This shows that $(S - \alpha I)$ is invertible on $\ell^2(\mathbb{C}^2)$ and completes the proof of Theorem 3.15. ∎

Next, we present a result which characterizes eigenvalues if any, of S in terms of the associated Fermi surface. It is analogous to Theorem 2.12.

Theorem 3.16. λ^{-1} *is an eigenvalue of* S *iff* $\mathrm{meas}\,(F_{\lambda^{-1}}) > 0$, *where the Fermi surface* $F_{\lambda^{-1}}$ *is defined by*

$$(3.102) \qquad F_{\lambda^{-1}} = \{w \in S^1 \times S^1 \mid \lambda^{-1} \text{ is an eigenvalue of } S(w)\}.$$

Proof. By definition, λ^{-1} is an eigenvalue of S iff there exists $\mathbf{a} \in \ell^2(\mathbb{C}^2)$, $\mathbf{a} \neq \mathbf{0}$ such that $S\mathbf{a} = \lambda^{-1}\mathbf{a}$. Let us decompose $\mathbf{a}$ as in (3.82):

$$\mathbf{a}_j = \int_{S^1} \int_{S^1} (\mathbf{a}_\#)_j(w)\,dw \quad \forall j \in \mathbb{Z}^2,$$

where $\mathbf{a}_\#(w) \in \ell_\#(w, \mathbb{C}^2)$. Since $\mathbf{a} \neq \mathbf{0}$, there exists a measurable subset $E \subset S^1 \times S^1$ such that

$$\mathrm{meas}\,(E) > 0, \quad \mathbf{a}_\#(w) \neq \mathbf{0} \quad \text{for a.e. } w \in E.$$

On the other hand, the equation $S\mathbf{a} = \lambda^{-1}\mathbf{a}$ is equivalent to (see (3.95))

$$S(w)\mathbf{a}_\#(w) = \lambda^{-1}\mathbf{a}_\#(w) \quad \text{for a.e. } w \in S^1 \times S^1.$$

This implies that λ^{-1} is an eigenvalue of $S(w)$ for a.e. $w \in E$. Therefore meas $(F_{\lambda^{-1}}) > 0$.

For the converse part, it suffices to retrace the above steps. ∎

With this result, we come to the end of our description of the spectrum of the operator S and this completes the second step of our non-standard homogenization procedure. The moral is that it is enough to solve the family of subproblems (3.42) which are parametrized by $w \in S^1 \times S^1$. As will be seen in numerical examples in Paragraph 3.5 this can be done without much difficulty.

3.4. Convergence Analysis

In this paragraph, we shall concentrate our efforts to the third step of the non-standard homogenization procedure. More precisely, we will prove that the spectral family of Green's operator S_n associated to problem (3.30) converges in a suitable sense to that of S. We shall apply Theorems 4.7 and 4.9 of Chapter I and deduce the convergence of the spectral family from the strong convergence of S_n towards S. Thus we come to the conclusion that $\sigma(S)$ describes the vibrations of the Laplace model for a large number of tubes.

Let us then begin identifying (3.30) as the spectral problem associated with an operator S_n acting on $\mathbb{C}^{2K_n}$ (recall $K_n = 4n^2$) and defined by the following rule:

$$(3.103) \qquad S_n \mathbf{a} = \left\{ \int_{\Gamma_j} \varphi_n \mathbf{n} d\gamma \right\}_{j \in Q_n} \qquad \forall \mathbf{a} \in \mathbb{C}^{2K_n},$$

where φ_n is the unique solution of

$$(3.104) \qquad \begin{cases} \varphi_n \in \dot{H}^1(\Omega_n), \\ \int_{\Omega_n} \nabla \varphi_n \cdot \nabla \bar{\psi} dx = \sum_{j \in Q_n} \mathbf{a}_j \cdot \int_{\Gamma_j} \bar{\psi} \mathbf{n} d\gamma \quad \forall \psi \in \dot{H}^1(\Omega_n). \end{cases}$$

Recall that the set Q_n of indices was defined in (3.31). The standard inner product in $\mathbb{C}^{2K_n}$ will be denoted by $(\cdot, \cdot)_n$ and the corresponding norm by $\| \cdot \|_n$. The operator S_n has been extensively studied in Paragraph 2.1 of Chapter II. In particular, we know that S_n is a self-adjoint operator and that it is positive definite. We arrange its eigenvalues as follows:

$$(3.105) \qquad 0 < \mu_1^{(n)} \leq \cdots \leq \mu_{2K_n}^{(n)}.$$

Then the spectral family of S_n is the following piece-wise constant family of projections:

(3.106) $\qquad E_\tau(S_n)\mathbf{a} = \begin{cases} 0 & \text{if } \tau \leq 0, \\ \displaystyle\sum_{\ell;\mu_\ell^{(n)} \leq \tau} (\mathbf{a}, \mathbf{s}_\ell^{(n)})_n \mathbf{s}_\ell^{(n)} & \text{if } \tau > 0, \end{cases}$

where $\{\mathbf{s}_\ell^{(n)}\}_{\ell=1}^{2K_n}$ are the eigenvectors of S_n.

$\Diamond$ **Exercise 3.6.** Adapt the proof Proposition of 3.6 and show that there exists $\alpha > 0$ independent of n such that

(3.107) $\qquad\qquad (S_n\mathbf{a}, \mathbf{a})_n \geq \alpha\|\mathbf{a}\|_n^2 \quad \forall \mathbf{a} \in \mathbf{C}^{2K_n}.$ ∎

In order to apply Theorem 4.7 of Chapter I, we need the strong convergence of S_n to S. Before announcing the main result, we fix some more notations. Given $\mathbf{a} \in \ell^2(\mathbf{C}^2)$, let $\mathbf{a}_n \in \mathbf{C}^{2K_n}$ denote the image of $\mathbf{a}$ under the canonical projection of $\ell^2(\mathbf{C}^2)$ onto $\mathbf{C}^{2K_n}$, i.e., $(\mathbf{a}_n)_j = \mathbf{a}_j$ for all $j \in Q_n$. On the other hand, given $\mathbf{a}_n \in \mathbf{C}^{2K_n}$, we denote by $\tilde{\mathbf{a}}_n \in \ell^2(\mathbf{C}^2)$ its extension by zero, i.e.,

$$(\tilde{\mathbf{a}}_n)_j = \begin{cases} (\mathbf{a}_n)_j & \text{if } j \in Q_n, \\ 0 & \text{if } j \notin Q_n. \end{cases}$$

Theorem 3.17. *Let* $\mathbf{a} \in \ell^2(\mathbf{C}^2)$ *be given. Then we have*

$$(S_n\mathbf{a}_n)^\sim \longrightarrow S\mathbf{a} \quad in \quad \ell^2(\mathbf{C}^2).$$

Proof. Let $\mathbf{a} \in \ell^2(\mathbf{C}^2)$ be arbitrary and $\mathbf{a}_n \in \mathbf{C}^{2K_n}$ be the projection of $\mathbf{a}$. Recall that $S_n\mathbf{a}_n$ is defined by (3.103), (3.104). Taking $\psi = \phi_n$ and using inequality (3.35), we see that there exists a constant $c > 0$ independent of n such that

(3.108) $\qquad\qquad |\varphi_n|_{1,\Omega_n} \leq c\|\mathbf{a}_n\|_n \leq c\|\mathbf{a}\|.$

Let us admit the following result which we prove later in this section.

Lemma 3.18. *There exists a sequence* $\{P_n\}$ *of continuous extension operators* P_n *from* $H^1(\Omega_n)$ *into* $BL(\mathcal{O})$ *enjoying the following properties: For all* $\psi \in H^1(\Omega_n)$,

(3.109) $\qquad\qquad P_n\psi(x) = \psi(x) \quad for\ a.e.\ x \in \Omega_n.$

There exists $c > 0$ *independent of* n *such that*

(3.110) $\qquad\qquad |P_n\psi|_{1,\Omega_n} \leq c|\psi|_{1,\Omega_n},$

(3.111) $P_n \psi(x) = 0 \quad \text{if} \quad x \in \Gamma_j, \; j \notin Q_n. \; \blacksquare$

From (3.108) and (3.110), it follows that $\{P_n \varphi_n\}$ remains bounded in $BL'(\mathcal{O})$. We can therefore extract a subsequence (still denoted by n) and pick up $\varphi \in BL'(\mathcal{O})$ such that

(3.112) $P_n \varphi_n \rightharpoonup \varphi \quad \text{in} \quad BL'(\mathcal{O}) \text{ weak.}$

Using classical arguments, one can easily pass to the limit in (3.104) and verify that φ is a solution of (3.34). Since the latter problem admits a unique solution, we conclude that the convergence (3.112) holds for the whole sequence. Next, using (3.111), we observe that

$$(S_n \mathbf{a}_n)^{\sim} = \left\{ \int_{\Gamma_j} P_n \varphi_n \mathbf{n} d\gamma \right\}_{j \in \mathbf{Z}^2}.$$

Now the weak convergence (3.112) and Proposition 3.5 imply that

(3.113) $(S_n \mathbf{a}_n)^{\sim} \rightharpoonup S\mathbf{a} \quad \text{in} \quad \ell^2(\mathbb{C}^2) \text{ weak.}$

It remains to prove that the above convergence is in fact strong. To achieve this, we first prove

(3.114) $|\varphi_n|_{1,\Omega_n}^2 \longrightarrow |\varphi|_{1,\mathcal{O}}^2.$

Taking $\psi = \varphi_n$ in (3.104), we see that

$$|\varphi_n|_{1,\Omega_n}^2 = (\mathbf{a}_n, S_n \mathbf{a}_n)_n = (\tilde{\mathbf{a}}_n, (S_n \mathbf{a}_n)^{\sim}).$$

Since $\tilde{\mathbf{a}}_n \longrightarrow \mathbf{a}$ in $\ell^2(\mathbb{C}^2)$, the above relation jointly with (3.113) yields

$$|\varphi_n|_{1,\Omega_n}^2 \longrightarrow (\mathbf{a}, S\mathbf{a}).$$

However $(\mathbf{a}, S\mathbf{a})$ is nothing but $|\varphi|_{1,\mathcal{O}}^2$ by the very definition of S. This proves (3.114).

We are now in a position to prove that the convergence in (3.113) is strong. We have

$$\|(S_n \mathbf{a}_n)^{\sim} - S\mathbf{a}\|^2 = \sum_{j \in Q_n} \left| \int_{\Gamma_j} (\varphi_n - \varphi) \mathbf{n} d\gamma \right|^2 + \sum_{j \notin Q_n} \left| \int_{\Gamma_j} \varphi \mathbf{n} d\gamma \right|^2.$$

We observe that the second term on the right side of the above equality goes to zero as $n \to \infty$ because $\varphi \in BL'(\mathcal{O})$. Regarding the first term, let us remark that the following estimate holds (see Proposition 3.5):

$$\sum_{j \in Q_n} \Big| \int_{\Gamma_j} (\varphi_n - \varphi) \mathbf{n} d\gamma \Big|^2 \leq c^2 |\varphi_n - \varphi|_{1,\Omega_n}^2$$

It is therefore sufficient to show that

(3.115) $$|\varphi_n - \varphi|_{1,\Omega_n}^2 \longrightarrow 0.$$

We can write

(3.116) $$|\varphi_n - \varphi|_{1,\Omega_n}^2 = |\varphi_n|_{1,\Omega_n}^2 + |\varphi|_{1,\Omega_n}^2 - 2\Re e \Big(\int_{\Omega_n} \nabla \varphi_n \cdot \nabla \bar{\varphi} dx \Big)$$

The last term is equal to

$$-2\Re e \Big(\sum_{j \in Q_n} \mathbf{a}_n \cdot \int_{\Gamma_j} \bar{\varphi} \mathbf{n} d\gamma \Big) = -2\Re e (\tilde{\mathbf{a}}_n, S\mathbf{a})$$

which converges to $-2\Re e(\mathbf{a}, S\mathbf{a}) = -2(\mathbf{a}, S\mathbf{a}) = -2|\varphi|_{1,\mathcal{O}}^2$. Using this information in (3.116) along with (3.114), we see that (3.115) holds. This shows that the convergence in (3.113) is strong. This finishes the proof of the theorem modulo Lemma 3.18. ∎

Proof of Lemma 3.18. We are not going into the details of the extension operator P_n because it is quite technical and classical. We remark that the construction of P_n consists of three major steps.

In the first step, given $\psi \in H^1(\Omega_n)$, we extend ψ to ψ^* in a neighbourhood of Γ_n by the well-known procedure of reflection shown in Figure 3.1: $\psi^*(B) = \psi(A)$ where $A \in \Omega_n$ is the image of $B \in (\Omega_{n+1} \setminus \Omega_n)$ under reflection on a side or at a corner point as the case may be. For details of this method, we refer the reader to BREZIS [1983], LIONS and MAGENES [1968] or NEČAS [1967].

The second step consists of truncating ψ^* by multiplying it by a suitable smooth cut-off function θ_n such that

(3.117) $$\begin{cases} \theta_n = 1 & \text{in} \quad \Omega_n, \\ \theta_n = 0 & \text{off} \quad \text{a small neighbourhood of } \Gamma_n. \end{cases}$$

Thus we get a modified extension operator

(3.118) $$\tilde{P}_n \psi = \theta_n \psi^*.$$

By a suitable choice of θ_n, one can arrange that the support of $\tilde{P}_n\psi$ is contained in a neighbourhood of Γ_n which can be chosen such that the support of $\tilde{P}_n\psi$ is contained in Ω_{n+1}.

We observe that $\tilde{P}_n$ satisfies the inequality

$$(3.119) \qquad \|\tilde{P}_n\psi\|_{1,\Omega_{n+1}} \leq c\|\psi\|_{1,\Omega_n} \quad \forall \psi \in H^1(\Omega_n),$$

where $c > 0$ is a constant independent of n. In fact, the above estimate is a consequence of a similar inequality which is valid on each cell adjoining the boundary Γ_n. More precisely, for each $j \in (Q_{n+1}\backslash Q_n)$, there exists $j' \in (Q_n\backslash Q_{n-1})$ such that

$$(3.120) \qquad \|\tilde{P}_n\psi\|_{1,Y_j^*} \leq c\|\psi\|_{1,Y_{j'}^*},$$

where the cells Y_j^* and $Y_{j'}^*$ are shown in Figure 3.1. The extension operator thus constructed has the properties (3.109) and (3.111) of Lemma 3.18. In order to have (3.110) also, we modify it further and that is the purpose of the third step. The idea involved is due to CIORANESCU and SAINT JEAN PAULIN [1979]. Indeed we define

$$(3.121) \qquad P_n\psi(x) = \tilde{P}_n(\psi - \mathcal{M}_{j'})(x) + \mathcal{M}_{j'}$$

for almost all $x \in Y_j^*$ and for every $j \in (Q_{n+1}\backslash Q_n)$. Here $\mathcal{M}_{j'}$ is the average of ψ on $Y_{j'}^*$, i.e.,

$$(3.122) \qquad \mathcal{M}_{j'} = \mathcal{M}_{j'}(\psi) = \frac{1}{|Y_{j'}^*|} \int\limits_{Y_{j'}^*} \psi(x)dx.$$

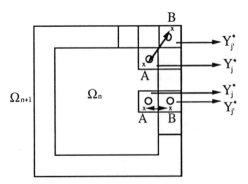

Figure 3.1.

Since P_n leaves constant functions invariant, it is a simple matter, using Exercise 1.1 (Bramble-Hilbert Lemma), to verify that P_n satisfies (3.110) also. This completes the proof of Lemma 3.18. ∎

The convergence of the spectral family $E_\tau(S_n)$ towards $E_\tau(S)$ is now an easy corollary of Theorem 4.7 of Chapter I. Indeed, we have

Theorem 3.19. *For all $\tau \in \mathbb{R}$ such that τ is not an eigenvalue of S, the following convergence holds: For each $\mathbf{a} \in \ell^2(\mathbb{C}^2)$,*

$$(3.123) \qquad \left(E_\tau(S_n)\mathbf{a}_n\right)^{\sim} \longrightarrow E_\tau(S)\mathbf{a} \quad in \quad \ell^2(\mathbb{C}^2). \quad \blacksquare$$

Since one does not know a priori whether S has eigenvalues or not, we cannot conclude the convergence (3.123) for all $\tau \in \mathbb{R}$. That is why the following result which ensures always the distributional convergence of the spectral family may be useful in applications.

Theorem 3.20. *The spectral family of S_n converges to that of S in the following sense: For every $\mathbf{a}, \mathbf{a}' \in \ell^2(\mathbb{C}^2)$,*

$$(3.124) \qquad \left(E_\tau(S_n)\mathbf{a}_n, \mathbf{a}'_n\right)_n \longrightarrow \left(E_\tau(S)\mathbf{a}, \mathbf{a}'\right)$$

in the space of tempered distributions on $\mathbb{R}$.

Proof. Observe that we cannot apply directly Theorem 4.9 of Chapter I to the sequence $\{\tilde{S}_n\}$ of operators defined by $\tilde{S}_n\mathbf{a} = (S_n\mathbf{a}_n)^{\sim}$ for all $\mathbf{a} \in \ell^2(\mathbb{C}^2)$, as they are not uniformly positive; they are only non-negative definite. However the operators $(I + \tilde{S}_n)$ are uniformly positive definite and so we can apply Theorem 4.9 of Chapter I to this new sequence and deduce the convergence of their spectral families in the space of tempered distributions. The convergence (3.124) is then an easy consequence. $\blacksquare$

Comments

1. The Laplace model, by definition, involves a Neumann boundary condition on Γ_n and the above convergence analysis was carried out in this case. A close look at the homogenized problem (i.e., the spectral problem associated to S) shows that this does not have a boundary condition explicitly. This suggests that the boundary condition on Γ_n does not play any essential role in the convergence analysis. This is indeed the case. Let us remark that the proof of Theorem 3.20 is based on the crucial properties of the operator S_n contained in Theorem 3.17 and Exercise 3.6. These results can be proved in a similar manner in the case of other boundary conditions as well, for instance, a homogeneous Dirichlet condition on Γ_n or a Ω_n-periodicity condition. In the case of Dirichlet condition on Γ_n, a natural extension operator P_n exists, namely the extension by zero outside Ω_n whereas the extension operator P_n of Lemma 3.18 serves the purpose in the case of a periodicity boundary condition. These operators can be used in establishing results analogous to Theorem 3.17. $\blacksquare$

2. As has already been mentioned in Paragraph 4.4 of Chapter I, convergence of spectral families or strong convergence of operators is not enough to ensure point-wise convergence of the eigenvalues. If we apply Theorem 4.10 of Chapter I along with Theorem 3.17 to S_n, we see that every element $\mu \in \sigma(S)$ can be attained as a limit of a sequence μ_n in $\sigma(S_n)$. There can exist, however, non-zero sequences $\mu_n \in \sigma(S_n)$ which converge to limits outside $\sigma(S)$. Therefore, one might think in the first place that S is perhaps not the "best possible" limit operator for the sequence S_n. Secondly, as $n \to \infty$, it is clear that the spectrum $\sigma(S_n)$ converges to a limiting set $\sigma_\infty \subset \mathbb{R}$ and a second question which arises naturally in this context is to find a characterization of this limiting spectrum σ_∞. In Paragraph 5.3 we will apply the usual homogenization procedure to a suitable renormalized version of problem (3.30) and this will lead us to a new homogenized operator (say S_c) for the sequence S_n. The spectral analysis of S_c will enlighten a little bit the above points to us. ∎

3.5. Numerical Results in Bloch Wave Decomposition

Several numerical experiments were carried out by F. Aguirre at the University of Chile at Santiago. A brief summary of his work can be found in AGUIRRE and CONCA [1988]. The goal of these experiments was to analyze the behaviour of the functions $\beta_\ell(w)$, $\ell = 1, 2$ as w varies over $S^1 \times S^1$.

The test problem on which we report here has a periodic tube bundle consisting of identical tubes each of which has a square cross section with side δ meters. (In this experiment δ is a parameter). The distance between the tubes is taken to be equal to one meter and the tube is placed symmetrically at the center of the reference cell.

Under these conditions, the eigenvalue problem (3.42) has been solved numerically varying δ. Lagrange finite elements of degree 1 on triangles were used to discretize (3.42). In the computations shown below, a uniform triangulation of Y^* consisting of 256 triangles has been used. More precisely, each side of Y^* contains 13 mesh points and that of the tube has 5 mesh points.

It is worth to point out that the corresponding linear systems (3.48) to compute $S(w)$ have been solved by using a conjugate-gradient method where at each step of the algorithm, the gradient is projected over the space of (w, Y^*)-periodic functions. This iterative method is preferred over a direct method because the (w, Y^*)-periodicity condition destroys the band structure of the matrices. The advantage of using a conjugate-gradient method is that it can be implemented in such a way that the matrices do not have to be stored.

For obvious technical reasons, the parameter $w = (w_1, w_2) \in S^1 \times S^1$ is represented by the equivalent form

$$w_\ell = e^{i\theta_\ell}, \quad \ell = 1, 2,$$

where $\theta_\ell \in [0, 2\pi[$. The subproblems (3.42) were solved for 1024 different values of (θ_1, θ_2) picked up from a uniform mesh of the square $[0, 2\pi[^2$, having a step size equal to $\pi/16$. Figure 3.2 given below shows a graphical view of the functions $\beta_1(\theta_1, \theta_2)$ and $\beta_2(\theta_1, \theta_2)$ with $\delta = 1/3$ m. Similar computations of these functions were carried out with other values of δ, the details of which can be found in CONCA, PLANCHARD, THOMAS and VANNINATHAN [1994]. We record here some of the characteristic features of β_ℓ, $\ell = 1, 2$ revealed by these computations. When $\delta \to 1$, the function β_1 tends to a constant and β_2 exhibits strong variations near the boundary of $[0, 2\pi[^2$. As $\delta \to 0$, we see that the picture is reversed, i.e., β_2 tends towards a constant and β_1 has strong gradients near the boundary.

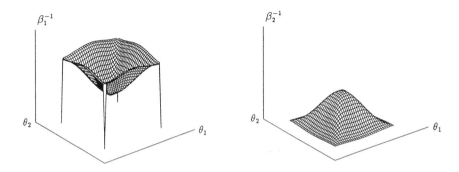

Figure 3.2: *Graphical views of the functions* $\beta_1^{-1}, \beta_2^{-1}$ ($\delta = 1/3$ m).

All numerical examples also show that the Fermi surface defined by (3.102) has always measure zero strongly suggesting that there is no eigen-value in the spectrum of S. Further, these results also show that

$$(3.125) \qquad \alpha_{\max}^{(1)} = \alpha_{\min}^{(2)}$$

for all values of δ. See (3.98) for the definition of these quantities. This implies thanks to Theorem 3.15, that

$$(3.126) \qquad \sigma(S) = [\alpha_{\min}^{(1)}, \alpha_{\max}^{(2)}].$$

This means that the band structure of the spectrum $\sigma(S)$ consists of two intervals with a common point.

The above properties of the functions β_1 and β_2 are simple reflections of the corresponding properties of the frequencies ω of the Laplace model. In particular, there is a band structure for the frequencies of the homogenized Laplace model. More exactly, these frequencies fill up the union of the intervals $[\omega_{\min}^{(1)}, \omega_{\max}^{(1)}]$ and $[\omega_{\min}^{(2)}, \omega_{\max}^{(2)}]$ which are nothing but the images of $[\alpha_{\min}^{(1)}, \alpha_{\max}^{(1)}]$ and $[\alpha_{\min}^{(2)}, \alpha_{\max}^{(2)}]$ under the transformation

$$(3.127) \qquad \omega = \left(\frac{k}{m + \rho_0 \beta^{-1}} \right)^{1/2}.$$

The union of these intervals, because of (3.125), gives one interval which we denote simply by $[\omega_{\min}, \omega_{\max}]$ and it is referred to as the *resonant interval* of the homogenized Laplace model. Table 3.3 presents the calculated values of $\omega_{\min}$ and $\omega_{\max}$ for different values of δ corresponding to the following data:

$$(3.128) \qquad \begin{cases} \rho_0 = 1000 \text{ kg/m}^3, \quad m = 0.22 \text{ kg/m}, \quad k = 27800 \text{ N/m}, \\ \text{natural vibration frequency of tubes} = \sqrt{k/m} = 56.6 \text{ Hz} \end{cases}$$

δ m	$\omega_{\min}$ [Hz]	$\omega_{\max}$ [Hz]
1/8	53.9	54.2
1/3	44.7	48.4
1/2	32.9	44.2
2/3	21.9	43.2

Table 3.3: *Resonant intervals for different values of δ.*

It would be desirable to compare the above numerical results with direct numerical calculations of frequencies with large number of tubes. However, the latter values are not available for square shaped tubes. To our knowledge, the only available numerical results concerning a large tube bundle are presented in IBNOU-ZAHIR and PLANCHARD [1983b]. The bundle contains 100 tubes whose cross-sections are identical and circular. On comparing these two results, we find that the orders of magnitude are the same. The difference in actual values is, of course, explained by the fact that they correspond to geometries which are not identical. For details of this comparison, the reader is referred to CONCA, PLANCHARD and VANNINATHAN [1990].

3.6. Analyticity Properties of the Eigenvalues

In this paragraph, we continue our study of the local problems (3.42) which are parametrized by $w = (w_1, w_2) \in S^1 \times S^1$ and investigate further the dependence of the eigenvalues $\beta_\ell(w)$ on w. Recall that we have already shown that they are continuous everywhere except at $w = (1,1)$. Our aim here is to examine their analyticity with respect to w. The usefulness of such results has been put into evidence in the case of Schrödinger equation in Paragraph 2.3. For instance, it can help one to deduce some properties of the limiting spectrum.

It will be convenient for us to work with parameters $\zeta = (\zeta_1, \zeta_2) \in]0, 1[^2$ where

$$w_\ell = e^{2\pi i \zeta_\ell} \quad \ell = 1, 2$$

and regard the operator S (see (3.47)) and its characteristic values β_ℓ as functions of ζ. Our results show that $S(\zeta)$ is real-analytic. The difficulty is due to rather non-standard fashion in which ζ enters into problem (3.42). Since ζ contains two parameters, there are some additional difficulties to analyze the behaviour of the eigenvalues. In the case of a single parameter, the classical Rellich's Theorem shows the existence of real-analytic branches of eigenvalues (see KATO [1966]). In the present case, we will be able to show that the eigenvalues are real-analytic functions of ζ except on a closed set of measure zero. (Ordering of the eigenvalues is respected here). The main tools used are the *Analytic Implicit Function Theorem* and the *Weierstrass Preparation Theorem* in the category of analytic functions for which the reference is made to BERGER [1977] and KRANTZ and PARKS [1992] (see Theorem 1.8 and Theorem 1.9 of Chapter I).

Let us begin by recalling that the operator $S = S(\zeta) : \mathbb{C}^2 \longrightarrow \mathbb{C}^2$ is given by

$$(3.129) \qquad Sa = \int_\Gamma \vartheta n d\gamma \quad \forall a \in \mathbb{C}^2,$$

where ϑ is the unique solution of

$$(3.130) \qquad \begin{cases} \Delta \vartheta = 0 \quad \text{in} \quad Y^*, \\ \dfrac{\partial \vartheta}{\partial n} = \mathbf{a} \cdot \mathbf{n} \quad \text{on} \quad \Gamma, \\ \vartheta \quad (\zeta, Y^*)\text{-periodic}. \end{cases}$$

In order to show that $S(\zeta)$ is real-analytic, we prove that $\zeta \longmapsto \vartheta$ is a real-analytic map in a suitable sense. To achieve this, we consider a more general problem, namely

$$(3.131) \qquad \begin{cases} \Delta \vartheta = f \quad \text{in} \quad Y^*, \\ \dfrac{\partial \vartheta}{\partial n} = g \quad \text{on} \quad \Gamma, \\ \vartheta \quad (\zeta, Y^*)\text{-periodic}. \end{cases}$$

For a given $f \in L^2(Y^*)$ and $g \in H^{1/2}(\Gamma)$, there is a unique weak solution $\vartheta \in H^1(Y^*)$ to the above problem. We assume that Γ is smooth enough so that the above solution is in fact in the space $H^2(Y^*)$. With this hypothesis we show that the map

(3.132)
$$\begin{cases}]0,1[^2 \longrightarrow H^2(Y^*) \\ \zeta \longmapsto \vartheta \end{cases}$$

is real-analytic in the sense defined is Paragraph 1.9 of Chapter I.

The first step is to make a transformation which takes problem (3.131) into one in which the parameter ζ appears in the coefficients in the usual way. Indeed, let us put

(3.133)
$$u(x) = e^{-i\zeta \cdot x}\vartheta(x).$$

Then (ζ, Y^*) periodicity condition on ϑ is equivalent to the usual Y^* periodicity condition on $u(x)$. Thus system (3.131) is transformed into

(3.134)
$$\begin{cases} \Delta u + 2i(\zeta \cdot \nabla)u - [\zeta_1^2 + \zeta_2^2]u = e^{-i\zeta \cdot x}f \quad \text{in} \quad Y^*, \\ \dfrac{\partial u}{\partial n} + i(\zeta \cdot \mathbf{n})u = e^{-i\zeta \cdot x}g \quad \text{on} \quad \Gamma, \\ u \quad Y^*\text{-periodic.} \end{cases}$$

It suffices to prove the real-analyticity of u and this is the purpose of the following result:

Theorem 3.21. *Let $f \in L^2(Y^*)$ and $g \in H^{1/2}(\Gamma)$. Suppose further that Γ is sufficiently smooth. Then the map $\zeta \longmapsto u(\zeta) \in H^2(Y^*)$ is real-analytic.*

Proof. The idea is to realize system (3.134) as a zero set of an analytic map. Indeed, let us introduce the spaces X, Y, Z and the map F as follows:

(3.135)
$$\begin{cases} X = \mathbb{C}^2, \quad Z = L^2(Y^*) \times H^{1/2}(\Gamma), \\ Y = \{w \in H^2(Y^*) \mid w \text{ is } Y^*\text{-periodic}\}, \\ F: X \times Y \longrightarrow Z, \\ F(\zeta, w) = \begin{pmatrix} \Delta w + 2i(\zeta \cdot \nabla)w - [\zeta_1^2 + \zeta_2^2]w - fe^{-i\zeta \cdot x} \\ \dfrac{\partial w}{\partial n} + i(\zeta \cdot \mathbf{n})w - ge^{-i\zeta \cdot x} \end{pmatrix}. \end{cases}$$

Let us observe that u solves (3.134) iff $F(\zeta, u) = 0$. Our plan is to apply *Implicit Function Theorem* to solve this equation in infinite dimensions. To this end, let us remark that $F(\zeta, w)$ is well-defined for complex $\zeta \in \mathbb{C}^2$ also and it is complex-analytic in the sense defined in Paragraph 1.9 of Chapter I.

Indeed F is easily seen to be Gateaux differentiable at every $(\zeta^0, u_0) \in X \times Y$ and its Gateaux derivative is given by the linear map

$$
\begin{cases}
DF(\zeta^0, u_0): X \times Y \longrightarrow Z, \\
DF(\zeta^0, u_0)(\zeta, w) = \begin{pmatrix} \Delta w + 2i(\zeta^0 \cdot \nabla)w + 2i(\zeta \cdot \nabla)u_0 - [(\zeta_1^0)^2 + (\zeta_2^0)^2]w \\ -2(\zeta^0 \cdot \zeta)u_0 + i(\zeta \cdot x)e^{-i\zeta^0 \cdot x}f \\ \dfrac{\partial w}{\partial n} + i(\zeta^0 \cdot n)w + i(\zeta \cdot n)u_0 + i(\zeta \cdot x)e^{-i\zeta^0 \cdot x}g \end{pmatrix}
\end{cases}
$$

From this expression, one can also calculate partial derivative $\frac{\partial F}{\partial w}(\zeta^0, u_0)$ as a linear bounded map from Y into Z. It is given by

$$
\frac{\partial F}{\partial w}(\zeta^0, u_0)(w) = \begin{pmatrix} \Delta w + 2i(\zeta^0 \cdot \nabla)w - [(\zeta_1^0)^2 + (\zeta_2^0)^2]w \\ \dfrac{\partial w}{\partial n} + i(\zeta^0 \cdot n)w \end{pmatrix}.
$$

Let us fix ζ^0 real and u_0 be the corresponding solution of (3.134) so that $F(\zeta^0, u_0) = 0$. We will prove the real-analyticity of $\zeta \longmapsto u(\zeta)$ at $\zeta = \zeta^0$. By our regularity hypothesis on Γ, the linear map $\frac{\partial F}{\partial w}(\zeta^0, u_0): Y \longrightarrow Z$ is an isomorphism onto. By Theorem 1.8 of Chapter I, we conclude, therefore, that there is a unique $u(\zeta)$ which is complex-analytic near ζ^0 such that $F(\zeta, u(\zeta)) = 0$. By uniqueness, $u(\zeta)$ must coincide with the solution of (3.134) for ζ real. This completes the proof of the theorem. ∎

As a consequence of the above result, we can deduce the following behaviour of the eigenvalues:

Theorem 3.22. *The eigenvalues $\beta_1(\zeta), \beta_2(\zeta)$ are real-analytic for $\zeta \in]0, 1[^2$ except on a closed set of (topological) dimension ≤ 1.*

Proof. To discuss the analyticity of the eigenvalues of $S(\zeta)$, let us introduce its characteristic polynomial:

$$
D(\zeta, \lambda) = \det(S(\zeta) - \lambda I),
$$

which can also be written as

$$
D(\zeta, \lambda) = \lambda^2 - a(\zeta)\lambda + b(\zeta),
$$

where $a(\zeta) = \mathrm{trace}(S(\zeta))$ and $b(\zeta) = \det(S(\zeta))$. Since $\zeta \longmapsto S(\zeta)$ is real-analytic, it follows that $a(\zeta)$ and $b(\zeta)$ are real-analytic functions as well. The eigenvalues of $S(\zeta)$ are given by

$$
\beta_1^{-1}(\zeta) = \frac{1}{2}[a(\zeta) + (a(\zeta)^2 - 4b(\zeta))^{1/2}],
$$

$$
\beta_2^{-1}(\zeta) = \frac{1}{2}[a(\zeta) - (a(\zeta)^2 - 4b(\zeta))^{1/2}].
$$

Since $S(\zeta)$ is self-adjoint, it follows that $a(\zeta)^2 \geq 4b(\zeta)$. From the above expressions, the real-analyticity of $\beta_\ell^{-1}(\zeta)$ is clear in the open region where $a(\zeta)^2 - 4b(\zeta) > 0$. Let us therefore consider a ζ^0 at which $a(\zeta^0)^2 - 4b(\zeta^0) = 0$. If $a(\zeta)^2 - 4b(\zeta) \equiv 0$ in a neighbourhood of ζ^0 then again the real-analyticity of $\beta_\ell^{-1}(\zeta)$ at ζ^0 is obvious because the above expressions then reduce to

$$\beta_1^{-1}(\zeta) = \beta_2^{-1}(\zeta) = \frac{1}{2}a(\zeta) \quad \text{near} \quad \zeta^0.$$

Thus we can assume that $a(\zeta)^2 - 4b(\zeta) \not\equiv 0$ in any neighbourhood of ζ^0. According to the following lemma, the dimension of the set consisting of such ζ^0's is ≤ 1 and this finishes the proof of Theorem 3.22. ∎

Lemma 3.23. *Let $F(\zeta) \not\equiv 0$ be a real-analytic function defined in a neighbourhood U of $0 \in \mathbb{R}^N$. Then the zero set of F is locally of dimension $\leq N-1$.*

Proof. We proceed by induction on N. If $N = 1$, then the zero set of F consists of an isolated sequence of points without accumulation point inside U (see KRANTZ and PARKS [1992] p.14). Thus the conclusion is obviously valid.

Assume the result for $N - 1$. Without loss of generality, we can assume that

$$F(0, 0, ..., 0, \zeta_N) \not\equiv 0.$$

According to *Weierstrass Preparation Theorem*, F may be written as

$$F(\zeta) = H(\zeta', \zeta_N)V(\zeta),$$

where $V(\zeta)$ is a real-analytic function which does not vanish near $0 \in \mathbb{R}^N$ and $H(\zeta', \zeta_N)$ is a polynomial of a certain degree, say m, with respect to the variable ζ_N and with coefficients $a_j(\zeta')$ being real-analytic in $\mathbb{R}^{N-1}$ with respect to $\zeta' = (\zeta_1, ..., \zeta_{N-1})$:

$$H(\zeta', \zeta_N) = \zeta_N^m + a_1(\zeta')\zeta_N^{m-1} + \cdots + a_m(\zeta').$$

Thus the zero set of F coincides with that of H.

Let us therefore analyze the zero set of H. Fix a zero (ζ_0', ζ_N^0) of H. If ζ_N^0 is a simple zero of the polynomial $H(\zeta_0', \zeta_N)$ then $\frac{\partial H}{\partial \zeta_N}(\zeta_0', \zeta_N^0) \neq 0$ and consequently by *Implicit Function Theorem*, one can parametrize all the zeroes of H lying near (ζ_0', ζ_N^0) by a real-analytic function $\zeta_N = \zeta_N(\zeta')$ and so the zero-set near (ζ_0', ζ_N^0) is a $(N - 1)$ dimensional real-analytic manifold.

On the other hand, if ζ_N^0 is a multiple zero of $H(\zeta_0', \zeta_N)$ then its *discriminant* (see VAN DER WAERDEN [1953] pp. 82–87) $R(\zeta')$ vanishes at $\zeta' = \zeta_0'$, i.e., ζ_0' belongs to the zero set of $R(\zeta')$ which, by induction hypothesis, is of dimension $\leq N - 2$. Thus multiple roots form a set of dimension $\leq N - 2$. This completes the proof. ∎

3.7. Behaviour of the Eigenvalues at the Origin

Let us now investigate the behaviour of $\beta_\ell(w)$ as $w \to (1,1)$. We do not expect their continuity because the spaces $H^1_\#(w, Y^*)$ vary continuously with respect to w but the values of β_ℓ at $w = (1,1)$ are defined in terms of the quotient space $\dot{H}^1_\#(Y^*)$ rather than $H^1_\#(Y^*)$. Further, as already noted in Paragraph 3.1.2 the constant which appears in Poincaré Inequality in the spaces $H^1_\#(w, Y^*)$ blows up as $w \to (1,1)$. Thus there are indications that some curious phenomenon takes place as w approaches $(1,1)$ and our aim now is to make it more transparent. For this purpose, it will be more convenient to work with variables $\zeta = (\zeta_1, \zeta_2) \in [0,1[^2$ rather than $w = (w_1, w_2) \in S^1 \times S^1$. Of course, these two are connected by $w_\ell = e^{2\pi i \zeta_\ell}, \ell = 1,2$. We will then denote the operator by $S(\zeta)$ and the corresponding solution of (3.48) by $\vartheta(\zeta)$. We will need to analyze their behaviour as $\zeta \to 0$.

Below, we will calculate the limit of $S(\zeta)$ as $\zeta \to 0$ at a given angle $\alpha \in S^1$. This will show, in particular, that $S(\zeta)$ is never continuous at $\zeta = 0$. The idea of the calculations that follows is taken from the article ALLAIRE and CONCA [1995].

Let us begin by recalling that there exists a constant $c > 0$ independent of $\zeta \in Y'$ such that

$$|\vartheta(\zeta)|_{1,Y^*} \le c \quad \text{and} \quad \|\vartheta(\zeta)\|_{0,Y^*} \le \frac{c}{|\zeta|}.$$

Actually the first inequality is a consequence of the weak formulation while the second one follows from the first and Poincaré Inequality in $H^1_\#(\zeta, Y^*)$ (see Exercise 3.1). Therefore if we introduce $v(\zeta)$ by

$$\vartheta(\zeta) = v(\zeta)e^{i\zeta \cdot x}$$

then $v(\zeta)$ admits the following bounds:

(3.136)
$$\|v(\zeta)\|_{0,Y^*} \le \frac{c}{|\zeta|} \quad \text{and} \quad |v(\zeta)|_{1,Y^*} \le c.$$

Moreover, it is the unique solution of the following variational problem:
(3.137)
$$\begin{cases} v(\zeta) \in H^1_\#(Y^*), \\ \int_{Y^*} (\nabla v(\zeta) + i\zeta v(\zeta)) \cdot (\nabla \bar{\psi} - i\zeta \bar{\psi}) dx = \mathbf{a} \cdot \int_{\Gamma} \bar{\psi} e^{-i\zeta \cdot x} n d\gamma \quad \forall \psi \in H^1_\#(Y^*). \end{cases}$$

In order to analyze the behaviour of $v(\zeta)$ we decompose it as follows:

(3.138)
$$v(\zeta) = \mathring{v}(\zeta) + \mathcal{M}(\zeta)$$

with

$$M(\zeta) = \frac{1}{|Y^*|} \int_{Y^*} v(\zeta) dx, \qquad \int_{Y^*} \mathring{v}(\zeta) dx = 0.$$

The above estimates on $v(\zeta)$ combined with Poincaré-Wirtinger Inequality imply that there exists $c > 0$ independent of ζ such that

(3.139) $$\|\mathring{v}(\zeta)\|_{0,Y^*} + \|\nabla \mathring{v}(\zeta)\|_{0,Y^*} \leq c.$$

In order to describe the behaviour of $\mathring{v}(\zeta)$ as ζ approaches 0 at a given angle $\alpha \in S^1$, we introduce the following problem wherein α is a parameter: Find $v \in H^1_\#(Y^*)$ with $\int_{Y^*} v dx = 0$ and such that

(3.140)
$$\begin{cases} \displaystyle\int_{Y^*} \nabla v \cdot \nabla \bar\psi dx - \frac{1}{|Y^*|}(\alpha \cdot \int_{Y^*} \nabla v dx)(\alpha \cdot \int_{Y^*} \nabla \bar\psi dx) = \\ \qquad\qquad = \left[\mathbf{a} + \frac{|T|}{|Y^*|}(\alpha \cdot \mathbf{a})\alpha \right] \cdot \int_{\Gamma} \bar\psi \mathbf{n} d\gamma \end{cases}$$

for all $\psi \in H^1_\#(Y^*)$ with $\int_{Y^*} \psi dy = 0$.

Our first concern is to establish the existence of a unique solution for the above problem. We can achieve this by a standard application of Lax-Milgram Lemma provided the bilinear form associated with (3.140) is coercive on the space

$$V \stackrel{\text{def}}{=} \{\psi \in H^1_\#(Y^*) \mid \int_{Y^*} \psi dx = 0\}.$$

This is not entirely obvious. First, observe that the bilinear form is given by

$$a(v, \psi) = \int_{Y^*} \nabla v \cdot \nabla \bar\psi dx - \frac{1}{|Y^*|}(\alpha \cdot \int_{Y^*} \nabla v dx)(\alpha \cdot \int_{Y^*} \nabla \bar\psi dx).$$

Theorem 3.10. *There is a constant $c > 0$ such that*

$$\mathfrak{Re}\, a(\psi, \psi) \geq c\|\psi\|^2_{1,Y^*} \quad \forall \psi \in V.$$

Consequently, there is one and only one solution to problem (3.140).

Proof. To establish the coercivity, we apply Exercise 2.1. Since we have:

$$|\alpha \cdot \int_{Y^*} \nabla\psi dx| \le \int_{Y^*} |\nabla\psi| dx \le |Y^*|^{1/2} \|\nabla\psi\|_{0,Y^*},$$

it is obvious that $\Re\, a(\psi,\psi) \ge 0 \; \forall\psi \in V$.

Next, if there is a $\psi \in V$ such that $\Re\, a(\psi,\psi) = 0$ then we see that equality holds in the above Cauchy-Schwarz's Inequality and as a consequence, we conclude that $\nabla\psi = c\alpha$ a.e. in Y^* for some constant c. Since Y^* is connected, this implies $\psi = c\alpha \cdot x + d$ where d is another constant. This contradicts the fact that ψ is Y^*-periodic unless $c = 0$. Because the average of ψ vanishes, it follows that $d = 0$ also. Thus $\psi = 0$.

Finally, to see that $\Re\, a(\psi,\psi)$ is coercive modulo compact terms, we deal with the following alternate expression for $a(\cdot,\cdot)$:

$$a(v,\psi) = \int_{Y^*} \nabla v \cdot \nabla\bar{\psi} dx - \frac{1}{|Y^*|}(\alpha \cdot \int_\Gamma v \mathbf{n} d\gamma)(\alpha \cdot \int_\Gamma \bar{\psi}\mathbf{n} d\gamma).$$

Hence

$$\Re\, a(\psi,\psi) = \int_{Y^*} |\nabla\psi|^2 dx - \frac{1}{|Y^*|}|\alpha \cdot \int_\Gamma \psi\mathbf{n} d\gamma|^2.$$

We note that the second term is continuous with respect to the norm in $L^2(\Gamma)$ and the trace map from V to $L^2(\Gamma)$ is compact. Thus all the conditions of Exercise 2.1 are fulfilled and hence the theorem. ∎

We are now ready for the analysis of the behaviour of $v(\zeta)$ as $\zeta \to 0$ at a given angle $\alpha \in S^1$.

Theorem 3.11. *Let $\zeta \in Y'$ tend to 0 at a given angle $\alpha \in S^1$, i.e., $\frac{\zeta}{|\zeta|} \to \alpha$. Then the solution $v(\zeta)$ of (3.137) can be decomposed as in (3.138) where $\mathring{v}(\zeta)$ converges in $H^1(Y^*)$ to v, the unique solution of (3.140) and $\mathcal{M}(\zeta)\zeta \longrightarrow m\alpha$ where m is a scalar which can be expressed in terms of v as follows:*

$$(3.141) \qquad m = \frac{i\alpha}{2\pi|Y^*|} \cdot \left[\int_{Y^*} \nabla v dx + |T|\mathbf{a} \right].$$

Proof. Let $\{\zeta_n\}$ be any sequence such that

$$\zeta_n \longrightarrow 0, \quad \frac{\zeta_n}{|\zeta_n|} \longrightarrow \alpha \in S^1.$$

We simplify our writing by posing $v_n = \hat{v}(\zeta_n)$ and $m_n = \mathcal{M}(\zeta_n)$. From the estimates (3.136) and (3.139), we see that we can extract a subsequence (denoted again by n) such that

$$\begin{cases} v_n \rightharpoonup v & \text{in } H^1_\#(Y^*)\text{-weak}, \\ \zeta_n m_n \longrightarrow m\alpha. \end{cases}$$

Using this convergence, we wish to pass to the limit in (3.137) and obtain (3.140).

To this end, let $\mu \in \mathbb{C}$ be arbitrary and $\psi \in H^1_\#(Y^*)$ be such that $\int_{Y^*} \psi dx = 0$. The test function in (3.137) is taken as $\psi(\zeta_n) = \psi + \mu_n$ where $\mu_n = \frac{\mu}{|\zeta_n|}$. Then $\zeta_n \bar{\mu}_n \longrightarrow \bar{\mu}\alpha$.

In order to pass to the limit on the right side of (3.137), we see that it can be expressed as

$$\mathbf{a} \cdot \int_\Gamma \bar{\psi} e^{-2\pi i \zeta_n \cdot x} \mathbf{n} d\gamma + \mathbf{a} \cdot \int_\Gamma \bar{\mu}_n e^{-2\pi i \zeta_n \cdot x} \mathbf{n} d\gamma.$$

The trick here is to express the second term as an integral over T and this gives

$$\mathbf{a} \cdot \int_\Gamma \bar{\psi} e^{-2\pi i \zeta_n \cdot x} \mathbf{n} d\gamma - \mathbf{a} \cdot \int_T \bar{\mu}_n \nabla(e^{-2\pi i \zeta_n \cdot x}) dx$$

which is the same as

$$(3.142) \qquad \mathbf{a} \cdot \int_\Gamma \bar{\psi} e^{-2\pi i \zeta_n \cdot x} \mathbf{n} d\gamma + \mathbf{a} \cdot \int_T 2\pi i \zeta_n \bar{\mu}_n e^{-2\pi i \zeta_n \cdot x} dx.$$

Because of the appearance of $\zeta_n \mu_n$, it is possible for us to pass to the limit in the second term also. Treatment of the left side of (3.137) is straightforward. At the limit, we thus obtain

$$(3.143) \quad \int_{Y^*} (\nabla v + 2\pi i m\alpha) \cdot (\nabla\bar{\psi} - 2\pi i \bar{\mu}\alpha) dx = \mathbf{a} \cdot \int_\Gamma \bar{\psi} \mathbf{n} d\gamma + \mathbf{a} \cdot \int_T 2\pi i \bar{\mu}\alpha dx.$$

Since μ and ψ are arbitrary, the above relation is equivalent to the following ones:

$$(3.144) \qquad 4\pi^2 m|Y^*| - 2\pi i \alpha \cdot \int_{Y^*} \nabla v dx = 2\pi i |T| \mathbf{a} \cdot \alpha,$$

(3.145)
$$\int_{Y^*} (\nabla v + 2\pi i m\alpha) \cdot \nabla \bar\psi dx = \mathbf{a} \cdot \int_\Gamma \bar\psi \mathbf{n} d\gamma.$$

We can eliminate m from the second relation using the first one. When we do so, we see that (3.145) coincides with (3.140). Further we observe that (3.144) is nothing but (3.141).

To conclude the proof, we must show the strong convergence of v_n in $H^1_\#(Y^*)$ which is done following the usual procedure. ∎

Let us remark that (3.140) provides a weak formulation characterizing v. From this, we can also write down the differential form of the system of equations satisfied by v:

(3.146)
$$\begin{cases} \Delta v = 0 \quad \text{in} \quad Y^*, \\ \dfrac{\partial v}{\partial n} = \left[\mathbf{a} + \dfrac{|T|}{|Y^*|} (\alpha \cdot \mathbf{a})\alpha \right] \cdot \mathbf{n} + \dfrac{1}{|Y^*|} \left(\alpha \cdot \int_{Y^*} \nabla v dx \right) \alpha \cdot \mathbf{n} \quad \text{on} \quad \Gamma, \\ v \quad Y^*\text{-periodic} \quad \text{and} \quad \int_{Y^*} v dx = 0. \end{cases}$$

The form (3.146) is comparable with problem (3.48) ($w = (1,1)$) which corresponds to $\zeta = 0$. To make this precise, let us recall the definition of $S(0)$:

(3.147)
$$S(0)\mathbf{a} = \int_\Gamma \eta(\mathbf{a})\mathbf{n} d\gamma$$

where $\eta = \eta(\mathbf{a})$ is the unique solution of

(3.148)
$$\begin{cases} \Delta \eta = 0 \quad \text{in} \quad Y^*, \\ \dfrac{\partial \eta}{\partial n} = \mathbf{a} \cdot \mathbf{n} \quad \text{on} \quad \Gamma, \\ \eta \quad Y^*\text{-periodic}, \quad \int_{Y^*} \eta dx = 0. \end{cases}$$

Comparison between (3.148) and (3.146) immediately yields

(3.149)
$$v = \eta(\mathbf{a}) + \rho\eta(\alpha)$$

where

$$\rho = \frac{|T|}{|Y^*|}(\alpha \cdot \mathbf{a}) + \frac{1}{|Y^*|} \left(\alpha \cdot \int_{Y^*} \nabla v dx \right).$$

We can eliminate v from the right side by using (3.149). We get finally that

$$(3.150) \qquad \rho = \frac{[|T|\mathbf{a} + S(0)\mathbf{a}] \cdot \alpha}{|Y^*| - S(0)\alpha \cdot \alpha}.$$

The above expression makes sense provided the denominator does not vanish, a fact which we prove now. First of all, using Cauchy-Schwarz's Inequality, we obtain

$$(3.151) \qquad \int_{Y^*} |\nabla \eta(\alpha)|^2 dx = \alpha \cdot \int_\Gamma \eta(\alpha)\mathbf{n}d\gamma = \alpha \cdot \int_{Y^*} \nabla \eta(\alpha)dx \leq$$

$$\leq |\alpha| \int_{Y^*} |\nabla \eta(\alpha)|dx \leq |Y^*|^{1/2}(\int_{Y^*} |\nabla \eta(\alpha)|^2 dx)^{1/2}.$$

Hence

$$\int_{Y^*} |\nabla \eta(\alpha)|^2 dx \leq |Y^*|.$$

On the other hand, if the denominator in (3.150) vanishes, then

$$|Y^*| = S(0)\alpha \cdot \alpha = \alpha \cdot \int_{Y^*} \nabla \eta(\alpha)dx \leq |Y^*|^{1/2}(\int_{Y^*} |\nabla \eta(\alpha)|^2 dx)^{1/2},$$

which shows that

$$|Y^*| \leq \int_{Y^*} |\nabla \eta(\alpha)|^2 dx.$$

This shows that equality is obtained in (3.151) and so there is a constant c such that

$$\nabla \eta(\alpha)(x) = c\alpha \quad \text{in} \quad Y^*.$$

Hence, it follows $\eta(\alpha)(x) = c\alpha \cdot x + d$ in Y^* where d is another constant. The conditions $\eta(\alpha) \in H^1_\#(Y^*)$ and $\int_{Y^*} \eta(\alpha)dx = 0$ then imply that $c = d = 0$ and so $\eta(\alpha) = 0$ which is clearly a contradiction. Thus we reach the conclusion that the expression (3.150) for ρ makes sense and the denominator in (3.150) is indeed positive, i.e.,

$$(3.152) \qquad S(0)\alpha \cdot \alpha < |Y^*|.$$

We are now in a position to obtain an expression for the limit A^α of $S(\zeta)$ as $\zeta \to 0$ at an angle $\alpha \in S^1$ and connect this limit with $S(0)$. Indeed, $S(\zeta)$ is given by

$$S(\zeta)\mathbf{a} = \int_\Gamma \vartheta(\zeta)\mathbf{n}d\gamma = \int_\Gamma v(\zeta)e^{2\pi i \zeta \cdot x}\mathbf{n}d\gamma.$$

which can be written, using the decomposition (3.138), as follows:

$$(3.153) \qquad S(\zeta)\mathbf{a} = \int_\Gamma \hat{v}(\zeta)e^{2\pi i\zeta\cdot x}\mathbf{n}d\gamma + \int_\Gamma \mathcal{M}(\zeta)e^{2\pi i\zeta\cdot x}\mathbf{n}d\gamma.$$

Let us now take the limit of the above relation as $\zeta \to 0$ at an angle $\alpha \in S^1$. It is easy to pass to the limit in the first term on the right side of (3.153) using Theorem 3.11. For the second term, we do exactly the same trick as in the proof of Theorem 3.11 (see (3.142) and (3.143)). At the limit, we obtain

$$(3.154) \qquad A^\alpha \mathbf{a} = \int_\Gamma v\mathbf{n}d\gamma - 2\pi i m|T|\alpha,$$

where v is the unique solution of (3.140) and m is the scalar given by (3.141).

Using (3.149) and the definition (3.147) of $S(0)$ we can give the following alternate expression for $A^\alpha \mathbf{a}$:

$$(3.155) \qquad A^\alpha \mathbf{a} = S(0)\mathbf{a} + \rho S(0)\alpha - 2\pi i m|T|\alpha.$$

We have the expression (3.150) for ρ in terms of $\mathbf{a}$. We also express m in terms of $\mathbf{a}$ as follows:

$$(3.156) \qquad m = \frac{i\alpha}{2\pi|Y^*|}\left[|T|\mathbf{a} + S(0)\mathbf{a} + \rho S(0)\alpha\right].$$

Plugging these expressions in (3.155), we find, after some computations, that A^α is given by

$$(3.157) \quad A^\alpha = S(0) + \frac{1}{|Y^*| - S(0)\alpha\cdot\alpha}\left[S(0)\alpha + |T|\alpha\right] \otimes \left[S(0)\alpha + |T|\alpha\right].$$

We announce the foregoing result in the form of a theorem:

Theorem 3.12. *Let $\zeta \in Y'$ tend to 0 at a given angle $\alpha \in S^1$, that is, $\frac{\zeta}{|\zeta|} \to \alpha$. Then the operator $S(\zeta)$ defined by (3.47) tends to A^α which is given in* (3.157). ∎

We can draw a few conclusions from the above result. We see that A^α is a perturbation from $S(0)$ by a symmetric, non-negative matrix of rank unity. In particular $A^\alpha \neq S(0)$ and hence $S(\zeta)$ is *never* continuous at $\zeta = 0$.

Secondly, if $S(0) = a_0 I$ (which is by definition the case of an *isotropic effective material*) then A^α is given by

$$A^\alpha = a_0 I + \frac{(a_0 + |T|)^2}{|Y^*| - a_0}[\alpha \otimes \alpha].$$

It is easy to check that the eigenvalues of A^{α} are $\{a_0, a_0 + \frac{(a_0 + |T|)^2}{|Y^*| - a_0}\}$ with corresponding eigenvectors $\{u, \alpha\}$ where u is orthogonal to α. This proves that the smallest eigenvalue of $S(\zeta)$ is continuous at $\zeta = 0$ whereas the largest one need not be. These conclusions have already been confirmed by the numerical experiments of AGUIRRE and CONCA [1988]. This result is after all expected because $S(\zeta)$ is not continuous at $\zeta = 0$ as shown by (3.157); the ellipticity constant of $S(\zeta)$ is not uniform as $\zeta \to 0$ (see (3.136)). This difficulty can be easily overcome in classical examples such as the treated in §2 by simply replacing $-\Delta$ by $(-\Delta + c)$ where c is a large positive constant.

§4. Bloch Wave Method in the Helmholtz Model

With our experience gained in the study of the Laplace model, we now consider the asymptotic behaviour of the Helmholtz model as the number of tubes goes to infinity. Of course, we propose to follow the method of non-standard homogenization procedure introduced in Paragraph 2.4 and its major steps were summarized in the beginning of Paragraph 3.2. Since this method has been presented in full details in the case of the Laplace model and its application to the present model is very similar, we shall not go into a detailed analysis. We shall concentrate our efforts to the description of the homogenized Helmholtz model and underline the main differences between these two models. In particular, it is to be noted that the usual Green's operator technique is not successful to describe the homogenized Helmholtz model which will be, however, identified with the spectral problem of an unbounded self-adjoint operator.

In Paragraph 4.1, we introduce the homogenized spectral problem and present its properties. Paragraph 4.2 is devoted to the decomposition of the homogenized problem in terms of Bloch waves and finally we conclude with the convergence analysis in Paragraph 4.3.

4.1. Description of the Limit Operator

We place ourselves under the geometric set-up described in Paragraph 3.2 and consider Helmholtz model in Ω_n with identical tubes in which case the corresponding equation can be written as follows (see (2.17) of Chapter II): Find $\omega(n) \in \mathbb{C}$ for which there is a function not identically zero $\phi^{(n)}$ and a vector $s^{(n)} \in \mathbb{C}^{2K_n}$ such that $(\omega(n), \phi^{(n)}, s^{(n)})$ satisfies

(4.1)
$$
\begin{cases}
\Delta\phi^{(n)} + \omega^2(n)\phi^{(n)} = 0 & \text{in} \quad \Omega_n, \\[2mm]
\dfrac{\partial\phi^{(n)}}{\partial n} = s_j^{(n)} \cdot \mathbf{n} & \text{on} \quad \Gamma_j, \ j \in Q_n, \\[2mm]
\dfrac{\partial\phi^{(n)}}{\partial n} = 0 & \text{on} \quad \Gamma_0, \\[2mm]
(k - m\omega^2(n))s_j^{(n)} = \rho_0\omega^2(n)\displaystyle\int_{\Gamma_j} \phi^{(n)}\mathbf{n}d\gamma & \forall j \in Q_n,
\end{cases}
$$

where, for simplicity, we have taken the constant c of the Helmholtz model to be unity.

We are interested in the behaviour of (4.1) as $n \to \infty$. Looking at (4.1) we see that the limit problem can formally be written as follows: Find $\omega \in \mathbb{C}$ for which there exist ϕ not identically zero and a sequence $s = \{s_j\}_{j \in \mathbb{Z}^2}$ such that

(4.2)
$$
\begin{cases}
\Delta\phi + \omega^2\phi = 0 & \text{in} \quad \mathcal{O}, \\[2mm]
\dfrac{\partial\phi}{\partial n} = s_j \cdot \mathbf{n} & \text{on} \quad \Gamma_j, \ j \in \mathbb{Z}^2, \\[2mm]
(k - m\omega^2)s_j = \rho_0\omega^2 \displaystyle\int_{\Gamma_j} \phi\mathbf{n}d\gamma & \forall j \in \mathbb{Z}^2.
\end{cases}
$$

We shall identify the above system as the spectral problem associated with an unbounded operator A with domain in $L^2(\mathcal{O}) \times \ell^2(\mathbb{C}^2)$. In accordance with the symmetrization procedure followed in the resolution of the Helmholtz model (see Paragraph 2.2 of Chapter II), we equip the above space with the following weighted inner product:

$$
\langle (\phi, s), (\psi, t) \rangle \overset{\text{def}}{=} \int_{\mathcal{O}} \phi\bar{\psi}dx + \rho_0^{-1}k \sum_{j \in \mathbb{Z}^2} s_j \cdot \bar{t}_j.
$$

Let the domain $D(A) \subset L^2(\mathcal{O}) \times \ell^2(\mathbb{C}^2)$ be defined as follows: $(\phi, s) \in D(A)$ iff

(4.3)
$$
\begin{cases}
\text{(i)} \quad \Delta\phi \in L^2(\mathcal{O}) \quad (\text{ i.e., } \phi \in H(\mathcal{O}; \Delta)). \\[2mm]
\text{(ii)} \begin{cases} \text{There exists } \mathbf{a} \in \ell^2(\mathbb{C}^2) \text{ such that for all } j \in \mathbb{Z}^2 \\[2mm] \text{(a) } \dfrac{\partial\phi}{\partial n} = \mathbf{a}_j \cdot \mathbf{n} \quad \text{on} \quad \Gamma_j, \\[2mm] \text{(b) } \rho_0^{-1}ks_j = \rho_0^{-1}m\mathbf{a}_j + \displaystyle\int_{\Gamma_j} \phi\mathbf{n}d\gamma. \end{cases}
\end{cases}
$$

It is worthwhile to recall that if $\phi \in H(\mathcal{O}; \Delta)$ then the traces $\{\phi, \frac{\partial\phi}{\partial n}\}|_{\Gamma_j}$ are well-defined elements of $H^{-1/2}(\Gamma_j)$ and $H^{-3/2}(\Gamma_j)$ respectively. Therefore, it is possible to interpret the integral in (4.3) as the duality bracket between $\phi \in H^{-1/2}(\Gamma_j)$ and $n_\ell \in H^{1/2}(\Gamma_j)$ for every $\ell = 1, 2$.

On the other hand, condition (4.3) (ii) (a) determines $\mathbf{a}_j$ on Γ_j in a unique way by ϕ and so $\mathbf{s}_j$ is also uniquely determined by ϕ. We can now define the operator by the following rule:

$$
(4.4) \qquad
\begin{cases}
A \colon D(A) \subset L^2(\mathcal{O}) \times \ell^2(\mathbb{C}^2) \longrightarrow L^2(\mathcal{O}) \times \ell^2(\mathbb{C}^2) \\
A(\phi, \mathbf{s}) = (-\Delta\phi, \mathbf{a}),
\end{cases}
$$

where $\mathbf{a} \in \ell^2(\mathbb{C}^2)$ is the unique element appearing in (4.3). We can easily verify that the spectral problem of A formally coincides with the system of equations (4.2). More precisely, $A(\phi, \mathbf{s}) = \omega^2(\phi, \mathbf{s})$ iff $(\frac{1}{\omega^2}\phi, \mathbf{s}, \omega^2)$ is a solution of (4.2). Indeed, the definition of the operator A is cooked up precisely to achieve this. The properties of A are summarized in the following result:

Theorem 4.1. *The unbounded operator A enjoys the following properties:*
(i) *The domain $D(A)$ is dense in $L^2(\mathcal{O}) \times \ell^2(\mathbb{C}^2)$.*
(ii) *A is closed.*
(iii) *A is self-adjoint.*
(iv) *A is non-negative definite in the following sense:*

$$
(4.5) \qquad \langle A(\phi, \mathbf{s}), (\phi, \mathbf{s}) \rangle \geq 0 \quad \forall (\phi, \mathbf{s}) \in D(A). \; \blacksquare
$$

To prove the above theorem, we shall need two density results.

Lemma 4.2. *Let $\mathbf{a} \in \ell^2(\mathbb{C}^2)$ be given. Consider the affine space*

$$
V(\mathbf{a}) = \left\{ \psi \in H(\mathcal{O}; \Delta) \; \Big| \; \frac{\partial \psi}{\partial n} = \mathbf{a}_j \cdot \mathbf{n} \quad on \quad \Gamma_j, \; j \in \mathbb{Z}^2 \right\}.
$$

Then $(H^2(\mathcal{O}) \cap V(\mathbf{a}))$ is dense in $V(\mathbf{a})$ with respect to $H(\mathcal{O}; \Delta)$ norm. $\blacksquare$

Lemma 4.3. *Let X be the subspace of $D(A)$ given by*

$$
(4.6) \qquad X = \{ (\phi, \mathbf{s}) \in D(A) \mid \phi \in H^2(\mathcal{O}) \}.
$$

Then X is dense in $D(A)$ with respect to the graph norm of A. Moreover, the following Green's formula holds for all $(\phi, \mathbf{s})$ and $(\phi', \mathbf{s}')$ in $D(A)$:

$$
(4.7) \qquad \int_{\mathcal{O}} [\Delta\phi\bar{\phi}' - \Delta\bar{\phi}'\phi]dx = \sum_{j \in \mathbb{Z}^2} \left[\mathbf{a}_j \cdot \int_{\Gamma_j} \bar{\phi}' \mathbf{n} d\gamma - \bar{\mathbf{a}}'_j \cdot \int_{\Gamma_j} \phi \mathbf{n} d\gamma \right]
$$

where $\mathbf{a}, \mathbf{a}'$ are the unique elements of $\ell^2(\mathbb{C}^2)$ associated to $(\phi, \mathbf{s})$ and $(\phi', \mathbf{s}')$ respectively as per (4.3) (ii). $\blacksquare$

The validity of the Green's formula (4.7) may seem strange at first sight because ϕ and ϕ' do not have sufficient regularity in $\mathcal{O}$. However as they satisfy smooth boundary conditions on $\partial\mathcal{O}$, we will be able to establish (4.7) by approximating ϕ and ϕ' by smooth functions.

Proof of Lemma 4.2. In the first step, we show that we are reduced to the case $\mathbf{a} = 0$. Indeed, we can pick up $\psi_0 \in H^2(\mathcal{O})$ such that

$$\frac{\partial \psi_0}{\partial n} = \mathbf{a}_j \cdot \mathbf{n} \quad \text{on} \quad \Gamma_j, \ j \in \mathbf{Z}^2.$$

To construct such a ψ_0, it suffices to work in the basic reference cell Y^* and use the surjectivity of the following trace map:

$$\begin{cases} \{\psi \in H^2(Y^*) \mid \psi, \dfrac{\partial \psi}{\partial n} = 0 \quad \text{on} \quad \partial Y\} \longrightarrow H^{1/2}(\Gamma) \\ \\ \qquad\qquad\qquad\qquad\qquad\qquad \psi \longmapsto \dfrac{\partial \psi}{\partial n}. \end{cases}$$

Once ψ_0 is constructed, it is enough to consider $(\psi - \psi_0)$ and we reduce to $\mathbf{a} = 0$.

As a second step, one can easily check that ϕ is in the orthogonal complement of $(H^2(\mathcal{O}) \cap V(0))$ iff ϕ satisfies

(4.8)
$$\begin{cases} \phi \in H(\mathcal{O}; \Delta), \ \Delta \phi \in H(\mathcal{O}; \Delta), \\ \Delta^2 \phi + \phi = 0 \quad \text{in} \quad \mathcal{O}, \\ \dfrac{\partial \phi}{\partial n} = \dfrac{\partial}{\partial n} \Delta \phi = 0 \quad \text{on} \quad \Gamma_j, \ j \in \mathbf{Z}^2. \end{cases}$$

Next, we multiply (4.8) by $\bar{\psi} \in H^4(\mathcal{O})$ and integrate by parts twice using formula (1.19) of Chapter I. We get

(4.9)
$$\int_{\mathcal{O}} \phi(\Delta^2 \bar{\psi} + \bar{\psi})dx - \sum_{j \in \mathbf{Z}^2} \int_{\Gamma_j} \left[\frac{\partial \Delta \bar{\psi}}{\partial n}\phi - \frac{\partial \bar{\psi}}{\partial n}\Delta\phi\right]d\gamma = 0$$

for all $\psi \in H^4(\mathcal{O})$.

The third step consists of applying the well-known *transposition method* of LIONS and MAGENES [1968] (see also GRISVARD [1985]) to prove $\phi = 0$. To this end, we consider the map $\Theta: H^4(\mathcal{O}) \longrightarrow L^2(\mathcal{O}) \times H^{5/2}(\partial\mathcal{O}) \times H^{1/2}(\partial\mathcal{O})$ defined by

$$\Theta\psi = (\Delta^2 \psi + \psi, \frac{\partial \psi}{\partial n}, \frac{\partial \Delta\psi}{\partial n}).$$

Using the classical regularity results, it is known that if $\mathcal{O}$ is sufficiently smooth then Θ is an isomorphism onto. In particular, this implies that Θ^* is injective and so for all $L \in H^4(\mathcal{O})'$ there exists a unique element $(u, u_0, u_1) \in L^2(\mathcal{O}) \times H^{-5/2}(\partial\mathcal{O}) \times H^{-1/2}(\partial\mathcal{O})$ such that

(4.10)
$$\int_{\mathcal{O}} u(\Delta^2 \bar{\psi} + \bar{\psi})dx + \int_{\partial\mathcal{O}} \left[u_0\frac{\partial \bar{\psi}}{\partial n} + u_1\frac{\partial \Delta\bar{\psi}}{\partial n}\right]d\gamma = L(\bar{\psi})$$

for all $\psi \in H^4(\mathcal{O})$. In particular for $L = 0$, the unique element satisfying (4.10) is nothing but $(\phi, \Delta\phi_{|\partial\mathcal{O}}, -\phi_{|\partial\mathcal{O}})$ because of (4.9). Thus $\phi = 0$ and this concludes the proof of the lemma. ∎

Proof of Lemma 4.3. Let $(\phi, \mathbf{s}) \in D(A)$ be orthogonal to X with respect to the inner product given by the graph of A, i.e.,

$$(4.11) \qquad \int_{\mathcal{O}} \phi \bar{\phi}' dx + \rho_0^{-1} k \sum_{j \in \mathbf{Z}^2} \left[\mathbf{s}_j \cdot \bar{\mathbf{s}}'_j + \mathbf{a}_j \cdot \bar{\mathbf{a}}'_j \right] + \int_{\mathcal{O}} \Delta \phi \Delta \bar{\phi}' dx = 0$$

for all $(\phi', \mathbf{s}') \in X$. To prove $(\phi, \mathbf{s}) = (0, 0)$, we make suitable choices of $(\phi', \mathbf{s}')$ in X and use (4.11).

The first one is that $\phi' \in \mathcal{D}(\mathcal{O})$ and so necessarily $\mathbf{s}' = 0$ (see (4.3) (ii)). From (4.11), we then get

$$\int_{\mathcal{O}} \phi \bar{\phi}' dx + \int_{\mathcal{O}} \Delta \phi \Delta \bar{\phi}' dx = 0.$$

This is equivalent to

$$(4.12) \qquad\qquad \Delta^2 \phi + \phi = 0 \quad \text{in} \quad \mathcal{O}$$

which implies, in particular, that $\Delta \phi \in H(\mathcal{O}; \Delta)$. Using now the classical Green's formula (see (1.19) of Chapter I) we obtain
(4.13)
$$\int_{\mathcal{O}} \Delta \phi \Delta \bar{\psi} dx = \int_{\mathcal{O}} (\Delta^2 \phi) \bar{\psi} dx - \sum_{j \in \mathbf{Z}^2} \int_{\Gamma_j} \left[\frac{\partial \Delta \phi}{\partial n} \bar{\psi} - \Delta \phi \frac{\partial \bar{\psi}}{\partial n} \right] d\gamma \quad \forall \psi \in H^2(\mathcal{O}).$$

We now make our second choice. More precisely, we take $(\phi', \mathbf{s}')$ in X with $\partial \phi' / \partial n = 0$ on Γ_j, for all $j \in \mathbf{Z}^2$. It then follows from (4.3) (ii) that $\mathbf{a}'_j = 0$ for all $j \in \mathbf{Z}^2$ and we have

$$\rho_0^{-1} k \mathbf{s}'_j = \int_{\Gamma_j} \phi' \mathbf{n} d\gamma \quad \forall j \in \mathbf{Z}^2.$$

On the other hand, if we take $\psi = \phi'$ in (4.13) and use (4.11), (4.12) we get

$$\sum_{j \in \mathbf{Z}^2} \left[\mathbf{s}_j \cdot \int_{\Gamma_j} \bar{\phi}' \mathbf{n} d\gamma - \int_{\Gamma_j} \frac{\partial \Delta \phi}{\partial n} \bar{\phi}' d\gamma \right] = 0.$$

Since ϕ' is arbitrary on Γ_j, we deduce that

$$(4.14) \qquad\qquad \frac{\partial}{\partial n} \Delta \phi = \mathbf{s}_j \cdot \mathbf{n} \quad \text{on} \quad \Gamma_j, \ j \in \mathbf{Z}^2.$$

Using this information in (4.11)–(4.13), we get, for all $(\phi', \mathbf{s}') \in X$

$$\rho_0^{-1} k \sum_{j \in \mathbf{Z}^2} \left[\mathbf{s}_j \cdot \bar{\mathbf{s}}'_j + \mathbf{a}_j \cdot \bar{\mathbf{a}}'_j \right] - \sum_{j \in \mathbf{Z}^2} \left[\mathbf{s}_j \cdot \int_{\Gamma_j} \bar{\phi}' \mathbf{n} d\gamma - \int_{\Gamma_j} \Delta \phi \frac{\partial \bar{\phi}'}{\partial n} d\gamma \right] = 0.$$

We observe that ϕ' can be eliminated from the above relation by a simple application of (4.3) (ii) and we obtain

$$\sum_{j \in \mathbb{Z}^2} \bar{\mathbf{a}}' \cdot \left[\rho_0^{-1} k \mathbf{a}_j + \rho_0^{-1} m \mathbf{s}_j + \int_{\Gamma_j} \Delta\phi \mathbf{n} d\gamma \right] = 0.$$

Finally, we make the third choice; namely, we take $(\phi', \mathbf{s}') \in X$ arbitrary. We observe, in such a case, that the associated element $\mathbf{a}' \in \ell^2(\mathbb{C}^2)$ is arbitrary. This is because of the fact that the trace map $\psi \mapsto \{\psi, \frac{\partial\psi}{\partial n}\}_{|\partial U}$ is surjective from $H^2(U)$ into $H^{3/2}(\partial U) \times H^{1/2}(\partial U)$ provided U is of class $\mathcal{C}^2$. Therefore, we deduce from the last relation that

$$(4.15) \qquad \rho_0^{-1} k \mathbf{a}_j + \rho_0^{-1} m \mathbf{s}_j + \int_{\Gamma_j} \Delta\phi \mathbf{n} d\gamma = \mathbf{0} \quad \forall j \in \mathbb{Z}^2.$$

Now, it remains to combine the information contained in (4.12), (4.14) and (4.15) to deduce that $(\phi, \mathbf{s}) = (0, \mathbf{0})$. To this end, we apply Lemma 4.2 and choose a sequence $\{\phi_\ell\}$ in $H^2(\mathcal{O})$ such that

$$(4.16) \qquad \begin{cases} \phi_\ell \longrightarrow \phi & \text{in} \quad H(\mathcal{O}; \Delta), \\ \dfrac{\partial\phi_\ell}{\partial n} = \mathbf{a}_j \cdot \mathbf{n} & \text{on} \quad \Gamma_j, \ \forall j \in \mathbb{Z}^2. \end{cases}$$

We multiply (4.12) by $\bar{\phi}_\ell$ and integrating by parts, we get

$$\int_{\mathcal{O}} \Delta\phi \cdot \Delta\bar{\phi}_\ell dx = -\int_{\mathcal{O}} \phi\bar{\phi}_\ell dx - \sum_{j \in \mathbb{Z}^2} \left[\mathbf{s}_j \cdot \int_{\Gamma_j} \bar{\phi}_\ell \mathbf{n} d\gamma - \int_{\Gamma_j} \Delta\phi \frac{\partial\bar{\phi}_\ell}{\partial n} d\gamma \right].$$

Thanks to (4.16), we can pass to the limit in the above relation as $\ell \to \infty$ and obtain

$$\int_{\mathcal{O}} |\Delta\phi|^2 dx = -\int_{\mathcal{O}} |\phi|^2 dx - \sum_{j \in \mathbb{Z}^2} \left[\mathbf{s}_j \cdot \int_{\Gamma_j} \bar{\phi} \mathbf{n} d\gamma - \bar{\mathbf{a}}_j \cdot \int_{\Gamma_j} \Delta\phi \mathbf{n} d\gamma \right].$$

Using now (4.3) (ii), we get

$$\int_{\mathcal{O}} |\Delta\phi|^2 dx = -\int_{\mathcal{O}} |\phi|^2 dx - \rho_0^{-1} k \left(\|\mathbf{s}\|^2 + \|\mathbf{a}\|^2 \right).$$

This implies immediately that $(\phi, \mathbf{s}) = (0, \mathbf{0})$ and this terminates the proof of the first part of Lemma 4.3.

To prove (4.7), it suffices to observe that this is nothing but the classical Green's formula for elements of X. Since each term in (4.7) is a continuous map with respect to the $H(\mathcal{O}; \Delta)$ norm, we conclude that (4.7) holds for all elements $(\phi, \mathbf{s})$ and $(\phi', \mathbf{s}')$ in $D(A)$ since X is dense in $D(A)$. ∎

Proof of Theorem 4.1. (i) Let $(\psi, \mathbf{t}) \in L^2(\mathcal{O}) \times \ell^2(\mathbf{C}^2)$ be such that

$$\int_{\mathcal{O}} \phi \bar{\psi} dx + \rho_0^{-1} k \sum_{j \in \mathbf{Z}^2} \mathbf{s}_j \cdot \bar{\mathbf{t}}_j = 0 \quad \forall (\phi, \mathbf{s}) \in D(A).$$

Observing that $\mathcal{D}(\mathcal{O}) \times \{0\} \subset D(A)$, we get

$$\int_{\mathcal{O}} \phi \bar{\psi} dx = 0 \quad \forall \phi \in \mathcal{D}(\mathcal{O})$$

which obviously implies that $\psi = 0$. Consequently, we have $(\mathbf{s}, \mathbf{t}) = 0$ for all $(\phi, \mathbf{s}) \in D(A)$. In order to show that $\mathbf{t} = \mathbf{0}$, we claim that for each $\mathbf{s} \in \ell^2(\mathbf{C}^2)$ with compact support, there exists a $\phi \in \mathcal{D}(\mathcal{O})$ such that $(\phi, \mathbf{s})$ belongs to $D(A)$. This claim can be easily established using the fact that the trace map $\phi \longmapsto \{\phi, \frac{\partial \phi}{\partial n}\}_{|\partial U}$ is surjective from $H^2(U)$ onto $H^{3/2}(\partial U) \times H^{1/2}(\partial U)$ for all domains U of class $\mathcal{C}^2$. Then we get $(\mathbf{s}, \mathbf{t}) = 0$, for all $\mathbf{s} \in \ell^2(\mathbf{C}^2)$ with compact support. This implies that $\mathbf{t} = \mathbf{0}$.

(ii) To prove that A is closed, let $\{(\phi_\ell, \mathbf{s}_\ell)\} \subset D(A)$ be a sequence such that

$$(\phi_\ell, \mathbf{s}_\ell) \longrightarrow (\phi, \mathbf{s}) \quad \text{in} \quad L^2(\mathcal{O}) \times \ell^2(\mathbf{C}^2),$$
$$(A\phi_\ell, \mathbf{a}_\ell) \longrightarrow (\psi, \mathbf{a}) \quad \text{in} \quad L^2(\mathcal{O}) \times \ell^2(\mathbf{C}^2),$$

where $\mathbf{a}_\ell$ is the unique element associated with ϕ_ℓ according to the definition of $D(A)$. Since $\phi_\ell \longrightarrow \phi$ in $H(\mathcal{O}; \Delta)$, $\psi = \Delta \phi \in L^2(\mathcal{O})$. Further, by using the continuity of the trace map (see Theorem 1.4 of Chapter I) we can pass to the limit in (4.3) (ii) and conclude simultaneously that $(\phi, \mathbf{s}) \in D(A)$ and $A(\phi, \mathbf{s}) = (\psi, \mathbf{a})$.

(iii) Let us first show that A is symmetric,. i.e.,

(4.17) $\langle A(\phi, \mathbf{s}), (\phi', \mathbf{s}') \rangle = \langle (\phi, \mathbf{s}), A(\phi', \mathbf{s}') \rangle$

for all $(\phi, \mathbf{s}), (\phi', \mathbf{s}') \in D(A)$. Let $\mathbf{a}$ and $\mathbf{a}' \in \ell^2(\mathbf{C}^2)$ be the unique elements associated with $(\phi, \mathbf{s})$ and $(\phi', \mathbf{s}')$ respectively as in the definition of $D(A)$. By the very definition of A, the left side of (4.17) is equal to

$$-\int_{\mathcal{O}} \Delta \phi \bar{\phi}' dx + \rho_0^{-1} k \sum_{j \in \mathbf{Z}^2} \mathbf{a}_j \cdot \bar{\mathbf{s}}'_j.$$

Upon using the relation (4.3) (ii), the above quantity can be rewritten as

$$-\int_{\mathcal{O}} \Delta \phi \bar{\phi}' dx + \sum_{j \in \mathbf{Z}^2} \mathbf{a}_j \cdot \left[\rho_0^{-1} m \bar{\mathbf{a}}'_j + \int_{\Gamma_j} \bar{\phi}' n d\gamma \right]$$

We now apply Green's formula (4.7) and the above expression becomes

$$-\int_{\mathcal{O}} \Delta\bar{\phi}'\phi dx + \rho_0^{-1}m \sum_{j\in\mathbb{Z}^2} \mathbf{a}_j \cdot \bar{\mathbf{a}}'_j + \sum_{j\in\mathbb{Z}^2} \bar{\mathbf{a}}'_j \cdot \int_{\Gamma_j} \phi n d\gamma.$$

It is easy to see that the right side of (4.17) coincides with the above identity and so this finishes the proof that A is symmetric.

To show that A is self-adjoint, it is now sufficient to prove that $D(A^*) = D(A)$. Let us therefore proceed to characterize the domain of A^*. By definition, $D(A^*)$ consists of $(\phi', \mathbf{s}') \in L^2(\mathcal{O}) \times \ell^2(\mathbb{C}^2)$ such that the map from $D(A)$ into $\mathbb{C}$ given by

(4.18) $$(\phi, \mathbf{s}) \longmapsto \langle A(\phi, \mathbf{s}), (\phi', \mathbf{s}')\rangle$$

is continuous with respect to the norm in $L^2(\mathcal{O}) \times \ell^2(\mathbb{C}^2)$.

In the first stage, we restrict (4.18) to elements of the form $(\phi, 0)$ with $\phi \in \mathcal{D}(\mathcal{O})$. The above map then becomes

$$(\phi, \mathbf{s}) \longmapsto -\int_{\mathcal{O}} \Delta\phi\bar{\phi}' dx.$$

The condition that this is continuous with respect to $L^2(\mathcal{O})$ norm on ϕ imposes that

(4.19) $$\Delta\phi' \in L^2(\mathcal{O}).$$

In the second step, we consider (4.18) with $(\phi, \mathbf{s}) \in X$ (see (4.6)). Using the definition of A and doing integration by parts which is permitted because of (4.19), we get

(4.20) $$\begin{cases} \langle A(\phi, \mathbf{s}), (\phi', \mathbf{s}')\rangle = -\int_{\mathcal{O}} \Delta\bar{\phi}'\phi dx + \sum_{j\in\mathbb{Z}^2}\int_{\Gamma_j} \frac{\partial\bar{\phi}'}{\partial n}\phi d\gamma - \\ -\sum_{j\in\mathbb{Z}^2}\left[km^{-1}\mathbf{s}_j - \rho_0 m^{-1}\int_{\Gamma_j}\phi n d\gamma\right]\cdot\left[\int_{\Gamma_j}\bar{\phi}' n d\gamma - \rho_0^{-1}k\bar{\mathbf{s}}'_j\right]. \end{cases}$$

If we restrict $(\phi, \mathbf{s}) \in X$ further by requiring $\mathbf{s} = 0$ and

(4.21) $$\int_{\Gamma_j}\phi n d\gamma = 0 \quad \forall j \in \mathbb{Z}^2,$$

then the expression (4.20) becomes

$$\langle A(\phi, 0), (\phi', \mathbf{s}')\rangle = -\int_{\mathcal{O}} \Delta\bar{\phi}'\phi dx + \sum_{j\in\mathbb{Z}^2}\int_{\Gamma_j} \frac{\partial\bar{\phi}'}{\partial n}\phi d\gamma.$$

It is obvious that the right side is continuous with respect $L^2(\mathcal{O})$ norm on ϕ iff

$$\int_{\Gamma_j} \frac{\partial \bar{\phi}'}{\partial n} \phi d\gamma = 0 \quad \forall \phi \in H^2(\mathcal{O}) \text{ such that } \int_{\Gamma_j} \phi n d\gamma = \mathbf{0}.$$

We observe that the above condition is equivalent to

(4.22) $$\frac{\partial \phi'}{\partial n} = \mathbf{a}'_j \cdot \mathbf{n} \quad \text{on} \quad \Gamma_j$$

for some $\mathbf{a}'_j \in \mathbb{C}^2$. Using this information in (4.20), we get

$$\langle A(\phi, \mathbf{s}), (\phi', \mathbf{s}') \rangle = -\int_{\mathcal{O}} \Delta \bar{\phi}' \phi dx + \sum_{j \in \mathbf{Z}^2} \int_{\Gamma_j} \phi n d\gamma \cdot \left[\bar{\mathbf{a}}'_j + \rho_0^{-1} m \int_{\Gamma_j} \bar{\phi}' n - k m^{-1} \bar{\mathbf{s}}'_j \right].$$

We observe here that the right side is continuous with respect to $L^2(\mathcal{O})$ norm on ϕ iff the coefficient corresponding to $\int_{\Gamma_j} \phi n d\gamma$ vanishes and this gives

(4.23) $$k m^{-1} \mathbf{s}'_j = \rho_0 m^{-1} \int_{\Gamma_j} \phi' n d\gamma + \mathbf{a}'_j \quad \forall j \in \mathbf{Z}^2.$$

Conditions (4.19), (4.22) and (4.23) deduced so far on $(\phi', \mathbf{s}')$ show that $(\phi', \mathbf{s}') \in D(A)$ provided that

(4.24) $$\mathbf{a}' \in \ell^2(\mathbb{C}^2).$$

To this end, we reconsider (4.20), this time with $\mathbf{s} \neq \mathbf{0}$. Using the above properties of $(\phi', \mathbf{s}')$, we see that the right side of (4.20) becomes

$$\int_{\mathcal{O}} \Delta \bar{\phi}' \phi dx + \rho_0^{-1} k \sum_{j \in \mathbf{Z}^2} \left[\mathbf{s}_j - \int_{\Gamma_j} \phi' n d\gamma \right] \cdot \bar{\mathbf{a}}'_j.$$

We observe that the first term is continuous with respect to $L^2(\mathcal{O})$ norm on ϕ. If we make the choice of ϕ such that (4.21) is satisfied, we see that $\mathbf{a}'$ has the property (4.24) because $\mathbf{s} \in \ell^2(\mathbb{C}^2)$ is arbitrary.

(iv) Since X is dense in $D(A)$ with respect to the graph norm of A and the map $(\phi, \mathbf{s}) \longmapsto \langle A(\phi, \mathbf{s}), (\phi, \mathbf{s}) \rangle$ is continuous with respect to the same norm, it suffices to prove the non-negative definiteness property (4.5) for $(\phi, \mathbf{s}) \in X$. This is easy. Indeed for $(\phi, \mathbf{s}) \in X$, we have

$$\langle A(\phi, \mathbf{s}), (\phi, \mathbf{s}) \rangle = -\int_{\mathcal{O}} \Delta \phi \bar{\phi} dx + \rho_0^{-1} k \sum_{j \in \mathbf{Z}^2} \mathbf{a}_j \cdot \bar{\mathbf{s}}_j.$$

By doing integration by parts in the first term and using (4.3) we obtain

(4.25) $$\langle A(\phi, \mathbf{s}), (\phi, \mathbf{s}) \rangle = \int_{\mathcal{O}} |\nabla \phi|^2 dx + \rho_0^{-1} m \|\mathbf{a}\|^2,$$

which clearly implies the desired result. This completes the proof. ∎

4.2. Bloch Wave Decomposition of the Limit Operator

After seeing the main properties of the operator A in the preceding paragraph, we turn now our attention to describe its spectrum. Following the procedure set-up in Paragraph 3.2 for the Laplace model, we introduce a continuous family of subproblems indexed by $w = (w_1, w_2) \in S^1 \times S^1$: Find $\lambda = \lambda(w) \in \mathbb{C}$ for which there exist $\psi = \psi(w)$ not identically zero and $\mathbf{b}(w) = \{\mathbf{b}_j(w)\}_{j \in \mathbb{Z}^2}$ such that

(4.26)
$$\begin{cases} \Delta\psi + \lambda^2\psi = 0 \quad \text{in} \quad \mathcal{O}, \\ \dfrac{\partial\psi}{\partial n} = \mathbf{b}_j(w) \cdot \mathbf{n} \quad \text{on} \quad \Gamma_j, \ j \in \mathbb{Z}^2, \\ (k - m\lambda^2)\mathbf{b}_j(w) = \rho_0\lambda^2 \displaystyle\int_{\Gamma_j} \psi\mathbf{n}d\gamma \quad \forall j \in \mathbb{Z}^2, \\ \psi \quad (w, Y^*)\text{-periodic.} \end{cases}$$

To solve these problems, we follow the same strategy as in Paragraph 2.2 of Chapter II. The idea is to introduce the operator $R(w)$ defined on $L^2_\#(w, Y^*) \times \ell_\#(w, \mathbb{C}^2)$. When $w = (1,1)$, we replace $L^2_\#(w, Y^*)$, as usual, by $L^2_\#(Y^*)/\mathbb{C}$. In the matrix form, $R(w)$ is defined as follows:

(4.27)
$$R(w) = \begin{bmatrix} R_{11}(w) & R_{12}(w) \\ \rho_0 k^{-1} R_{21}(w) & R_{22}(w) \end{bmatrix},$$

where $R_{11}(w), R_{12}(w)$ and $R_{22}(w)$ are defined by the following rules:

(4.28)
$$R_{11}(w) \colon L^2_\#(w, Y^*) \longrightarrow L^2_\#(w, Y^*), \quad R_{11}(w)\vartheta = \vartheta_1,$$

(4.29)
$$R_{12}(w) \colon \ell_\#(w, \mathbb{C}^2) \longrightarrow L^2_\#(w, Y^*), \quad R_{12}(w)\mathbf{a} = \vartheta_2,$$

(4.30)
$$R_{21}(w) \colon L^2_\#(w, Y^*) \longrightarrow \ell_\#(w, \mathbb{C}^2), \quad R_{21}(w)\vartheta = \left\{\int_{\Gamma_j} \vartheta_1\mathbf{n}d\gamma\right\}_{j \in \mathbb{Z}^2},$$

(4.31)
$$\begin{cases} R_{22}(w) \colon \ell_\#(w, \mathbb{C}^2) \longrightarrow \ell_\#(w, \mathbb{C}^2), \\ R_{22}(w)\mathbf{a} = \rho_0 k^{-1}\left\{\displaystyle\int_{\Gamma_j} \vartheta_2\mathbf{n}d\gamma\right\}_{j \in \mathbb{Z}^2} + mk^{-1}\mathbf{a}. \end{cases}$$

Here ϑ_1 and ϑ_2 are the unique solutions of the following (w, Y^*)-periodic problems:

(4.32)
$$\begin{cases} \vartheta_1 \in H^1_\#(w, Y^*), \\ \displaystyle\int_{Y^*} \nabla\vartheta_1 \cdot \nabla\bar\chi dx = \int_{Y^*} \vartheta\bar\chi dx \quad \forall \chi \in H^1_\#(w, Y^*) \end{cases}$$

and

$$(4.33) \quad \begin{cases} \vartheta_2 \in H^1_\#(w, Y^*), \\ \displaystyle\int_{Y^*} \nabla\vartheta_2 \cdot \nabla\bar\chi \, dx = \mathbf{a}_0(w) \cdot \int_\Gamma \bar\chi \, \mathbf{nd}\gamma \quad \forall \chi \in H^1_\#(w, Y^*). \end{cases}$$

We have the usual understanding when $w = (1,1)$.

Following the method of Paragraph 2.2 of Chapter II, it can be shown that $R(w)$ is a compact, positive and self-adjoint operator in $L^2_\#(w, Y^*) \times \ell_\#(w, \mathbb{C}^2)$ which is equipped with the weighted inner product defined below:

$$(4.34) \quad \langle (\phi, \mathbf{a}), (\phi', \mathbf{a}') \rangle_\# = \int_{Y^*} \phi\bar\phi' \, dx + \rho_0^{-1} k \mathbf{a}_0 \cdot \bar{\mathbf{a}}_0'.$$

We then conclude that the spectrum of $R(w)$ consists of a sequence of positive numbers converging to zero. We denote their inverses by $\lambda_\ell^2(w)$ and arrange them as follows:

$$(4.35) \quad 0 < \lambda_1^2(w) \le \cdots \le \lambda_\ell^2(w) \le \cdots \longrightarrow \infty.$$

Let the corresponding eigenvectors be denoted as

$$(\psi_1(w), \mathbf{b}^1(w)), ..., (\psi_\ell(w), \mathbf{b}^\ell(w)), ...$$

We can now state the following result:

Theorem 4.4. *The triplets* $\{(\lambda_\ell^2(w), \psi_\ell(w), \mathbf{b}^\ell(w))\}_\ell$ *satisfy the following conditions*:
(i) *For each* $\ell \ge 1$, *they are solutions of* (4.26).
(ii) *The set* $\{(\psi_\ell(w), \mathbf{b}^\ell(w))\}_\ell$ *forms a basis of* $L^2_\#(w, Y^*) \times \ell_\#(w, \mathbb{C}^2)$ *which can be orthonormalized with respect to the inner product* (4.34).
(iii) *They exhaust all the solutions of* (4.26). ∎

Our next goal is to establish some minimal regularity properties of the solutions provided by Theorem 4.4 with respect to $w \in S^1 \times S^1$. We will use these later in obtaining the spectral decomposition of A. We begin by getting bounds on the eigenvalues $\lambda_\ell^2(w)$ independent of w.

◇ **Exercise 4.1.** (Bounds on the eigenvalues $\lambda_\ell^2(w)$) The object of this exercise is to establish interlacing inequalities for $\lambda_\ell^2(w)$ similar to the ones obtained in Paragraph 3.2 of Chapter II for the Helmholtz model. One again the idea consists in comparing the functions $\lambda_\ell^2(w)$ with the eigenvalues of the Laplace model and Neumann problem always conserving the (w, Y^*)-periodicity condition on ∂Y. Bounds on $\lambda_\ell^2(w)$ will then be deduced.

(i) Show that the eigenvalues $\lambda_\ell^2(w)$ satisfy

(4.36)
$$\begin{cases} \nu_{\ell-2}^2 \leq \lambda_\ell^2 \leq \nu_{\ell+2}^2 & \forall \ell \geq 3, \\ \quad\quad \lambda_\ell^2 \leq \nu_{\ell+2}^2 & \forall \ell = 1, 2, \end{cases}$$

where $\{\nu_\ell^2(w)\}$ are the characteristic values of the operator $R_0(w)$ acting on the product space $L_\#^2(w, Y^*) \times \ell_\#(w, \mathbb{C}^2)$ whose matrix form is given by

$$R_0(w) = \begin{bmatrix} R_{11}(w) & 0 \\ 0 & R_{22}(w) \end{bmatrix}.$$

Of course, it is understood that the ν_ℓ^2 are arranged in the increasing order.

(ii) Show that the ℓth characteristic value of R_{11} lies in the closed interval $[(\nu_\ell^N)^2, (\nu_\ell^D)^2]$ where $\{(\nu_\ell^N)^2\}$ and $\{(\nu_\ell^D)^2\}$ are the eigenvalues of $(-\Delta)$ in Y^* with Neumann and Dirichlet boundary condition on ∂Y, respectively; in both cases we have Neumann condition on Γ.

(iii) Deduce that there exist constants $\alpha^{(\ell)}$ independent of $w \in S^1 \times S^1$ such that

(4.37)
$$0 \leq \lambda_\ell^2(w) \leq \alpha^{(\ell)} \quad \forall \ell \geq 1, \ \forall w \in S^1 \times S^1. \ ∎$$

We are now in a position to adapt the proof of Theorem 3.9 using Lemma 3.8 to deduce the following result:

Theorem 4.5. *For each $\ell \geq 1$, the eigenvalue $\lambda_\ell^2(w)$ as a function of w is continuous on $S^1 \times S^1 \backslash \{(1,1)\}$.* ∎

Now, we come to a difficult point of the theory, namely, making a selection of eigenvectors $\{(\psi_\ell(w), \mathbf{b}^\ell(w))\}_\ell$ such that they are measurable functions of $w \in S^1 \times S^1$. Such questions have been treated by WILCOX [1973], in a very general framework. His method calls for applications of the theory of *Fredholm minors* which are beyond the scope of this book. Without using this theory, we were able to give an explicit measurable selection in the case of the Laplace model. Since now we have no apriori control on the multiplicities, our earlier method does not seem to extend to cover the Helmholtz model. Hence we shall assume the existence of a measurable selection of eigenvectors and work with it in the sequel.

Definition. Any measurable selection of eigenvectors $\{(\psi_\ell(w), \mathbf{b}^\ell(w))\}_\ell$, $w \in S^1 \times S^1$ enjoying the properties stated in Theorem 4.4 is called *Bloch waves* associated with the Helmholtz model. ∎

To obtain the spectral decomposition of A in terms of the Bloch waves introduced above, we will be needing the so-called Bloch wave decomposition of the space $L^2(\mathcal{O}) \times \ell^2(\mathbb{C}^2)$. We now announce a result which achieves this and which can be proved along the lines of Theorem 3.11.

Theorem 4.6. (Bloch Wave Decomposition of $L^2(\mathcal{O}) \times \ell^2(\mathbb{C}^2)$) *Let* $(\phi, \mathbf{a})$ *in* $L^2(\mathcal{O}) \times \ell^2(\mathbb{C}^2)$ *be arbitrary. For almost every* $w \in S^1 \times S^1$ *we define a pair* $(\phi_\#(w), \mathbf{a}_\#(w)) \in L^2_\#(w, Y^*) \times \ell_\#(w, \mathbb{C}^2)$ *as follows:*

$$\phi_\#(w, x) = \frac{1}{(2\pi)^2} \sum_{j \in \mathbb{Z}^2} \phi(x + 2\pi j) w_1^{-j_1} w_2^{-j_2} \quad \text{for a.e. } x \in \mathcal{O},$$

$$(\mathbf{a}_\#)_{j'}(w) = \frac{1}{(2\pi)^2} \sum_{j \in \mathbb{Z}^2} \mathbf{a}_{j'+j} w_1^{-j_1} w_2^{-j_2} \quad \forall j' \in \mathbb{Z}^2.$$

Then we have the following decomposition:

$$(\phi, \mathbf{a}) = \int_{S^1} \int_{S^1} (\phi_\#(w), \mathbf{a}_\#(w)) dw.$$

Furthermore, the following Parseval's Identity holds:

$$(4.38) \qquad \|\phi\|_{0,\mathcal{O}}^2 + \rho_0^{-1} k \|\mathbf{a}\|^2 = (2\pi)^2 \int_{S^1} \int_{S^1} \sum_{\ell=1}^\infty |\alpha_\ell(w)|^2 dw,$$

where the complex constants $\alpha_\ell(w)$ *are given by*

$$(4.39) \qquad \alpha_\ell(w) = \langle (\phi_\#(w), \mathbf{a}_\#(w)), (\psi_\ell(w), \mathbf{b}^\ell(w)) \rangle_\# \quad \forall \ell \geq 1.$$

Finally, we have the decomposition of the identity:

$$(\phi, \mathbf{a}) = \int_{S^1} \int_{S^1} \sum_{\ell=1}^\infty \alpha_\ell(w) (\psi_\ell(w), \mathbf{b}^\ell(w)) dw. \quad \blacksquare$$

Definition. The complex constant $\alpha_\ell(w)$ defined by (4.39) is called *Bloch coefficient* of $(\phi, \mathbf{a}) \in L^2(\mathcal{O}) \times \ell^2(\mathbb{C}^2)$ with respect to the Bloch wave $(\psi_\ell(w), \mathbf{b}^\ell(w))$. $\blacksquare$

$\lozenge$ **Exercise 4.2.** Show that the Parseval's Identity (4.38) can be generalized to the following *Plancherel's Identity*:

$$\int_{\mathcal{O}} \phi \bar{\phi}' dx + \rho_0^{-1} k(\mathbf{a}, \mathbf{a}') = (2\pi)^2 \int_{S^1} \int_{S^1} \left(\sum_{\ell=1}^\infty \alpha_\ell(w) \bar{\alpha}'_\ell(w) \right) dw$$

for all $(\phi, \mathbf{a}), (\phi', \mathbf{a}')$ in $L^2(\mathcal{O}) \times \ell^2(\mathbb{C}^2)$ where $\alpha_\ell(w)$ and $\alpha'_\ell(w)$ are the Bloch coefficients of $(\phi, \mathbf{a})$ and $(\phi', \mathbf{a}')$ respectively. $\blacksquare$

The spectral decomposition of A follows now easily from Theorem 4.6 and it is obtained exactly as in Laplace model (see Theorem 3.13).

Theorem 4.7. *Let A be the unbounded operator defined by* (4.3) *and* (4.4). *Then the spectral resolution of A can be written as follows: For all* $(\phi, \mathbf{a})$ *in* $D(A)$, *we have*

$$A(\phi, \mathbf{a}) = \int_{S^1} \int_{S^1} \sum_{\ell=1}^{\infty} \alpha_\ell(w) \lambda_\ell^2(w) \big(\psi_\ell(w), \mathbf{b}^\ell(w)\big) \, dw$$

where $(\lambda_\ell^2(w), \psi_\ell(w), \mathbf{b}^\ell(w))$ *are as in Theorem 4.4.* ∎

An immediate consequence of the above result is that the spectral family of the operator A is given by

$$(4.40) \quad E_\tau(A)(\phi, \mathbf{a}) = \begin{cases} (0,0) & \text{if } \tau \leq 0, \\ \int_{S^1} \int_{S^1} \sum_{\ell;\lambda_\ell^2(w)\leq\tau} \alpha_\ell(w)\big(\psi_\ell(w), \mathbf{b}^\ell(w)\big) \, dw & \text{if } \tau > 0. \end{cases}$$

We add that one can also prove other corollaries as well. In particular, the results analogous to Theorem 3.15 showing the *band structure* of $\sigma(A)$ and to Theorem 3.16 which gives a criterion for the existence of eigenvalues for A.

With this, we come to the end of the second step of our non-standard homogenization procedure. In the next paragraph, we take up the last step.

4.3. Convergence Analysis

In this paragraph, we show that the spectral family associated with problem (4.1) converges to that of A. We cannot do this as in the Laplace model (see Paragraph 3.4) because we are now dealing with unbounded operators which are not uniformly positive definite. To overcome this difficulty, we require certain modifications in the analysis of convergence and we propose to go through the details.

4.3.1. Main Convergence Result

To formulate the theorem on convergence, we need several notations. First, let us remark that one can introduce a sequence $\{A_n\}$ of unbounded self-adjoint operators which are non-negative definite and with domains $D(A_n) \subset L_0^2(\Omega_n) \times \mathbb{C}^{2K_n}$ in such a way that the spectral problem of A_n coincides with (4.1). Of course, the space $L_0^2(\Omega_n) \times \mathbb{C}^{2K_n}$ is equipped with the following inner product:

$$\langle(\phi, \mathbf{s}), (\psi, \mathbf{t})\rangle \overset{\text{def}}{=} \int_{\Omega_n} \phi\bar{\psi}\,dx + \rho_0^{-1}k \sum_{j\in Q_n} \mathbf{s}_j \cdot \bar{\mathbf{t}}_j.$$

Its definition is analogous to that of A replacing $L^2(\mathcal{O}) \times \ell^2(\mathbb{C}^2)$ by $L_0^2(\Omega_n) \times \mathbb{C}^{2K_n}$. In the sequel, we will be using the fact that A_n's are uniformly bounded below (see Proposition 2.7 of Chapter II):

$$(4.41) \qquad \langle A_n(\phi, \mathbf{s}), (\phi, \mathbf{s}) \rangle \geq 0 \quad \forall (\phi, \mathbf{s}) \in D(A_n).$$

Let $E_\tau(A_n)$ denote the spectral family associated with A_n. Then, for $\tau \in \mathbb{R}$, $E_\tau(A_n)$ is a bounded operator on $L_0^2(\Omega_n) \times \mathbb{C}^{2K_n}$ given by

$$(4.42) \qquad E_\tau(A_n)(\phi, \mathbf{s}) = \sum_{\ell; \omega_\ell^2(n) \leq \tau} \langle (\phi, \mathbf{s}), (\phi_\ell^{(n)}, \mathbf{s}_\ell^{(n)}) \rangle (\phi_\ell^{(n)}, \mathbf{s}_\ell^{(n)})$$

where $\{\omega_\ell^2(n)\}_\ell$ are the eigenvalues of problem (4.1) with the corresponding orthonormalized eigenvectors $\{(\phi_\ell^{(n)}, \mathbf{s}_\ell^{(n)})\}_\ell$. Next, given $\mathbf{a} \in \ell^2(\mathbb{C}^2)$, we denote by $\mathbf{a}_n \in \mathbb{C}^{2K_n}$ the restriction of $\mathbf{a}$ to Q_n. In the same way, given $f \in L^2(\mathcal{O})$, we associate an element $f_n \in L_0^2(\Omega_n)$ defined as follows:

$$f_n(x) = f(x) - \frac{1}{\text{meas}(\Omega_n)} \int_{\Omega_n} f \, dx \quad \text{for a.e. } x \in \Omega_n.$$

The operators given by extension by zero from $\mathbb{C}^{2K_n}$ to $\ell^2(\mathbb{C}^2)$ and from $L^2(\Omega_n)$ to $L^2(\mathcal{O})$ are denoted by tildas: $\sim$. Observe that $\tilde{\mathbf{a}}_n \longrightarrow \mathbf{a}$ in $\ell^2(\mathbb{C}^2)$ for all $\mathbf{a} \in \ell^2(\mathbb{C}^2)$ and $\tilde{f}_n \longrightarrow f$ in $L^2(\mathcal{O})$ weak, for all $f \in L^2(\mathcal{O})$.

Since A_n and A are only negative definite, we introduce uniformly positive definite operators B_n and B as follows:

$$B_n = A_n + I, \quad B = A + I.$$

We add that this modification was not necessary in the Laplace model.

We now state the main result of this paragraph.

Theorem 4.8. *The spectral family of B_n converges to that of B in the following sense: For all $(\phi, \mathbf{s}), (\psi, \mathbf{t}) \in L^2(\mathcal{O}) \times \ell^2(\mathbb{C}^2)$,*

$$\langle E_\tau(B_n^{1/2})(\phi_n, \mathbf{s}_n), (\psi_n, \mathbf{t}_n) \rangle \longrightarrow \langle E_\tau(B^{1/2})(\phi, \mathbf{s}), (\psi, \mathbf{t}) \rangle$$

in the space of tempered distributions on $\mathbb{R}$. ∎

Some remarks are in order about this result. If we make a change of variables $\tau \longmapsto \tau^{1/2}$, we get the convergence of $E_\tau(B_n)$ because

$$E_\tau(B_n) = E_{\tau^{1/2}}(B_n^{1/2}), \quad E_\tau(B) = E_{\tau^{1/2}}(B^{1/2}) \quad \forall \tau > 0.$$

Next, we note that the spectral measures of A_n, B_n and A, B are same up to translation:

$$E_\tau(B_n) = E_{\tau-1}(A_n), \quad E_\tau(B) = E_{\tau-1}(A).$$

Thus Theorem 4.8 implies the convergence of the spectral families $E_\tau(A_n)$ towards that of A.

The proof of Theorem 4.8 involves two major steps. In the first one, one shows that B_n converges to B in the following sense: The solution sequence of the corresponding wave equation converges. In the second step, we translate this result in terms of the convergence of the spectral family of $B_n^{1/2}$ to that of $B^{1/2}$ by means of Fourier transform. As is transparent, this method is the same as the one followed in proving Theorem 4.9 of Chapter I.

4.3.2. The Wave Equation associated with B_n

For ease of writing, we denote by $u, v, \dots$ the elements in $L_0^2(\Omega_n) \times \mathbb{C}^{2K_n}$ (or in $L^2(\mathcal{O}) \times \ell_2(\mathbb{C}^2)$):

$$u = (\phi, \mathbf{s}) \in L_0^2(\Omega_n) \times \mathbb{C}^{2K_n}.$$

We consider the following Cauchy problem for the wave equation associated with B_n: Find $u_n : \mathbb{R} \to L_0^2(\Omega_n) \times \mathbb{C}^{2K_n}$ such that $u_n \in L^\infty(\mathbb{R}; H^1(\Omega_n) \times \mathbb{C}^{2K_n})$ and $u'_n \in L^\infty(\mathbb{R}; L_0^2(\Omega_n) \times \mathbb{C}^{2K_n})$ satisfying

(4.43)
$$\begin{cases} u''_n(t) + B_n u_n(t) = 0, \\ u_n(0) = 0, \\ u'_n(0) = u_n^1, \end{cases}$$

where u_n^1 is an element of $L_0^2(\Omega_n) \times \mathbb{C}^{2K_n}$ to be chosen later.

To see the existence of a weak solution to (4.43) we introduce the bilinear form associated with the operators A_n and B_n:

$$\begin{cases} a_n\big((\phi,\mathbf{s}),(\psi,\mathbf{t})\big) = \\ = \displaystyle\int_{\Omega_n} \nabla\phi \cdot \nabla\bar\psi\, dx + m^{-1}\rho \sum_{j \in Q_n} \left(\rho_0^{-1} k s_j - \int_{\Gamma_j} \phi n\, d\gamma\right) \cdot \left(\rho_0^{-1} k \bar t_j - \int_{\Gamma_j} \bar\psi n\, d\gamma\right), \\ b_n\big((\phi,\mathbf{s}),(\psi,\mathbf{t})\big) = a_n\big((\phi,\mathbf{s}),(\psi,\mathbf{t})\big) + \big\langle(\phi,\mathbf{s}),(\psi,\mathbf{t})\big\rangle. \end{cases}$$

These bilinear forms are defined for $\phi, \psi \in H^1(\Omega_n)$ and $\mathbf{s}, \mathbf{t} \in \mathbb{C}^{2K_n}$ and are continuous. These forms are cooked up so that we have

$$a_n\big((\phi,\mathbf{s}),(\psi,\mathbf{t})\big) = \big\langle A_n(\phi,\mathbf{s}),(\psi,\mathbf{t})\big\rangle,$$
$$b_n\big((\phi,\mathbf{s}),(\psi,\mathbf{t})\big) = \big\langle B_n(\phi,\mathbf{s}),(\psi,\mathbf{t})\big\rangle,$$

for all $(\phi,\mathbf{s}) \in D(A_n), (\psi,\mathbf{t}) \in H^1(\Omega_n) \times \mathbb{C}^{2K_n}$. The main properties of these forms are summarized in the following result:

Theorem 4.9. *The following statements hold:*
(i) $a_n(\cdot, \cdot), b_n(\cdot, \cdot)$ *are sesquilinear and Hermitian.*
(ii) *For some $\alpha > 0$ independent of n, we have*

$$a_n((\phi, \mathbf{s}), (\phi, \mathbf{s})) \geq \alpha \left[\int_{\Omega_n} |\nabla \phi|^2 dx + \|\mathbf{s}\|^2 \right],$$

$$b_n((\phi, \mathbf{s}), (\phi, \mathbf{s})) \geq \alpha \left[\|\phi\|_{1,\Omega_n}^2 + \|\mathbf{s}\|^2 \right].$$

Proof. (i) is evident to prove. (ii) We see first that

$$a_n((\phi, \mathbf{s}), (\phi, \mathbf{s})) = \int_{\Omega_n} |\nabla \phi|^2 dx + m^{-1} \rho_0 \sum_{j \in Q_n} \|\rho_0^{-1} k\mathbf{s} - \int_{\Gamma_j} \phi \mathbf{n} d\gamma\|^2.$$

On the other hand, we have always (see (3.35))

$$\left\| \left\{ \int_{\Gamma_j} \phi \mathbf{n} d\gamma \right\}_{j \in Q_n} \right\|^2 \leq c \int_{\Omega_n} |\nabla \phi|^2 dx,$$

with c independent of n. We now write $\rho_0^{-1} k\mathbf{s} = \rho_0^{-1} k\mathbf{s} - \int_{\Gamma_j} \phi \mathbf{n} d\gamma + \int_{\Gamma_j} \phi \mathbf{n} d\gamma$
and deduce that

$$\rho_0^{-2} k^2 \|\mathbf{s}\|^2 \leq 2 \left[\left\| \rho_0^{-1} k\mathbf{s} - \left\{ \int_{\Gamma_j} \phi \mathbf{n} d\gamma \right\}_{j \in \mathbb{Q}^N} \right\|^2 + \left\| \left\{ \int_{\Gamma_j} \phi \mathbf{n} d\gamma \right\}_{j \in \mathbb{Q}^N} \right\|^2 \right].$$

Combining the above inequalities, we get

$$\rho_0^{-2} k^2 \|\mathbf{s}\|^2 \leq 2 a_n((\phi, \mathbf{s}), (\phi, \mathbf{s})) + 2c \int_{\Omega_n} |\nabla \phi|^2 dx$$

$$\leq 2(1 + c) a_n((\phi, \mathbf{s}), (\phi, \mathbf{s})).$$

This completes the proof of Theorem 4.9. ∎

Thanks to the above properties of the bilinear form $b_n(\cdot, \cdot)$, we can easily deduce the following result (see SÁNCHEZ-PALENCIA [1980] pp. 37–38):

Corollary 4.10. *There is a unique weak solution u_n to the Cauchy problem* (4.43) *such that*

$$\begin{cases} u_n \in L^\infty(\mathbb{R}; V(\Omega_n) \times \mathbb{C}^{2K_n}), \\ u_n' \in L^\infty(\mathbb{R}; L_0^2(\Omega_n) \times \mathbb{C}^{2K_n}), \end{cases}$$

where $V(\Omega_n) \overset{\text{def}}{=} H^1(\Omega_n) \cap L_0^2(\Omega_n)$.

4.3.3. Passage to the Limit

In this section, we show that the sequence $\{u_n\}$ defined by (4.43) converges to the solution u of the wave equation associated with B under suitable hypothesis on u_n^1. As the first step towards this, we obtain bounds on u_n. Multiplying (4.43) by $u_n'(t)$, integrating by parts and using Theorem 4.9, we get a constant $M > 0$ independent of n such that

$$\begin{cases} \|u_n\|_{L^\infty(\mathbb{R};V(\Omega_n)\times\mathbb{C}^{2K_n})} \leq M, \\ \|u_n'\|_{L^\infty(\mathbb{R};L_0^2(\Omega_n)\times\mathbb{C}^{2K_n})} \leq M, \end{cases}$$

provided that u_n^1 is bounded in $L_0^2(\Omega_n) \times \mathbb{C}^{2K_n}$ independent of n.

In order to get a converging subsequence, we use the extension operator P_n constructed in Lemma 3.18. Recalling that u_n has two components ϕ_n, $\mathbf{a}_n$ in $V(\Omega_n)$ and $\mathbb{C}^{2K_n}$ respectively, we extend the first component using P_n to get an element of $BL(\mathcal{O})$ and the second component by zero outside Q_n to get an element of $\ell^2(\mathbb{C}^2)$. For ease of notation, we denote again $P_n u_n$, the extension of u_n thus obtained. Both components of u_n' are extended by zero and this yields an element $\tilde{u}_n'$ of $L^2(\mathcal{O}) \times \ell^2(\mathbb{C}^2)$. Using Lemma 3.18 and the bounds obtained just above, we can extract subsequences such that

(4.44)
$$\begin{cases} P_n u_n \rightharpoonup u \quad \text{in} \quad L^\infty(\mathbb{R}; H^1(\mathcal{O}) \times \ell^2(\mathbb{C}^2)) \text{ weak*}, \\ \tilde{u}_n' \rightharpoonup u' \quad \text{in} \quad L^\infty(\mathbb{R}; L^2(\mathcal{O}) \times \ell^2(\mathbb{C}^2)) \text{ weak*}. \end{cases}$$

Intuitively, we feel that u is the solution of the wave equation associated with B. This is the aim of our next result. In order to formulate it, we introduce the following problem: Find $u: \mathbb{R} \longrightarrow L^2(\mathcal{O}) \times \ell^2(\mathbb{C}^2)$ such that $u \in L^\infty(\mathbb{R}; H^1(\mathcal{O}) \times \ell^2(\mathbb{C}^2))$ and $u' \in L^\infty(\mathbb{R}; L^2(\mathcal{O}) \times \ell^2(\mathbb{C}^2))$ satisfying

(4.45)
$$\begin{cases} u'' + Bu(t) = 0, \\ u(0) = 0, \quad u'(0) = u^1. \end{cases}$$

Then we have

Theorem 4.11.
(i) *Given $u^1 \in L^\infty(\mathcal{O}) \times \ell^2(\mathbb{C}^2)$, there is a unique solution u to problem (4.45).*
(ii) *Given $u^1 \in L^2(\mathcal{O}) \times \ell^2(\mathbb{C}^2)$, define $u_n^1 = (\phi_n, \mathbf{s}_n) \in L_0^2(\Omega_n) \times \mathbb{C}^{2K_n}$ by the restriction process defined in Subparagraph 4.3.1. Then the solution u_n of problem (4.43) converges to the unique solution u of (4.45) in the sense of (4.44).*

Proof. (i) The proof is analogous to Corollary 4.10. Indeed, the bilinear form corresponding to B is given by

$$b((\phi, \mathbf{s}), (\psi, \mathbf{t})) = \langle (\phi, \mathbf{s}), (\psi, \mathbf{t}) \rangle +$$

$$+ \int_{\mathcal{O}} \nabla \phi \cdot \nabla \bar{\psi} dx + \rho_0 m^{-1} \sum_{j \in \mathbb{Z}^2} \left(\rho_0^{-1} k s_j - \int_{\Gamma_j} \phi \mathbf{n} d\gamma \right) \cdot \left(\rho_0^{-1} k \bar{t}_j - \int_{\Gamma_j} \bar{\psi} \mathbf{n} d\gamma \right).$$

It obviously satisfies the coercive inequality

$$b((\phi, \mathbf{s}), (\phi, \mathbf{s})) \geq \beta \left[\|\phi\|_{1,\mathcal{O}}^2 + \|\mathbf{s}\|^2 \right]$$

for all $(\phi, \mathbf{s}) \in H^1(\mathcal{O}) \times \ell^2(\mathbb{C}^2)$ with a constant $\beta > 0$. Moreover, we observe that $b(\cdot, \cdot)$ is Hermitian. We can then appeal to the result of SÁNCHEZ-PALENCIA [1980] pp. 37–38 to conclude that there is a unique solution to problem (4.45).

(ii) In order to pass to the limit in (4.43), we invoke its weak formulation which is as follows: Find $u_n \in L^\infty(\mathbb{R}; V(\Omega_n) \times \mathbb{C}^{2K_n})$ such that $u_n' \in L^\infty(\mathbb{R}; L_0^2(\Omega_n) \times \mathbb{C}^{2K_n})$ and

$$(4.46) \quad \begin{cases} - \int_0^\infty \langle u_n'(t), v'(t) \rangle dt + \int_0^\infty b_n(u_n(t), v(t)) dt - \langle u_n^1, v(0) \rangle = 0, \\ u_n(0) = 0, \quad u_n(t) = -u_n(-t) \quad \forall t \in \mathbb{R}, \end{cases}$$

where the test function $v(t)$ is of the form $v(t) = \eta(t)v$ where $\eta(t)$ varies in $\mathcal{C}_0^1[0, +\infty)$ and $v = (\psi, \mathbf{t})$ varies over $\mathcal{D}(\bar{\Omega}_n) \times \mathbb{C}^{2K_n}$.

We fix one function $v(t) \in \mathcal{D}(\bar{\mathcal{O}}) \times \ell_0^2(\mathbb{C}^2)$ where $\ell_0^2(\mathbb{C}^2)$ is the subspace of $\ell^2(\mathbb{C}^2)$ consisting of elements with compact support. For sufficiently large n, we can take this $v(t)$ as a test function in (4.46) and pass to the limit. There is no difficulty with the first and the third terms in (4.46). To handle the middle term, we observe that

$$b_n(u_n(t), v(t)) = b(P_n u_n(t), v(t))$$

for large n. Thus the weak convergence (4.44) is sufficient to pass to the limit in (4.46). At the limit, we find the weak formulation of problem (4.45). Since the solution of (4.45) is unique, the whole sequence $\{u_n\}$ converges to u in the sense of (4.44). This completes the proof of Theorem 4.11. ∎

4.3.4. Proof of the Main Result

Following the arguments in Theorem 4.9 of Chapter I, we prove in this section that the convergence (4.44) implies that of the spectral families. We use Fourier transformation in this process. First, we remark that the solutions of problems (4.43), (4.45) have the following representations via operator calculus:

$$u'_n(t) = \cos(B_n^{1/2}t)u_n^1, \quad u'(t) = \cos(B^{1/2}\,t)u^1.$$

The convergence (4.44) now implies that

$$\langle \cos(B_n^{1/2}t)u_n^1, v_n^1 \rangle \rightharpoonup \langle \cos(B^{1/2}t)u^1, v^1 \rangle \quad \text{in} \quad L^\infty(\mathbb{R}) \text{ weak*},$$

where u_n^1, v_n^1 represent as usual the restriction of u^1, v^1 to Ω_n as prescribed in Sub-paragraph 4.3.1. Let us remark that the above convergence implies the convergence in $\mathcal{S}'$, the space of tempered distributions. Therefore by Fourier transform, we get

$$(4.47) \qquad \langle \cos(B_n^{1/2}t)u_n^1, v_n^1 \rangle^\wedge(\tau) \longrightarrow \langle \cos(B^{1/2}t)u^1, v^1 \rangle^\wedge(\tau) \quad \text{in} \quad \mathcal{S}'.$$

Here, τ denotes the variable dual to t in the Fourier transform.

On the other hand, we have the following representation of $\cos(B_n^{1/2}t)$ in terms of spectral families:

$$\cos(B_n^{1/2}t) = \frac{1}{2}\int_0^\infty (e^{i\tau t} + e^{-i\tau t})dE_\tau(B_n^{1/2})$$

or equivalently

$$\cos(B_n^{1/2}t) = \frac{1}{2}\int_{-\infty}^\infty e^{i\tau t}d\big[E_\tau(B_n^{1/2}) - E_{-\tau}(B_n^{1/2})\big].$$

We note that the right hand side of this identity is nothing but the inverse Fourier transform of

$$\frac{d}{d\tau}\big[E_\tau(B_n^{1/2}) - E_{-\tau}(B_n^{1/2})\big].$$

Certainly, an analogous result holds for the operator B. Thus the convergence (4.47) is equivalent to

$$\frac{d}{d\tau}\langle [E_\tau(B_n^{1/2}) - E_{-\tau}(B_n^{1/2})]u_n^1, v_n^1 \rangle \longrightarrow \frac{d}{d\tau}\langle [E_\tau(B^{1/2}) - E_{-\tau}(B^{1/2})]u^1, v^1 \rangle$$

in $\mathcal{S}'$. Since the distributions $E(B_n^{1/2}, \cdot)$ have supports in $[\alpha, +\infty)$ with $\alpha > 0$, we conclude that

$$\langle E_\tau(B_n^{1/2})u_n^1, v_n^1 \rangle \longrightarrow \langle E_\tau(B^{1/2})u^1, v^1 \rangle \quad \text{in} \quad \mathcal{S}'.$$

This is nothing but the convergence mentioned in Theorem 4.8 and hence the proof is finished. ∎

§5. Two-Scale Convergence Method

The study of the asymptotic behaviour of vibrations of fluid-solid structures has been our aim in earlier sections. This concerns the case when the number of tubes immersed in the fluid is large (it can be of several hundreds or even thousands in practice). As mentioned in the Preface, the reasons for having a large number of tubes are quite clear. In the context of nuclear reactors, this is done to increase the heat exchanges. Direct numerical simulations in such complicated geometries pose enormous difficulties starting from the discretization of the domain. Hence an asymptotic analysis is called for as the number of tubes goes to infinity. This led us naturally to a problem of homogenization which, as mentioned earlier, can be tackled in two different ways. In §2, §3 and §4, we followed what we called a *non-standard homogenization* procedure which consisted of fixing the size of the tubes and the distance between them and letting the size of the domain go to infinity. Our purpose in this section and in the following one is to return to the classical framework of *standard homogenization* procedure found, for instance, in BENSOUSSAN, LIONS and PAPANICOLAOU [1987], MURAT [1977-78], SÁNCHEZ-PALENCIA [1980] and TARTAR [1977]. Thus in this section, we fix the size of the domain and assume the size of the tubes and the distance between them to be of same order; we represent them by a single parameter $\varepsilon > 0$ destined to tend to zero. While in the next section, we shall apply the classical *multiple scale expansions* to obtain the *homogenized problem* describing the asymptotic behaviour, our aim here is to justify these formal expansions by proving convergence results.

To this end, we plan to use a recent technique called *two-scale convergence analysis* which is an alternative to the so-called *energy method* of L. Tartar (see TARTAR [1977], [1980]) found in the works cited above. This new technique is based on original ideas of G. Allaire (see ALLAIRE [1991], [1992]) and of G. Nguetseng (see NGUETSENG [1989]). Classically, one first finds the homogenized problem by using a two-scale asymptotic expansion and then one applies the energy method to justify the result obtained in the first step. In our study of models of fluid-solid structures, we prefer to invert this process mainly for two reasons. First, the usual two-scale expansion has to be combined with a suitable *Taylor expansion* for the velocity vector of the tubes. This results in calculations which are not so obvious and not found in classical cases. On the other hand, the two-scale convergence analysis is not influenced by these changes and directly gives the homogenized problem along with the justification. Thus as a net-result of this method, we see that our original problem is split into two: one in the basic periodic cell containing just one tube and a homogenized problem without tubes at all.

To carry out the goals listed above, we plan to follow the program below: In Paragraph 5.1, we present following ALLAIRE [1992] the basic results of the two-scale convergence analysis. It is then applied in Paragraph 5.2 to a classical problem of homogenization involving periodically oscillating coefficients. Next, we employ this technique in Paragraph 5.3 to find the homogenized operator corresponding to the Laplace model and deduce the convergence of the associated spectral families. Finally, Paragraph 5.4 is devoted to the study of Helmholtz model and follows the same pattern as the preceding paragraph.

For a better understanding of this section it is *highly desirable* that the reader be acquainted with the classical works in the theory of homogenization cited above.

We end this introduction by pointing out an essential difference in the conclusions drawn out of the present approach over those drawn out of the non-standard homogenization procedure presented in the previous sections. This concerns the nature of the limiting spectrum. Recall that we proved the absence of isolated eigenvalues of finite multiplicity by means of the Bloch wave method (see Theorem 3.15). Moreover, we provided numerical evidence to show that the limiting spectrum is continuous. This time, however, we will be able to show, in some special cases, that the limiting vibrations have their energies concentrated at a finite number of frequencies which are all of infinite multiplicities. Though these conclusions differ from the ones of the previous paragraphs, they are not contradictory for the simple reason that they are obtained at different limits. Moreover, recent results of G. Allaire and C. Conca (see ALLAIRE and CONCA [1995]) show that all these conclusions can be reached using a single "unified" homogenization theory which includes both the above non-standard procedure and the classical one. Since the results of both approaches are possible in applications, it is hoped that our study will have practical utility.

5.1. Basic Results

We try to motivate the results of the two-scale convergence method through the classical problem of homogenization. We consider, therefore, a family of differential operators A_ε with oscillating coefficients with period ε and a family of solutions u_ε which satisfy the following equation in a bounded domain $\Omega \subset \mathbb{R}^N$:

$$(5.1) \qquad A_\varepsilon u_\varepsilon = f \quad \text{in} \quad \Omega.$$

Here f is the forcing term and this equation is usually supplemented by appropriate boundary conditions. The problem in which one is interested consists of finding the limit u of u_ε (in a suitable topology) as $\varepsilon \to 0$ and to find, if possible, an asymptotic expansion for u_ε. Usually, one is led to look for the so-called *homogenized operator* A_0 such that u is a solution of

$$(5.2) \qquad A_0 u = f \quad \text{in} \quad \Omega$$

supplemented, of course, with appropriate boundary conditions. Passage from (5.1) to (5.2) is the standard homogenization procedure. One of the well-known methods to find the operator A_0 is the *two-scale asymptotic expansion* which consists in proposing the following ansatz for u_ε:

$$(5.3) \qquad u_\varepsilon(x) = u_0(x, \frac{x}{\varepsilon}) + \varepsilon u_1(x, \frac{x}{\varepsilon}) + \varepsilon^2 u_2(x, \frac{x}{\varepsilon}) + \cdots,$$

where each term $u_j(x, y)$ is periodic in y. One then proceeds with formal manipulations of (5.3) to get the *homogenized operator A_0* as illustrated in §6. The so-called *energy method* of L. Tartar was developed to justify that the expansion (5.3) is correct at least to the zero order term. Loosely speaking, his method consists of multiplying equation (5.1) by special test functions and then pass to the limit. The construction of these test functions calls for the resolution of suitable local problems in the basic reference cell.

This method is very general and it works in a variety of contexts. It is therefore not surprising that it does not take full advantage of the periodic structure of the problem and, in particular, it uses very little information coming from (5.3); it was not conceived by L. Tartar for periodic problems but rather in the more general context of H-convergence. Thus, there is some room for improvements in the periodic case. The works of G. Allaire and G. Nguetseng cited above served this purpose introducing a new technique called *two- scale convergence method*.

The starting point is to consider a sequence $\{u_\varepsilon\}$ that is bounded in $L^2(\Omega)$ where Ω is an open bounded subset of $\mathbf{R}^N (N \geq 1)$. By the classical result of Banach-Alaoglu, we can extract a subsequence (which we still denoted by ε for simplicity) for which there exists $u \in L^2(\Omega)$ such that $u_\varepsilon \rightharpoonup u$ in $L^2(\Omega)$ weak, i.e., we have

$$(5.4) \qquad \lim_{\varepsilon \to 0} \int_\Omega u_\varepsilon(x)\phi(x)dx = \int_\Omega u(x)\phi(x)dx$$

for all test functions $\phi \in \mathcal{D}(\Omega)$. The basic question is to know when the above convergence becomes strong. It is well-known that one of the obstacles is the presence of oscillations in the sequence. In order to see them, we need an oscillatory microscope in the form of test functions. The idea of the two-scale convergence is therefore to generalize (5.4) for a "larger" class of test functions which involve two scales. Looking at (5.3), it is quite clear that the best candidates for such test functions are functions of two variables, periodic with respect to one of them. More precisely, let $Y =]0, 2\pi[^N$ be the cube representing the reference cell and denote by $\mathcal{C}_\#^\infty(Y)$ the space of infinitely differentiable functions in $\mathbf{R}^N$ which are Y-periodic.

Definition. (Weak two-scale convergence) A sequence $\{u_\varepsilon\}$ in $L^2(\Omega)$ is said to *converge weakly* to a function $u_0(x, y)$ in $L^2(\Omega \times Y)$ in the sense of *two-scales* if for all test functions $\psi(x, y)$ in $\mathcal{D}(\Omega; C_\#^\infty(Y))$, we have

$$(5.5) \qquad \lim_{\varepsilon \to 0} \int_\Omega u_\varepsilon(x)\bar{\psi}(x, \frac{x}{\varepsilon})dx = \frac{1}{|Y|} \int_\Omega \int_Y u_0(x, y)\bar{\psi}(x, y)dydx,$$

where $|Y|$ denotes the Lebesgue measure of Y. ∎

The basic result with regard to the above definition is the following theorem due to G. Nguetseng:

Theorem 5.1. *For each sequence $\{u_\varepsilon\}$ bounded in $L^2(\Omega)$, one can extract a subsequence and there exists a function $u_0(x, y)$ in $L^2(\Omega \times Y)$ such that the subsequence converges weakly to $u_0(x, y)$ in the sense of two-scales.* ∎

The weak two-scale convergence implies the classical weak convergence and the weak limit is nothing but the average of $u_0(x, y)$ with respect to y. We denote this average by

$$\mathcal{M}_Y(u_0)(x) = \frac{1}{|Y|} \int_Y u_0(x, y)dy \quad \text{for a.e. } x \in \Omega.$$

Indeed we have the following result due to G. Allaire:

Theorem 5.2. *Let $\{u_\varepsilon\}$ be a sequence in $L^2(\Omega)$ which converges weakly to $u_0(x, y)$ in the sense of two-scales. Then $\{u_\varepsilon\}$ converges weakly in $L^2(\Omega)$ to $u(x) = \mathcal{M}_Y(u_0)$ and one has*

$$(5.6) \qquad \liminf_{\varepsilon \to 0} \|u_\varepsilon\|_{0,\Omega} \geq \frac{1}{|Y|^{1/2}} \|u_0\|_{0,\Omega \times Y} \geq \|u\|_{0,\Omega}.$$

Furthermore, if

$$\lim_{\varepsilon \to 0} \|u_\varepsilon\|_{0,\Omega} = \frac{1}{|Y|^{1/2}} \|u_0\|_{0,\Omega \times Y}$$

and $\{v_\varepsilon\}$ is another sequence which converges weakly to $v_0(x, y) \in L^2(\Omega \times Y)$ in the sense of two-scales then

$$(5.7) \qquad u_\varepsilon(x)v_\varepsilon(x) \to \mathcal{M}_Y(u_0(x, \cdot)v_0(x, \cdot)) \quad \text{in} \quad \mathcal{D}'(\Omega).$$

Under the same assumption as above, if $u_0(x, y) \in L^2(\Omega; C_\#^0(Y))$, i.e., if u_0 is measurable and square integrable with respect to x with values in the space $C_\#^0(Y)$ of continuous functions in $\mathbb{R}^N$ that are Y-periodic then one has

$$(5.8) \qquad \lim_{\varepsilon \to 0} \|u_\varepsilon(x) - u_0(x, \frac{x}{\varepsilon})\|_{0,\Omega} = 0. \quad ∎$$

Definition. If $\{u_\varepsilon\}$ is a sequence in $L^2(\Omega)$ satisfying (5.8) with u_0 in $L^2(\Omega; \mathcal{C}^0_\#(Y))$, then we say that $\{u_\varepsilon\}$ *converges strongly* to u_0 in the sense of *two-scales.* ∎

Roughly speaking, Theorem 5.1 provides a rigorous justification of the first term in the ansatz (5.3) while Theorem 5.2 provides the corresponding corrector result. We will give a detailed explanation in the next paragraph as to how both theorems stated above yield the classical homogenization results. For the time being, we shall present the proof of Theorem 5.1 which is based on the following convergence result:

Lemma 5.3. *For all $\psi \in L^2(\Omega; \mathcal{C}^0_\#(Y))$ and $\varepsilon > 0$ the map $x \longmapsto \psi(x, x/\varepsilon)$ is a measurable function on Ω and square integrable on Ω. Moreover, we have*

$$(5.9) \qquad \lim_{\varepsilon \to 0} \int_\Omega |\psi(x, \frac{x}{\varepsilon})|^2 dx = \frac{1}{|Y|} \int_\Omega \int_Y |\psi(x,y)|^2 dy dx. \quad \blacksquare$$

Remark. At first sight, one might think that (5.9) is valid once ψ is in $L^2(\Omega; L^2_\#(Y))$. However, it is classically known that the measurability of the function $x \longmapsto \psi(x, x/\varepsilon)$ is not guaranteed even if $\psi \in \mathcal{C}^\infty(\bar{\Omega}; L^2_\#(Y))$. In particular, the elements of this space are not usually of *Carathéodory class* (i.e., measurable in x and continuous in y for almost all x). The significance of Carathéodory conditions is that they ensure the measurability of the function $x \longmapsto \psi(x, x/\varepsilon)$. ∎

Proof. We observe that the elements of $L^2(\Omega; \mathcal{C}^0_\#(Y))$ are of Carathéodory class and hence the measurability of the function $x \longmapsto \psi(x, x/\varepsilon)$ (see, for instance, EKELAND and TEMAM [1973] pp. 216–218 or BERGER [1977] p.76). Moreover, we have the inequality

$$\int_\Omega |\psi(x, \frac{x}{\varepsilon})|^2 dx \leq \int_\Omega \sup_{y \in Y} |\psi(x, y)|^2 dx = \|\psi\|^2_{L^2(\Omega; \mathcal{C}^0_\#(Y))} < \infty$$

which shows that $\psi(x, x/\varepsilon) \in L^2(\Omega)$.

It remains to prove (5.9) which will be done by density arguments. Suppose therefore that $\psi \in L^2(\Omega; \mathcal{C}^0_\#(Y))$. Since $L^2(\Omega) \otimes \mathcal{C}^0_\#(Y)$ is dense in $L^2(\Omega; \mathcal{C}^0_\#(Y))$ (see BOURBAKI [1965] p.46 or SCHWARTZ [1966] pp. 108–109), given $\delta > 0$, there exists a function

$$\psi_\delta(x,y) = \sum_{j=1}^n f_j(x) \phi_j(y)$$

with $f_j \in L^2(\Omega)$ and $\phi_j \in \mathcal{C}^0_\#(Y)$, $j = 1, ..., n$ such that

$$\|\psi - \psi_\delta\|_{L^2(\Omega;\mathcal{C}^0_\#(Y))} \leq \delta.$$

It is easily seen that (5.9) holds for ψ_δ. Indeed, we can express the left side of (5.9) as

$$\int_\Omega \sum_{j,k=1}^n f_j(x)\bar{f}_k(x)\phi_j(\frac{x}{\varepsilon})\bar{\phi}_k(\frac{x}{\varepsilon})dx$$

and we can pass to the limit as $\varepsilon \to 0$ in this integral since

$$\phi_j\bar{\phi}_k(x/\varepsilon) \to \mathcal{M}_Y(\phi_j\bar{\phi}_k) \quad \text{in } L^\infty(\Omega)\text{-weak*}$$

and obtain at the limit the following integral

$$\int_\Omega \sum_{j,k=1}^n f_j(x)\bar{f}_k(x)\mathcal{M}_Y(\phi_j\bar{\phi}_k)dx$$

which is equal to

$$\frac{1}{|Y|} \int_\Omega \int_Y |\psi_\delta(x,y)|^2 dy dx.$$

Thus (5.9) holds for ψ_δ.

Finally, observe that we have

$$\left| \int_\Omega |\psi(x,\frac{x}{\varepsilon})|^2 - \frac{1}{|Y|}\int_\Omega\int_Y |\psi(x,y)|^2 \right| \leq \left| \int_\Omega |\psi_\delta(x,\frac{x}{\varepsilon})|^2 - \frac{1}{|Y|}\int_\Omega\int_Y |\psi_\delta(x,y)|^2 \right| + O(\delta)$$

which clearly implies (5.9) since δ is arbitrary. ∎

Before the start of the proofs of Theorems 5.1 and 5.2, let us give a few examples in the form of exercises.

◇ **Exercise 5.1.** Let $\psi \in L^2(\Omega;\mathcal{C}^0_\#(Y))$ be given. Show that the sequence $\{\psi(x,x/\varepsilon)\}_\varepsilon$ is bounded in $L^2(\Omega)$ and

$$\psi(x,\frac{x}{\varepsilon}) \to \mathcal{M}_Y(\psi(x,\cdot)) \quad \text{in} \quad L^2(\Omega) \text{ weak.}$$

Deduce that if $\psi \in L^\infty(\Omega;\mathcal{C}^0_\#(Y))$ then

$$\psi(x,\frac{x}{\varepsilon}) \to \mathcal{M}_Y(\psi(x,\cdot)) \quad \text{in} \quad L^\infty(\Omega) \text{ weak*.}$$

Deduce further that if $\psi \in L^2(\Omega;\mathcal{C}^0_\#(Y))$ then

$$\psi(x,\frac{x}{\varepsilon}) \to \psi(x,y)$$

weakly in the sense of two-scales. ∎

Proof of Theorem 5.1. Thanks to the previous lemma, there is a constant c independent of ψ and a subsequence of ε which may depend on ψ such that

$$(5.10) \qquad |\lim_{\varepsilon \to 0} \int_{\Omega} u_\varepsilon \bar{\psi}(x, \frac{x}{\varepsilon}) dx| \le c \|\psi\|_{L^2(\Omega \times Y)} \quad \forall \psi \in L^2(\Omega; \mathcal{C}^0_\#(Y)).$$

We vary ψ over a countable subset of $L^2(\Omega; \mathcal{C}^0_\#(Y))$ which is dense in $L^2(\Omega \times Y)$. Since the subset is countable, by a standard diagonalization process, we can extract a subsequence of ε such that (5.10) is valid for all functions ψ in this subset. Then, by density, we conclude that the limit in the left side of (5.10) as a function of ψ defines a continuous linear form on $L^2(\Omega \times Y)$. Then the classical *Riesz Representation Theorem* immediately implies the existence of a function $u_0(x, y)$ which satisfies (5.5). This finishes the proof of Theorem 5.1. ∎

Proof of Theorem 5.2. By choosing test functions $\psi(x) \in \mathcal{D}(\Omega)$ which are independent of y in (5.5), we immediately obtain that $\{u_\varepsilon\}$ converges weakly in $L^2(\Omega)$ to $u(x) = \mathcal{M}_Y(u_0)(x)$. To obtain (5.6) we remark first that the second inequality is evident by Cauchy-Schwarz's Inequality. To prove the first one, we take $\psi \in L^2(\Omega; \mathcal{C}^0_\#(Y))$ and expand

$$(5.11) \quad \left\{ \begin{aligned} 0 \le \int_{\Omega} |u_\varepsilon(x) - \psi(x, \frac{x}{\varepsilon})|^2 dx &= \int_{\Omega} |u_\varepsilon(x)|^2 dx - \int_{\Omega} u_\varepsilon(x) \bar{\psi}(x, \frac{x}{\varepsilon}) dx - \\ & - \int_{\Omega} \bar{u}_\varepsilon(x) \psi(x, \frac{x}{\varepsilon}) dx + \int_{\Omega} |\psi(x, \frac{x}{\varepsilon})|^2 dx. \end{aligned} \right.$$

Using the weak two-scale convergence of u_ε and Lemma 5.3, we can pass to the limit in the above inequality and obtain

$$\liminf_{\varepsilon \to 0} \int_{\Omega} |u_\varepsilon(x)|^2 \ge \frac{1}{|Y|} \int_{\Omega} \int_Y u_0(x, y) \bar{\psi}(x, y) dy dx +$$

$$+ \frac{1}{|Y|} \int_{\Omega} \int_Y \bar{u}_0(x, y) \psi(x, y) dy dx - \frac{1}{|Y|} \int_{\Omega} \int_Y |\psi(x, y)|^2 dy dx.$$

Choosing a sequence ψ_n of smooth functions approximating u_0 in $L^2(\Omega \times Y)$, we get the required inequality (5.6).

To prove (5.7), let us first observe that if $\{\psi_n\}$ is a sequence in $L^2(\Omega; \mathcal{C}^0_\#(Y))$ approximating u_0 in $L^2(\Omega \times Y)$ then we have from (5.11) that

$$\lim_{\varepsilon \to 0} \int_{\Omega} |u_\varepsilon(x) - \psi_n(x, \frac{x}{\varepsilon})|^2 dx = \frac{1}{|Y|} \int_{\Omega} \int_Y |u_0(x, y) - \psi_n(x, y)|^2 dy dx.$$

Taking now the limit as $n \to \infty$, we obtain

$$(5.12) \qquad \lim_{n \to \infty} \lim_{\varepsilon \to 0} \int_{\Omega} |u_\varepsilon(x) - \psi_n(x, \frac{x}{\varepsilon})|^2 dx = 0.$$

Now let $\{v_\varepsilon\}$ be a sequence converging weakly to v_0 in the sense of two-scales. For any $\phi \in \mathcal{D}(\Omega)$, we then have

$$\int_{\Omega} \phi(x)u_\varepsilon(x)v_\varepsilon(x)dx = \int_{\Omega} \phi(x)\psi_n(x, \frac{x}{\varepsilon})v_\varepsilon(x)dx + \int_{\Omega} \phi(x)\Big(u_\varepsilon(x) - \psi_n(x, \frac{x}{\varepsilon})\Big)v_\varepsilon(x)$$

Passing to the limit, first as $\varepsilon \to 0$, we obtain

$$\lim_{\varepsilon \to 0} \int_{\Omega} \phi(x)u_\varepsilon(x)v_\varepsilon(x)dx - \frac{1}{|Y|} \int_{\Omega} \int_{Y} \phi(x)\psi_n(x, y)v_0(x, y)dydx \le$$

$$\le c \lim_{\varepsilon \to 0} \|u_\varepsilon(x) - \psi_n(x, \frac{x}{\varepsilon})\|_{0,\Omega}.$$

Next, taking the limit as $n \to \infty$ and using (5.12), we get

$$\lim_{\varepsilon \to 0} \int_{\Omega} \phi(x)u_\varepsilon(x)v_\varepsilon(x)dx = \frac{1}{|Y|} \int_{\Omega} \int_{Y} \phi(x)u_0(x, y)v_0(x, y)dydx.$$

This is nothing but (5.7).

Finally, if $u_0 \in L^2(\Omega; \mathcal{C}^0_{\#}(Y))$, then taking directly $\psi_n = u_0$ in (5.12), we get (5.8). This completes the proof of Theorem 5.2. ∎

From the last part of Theorem 5.2, we see that for a given bounded sequence $\{u_\varepsilon\}$ in $L^2(\Omega)$ converging weakly to u_0 in the sense of two-scales, u_0 contains more information than its weak limit u. Indeed, u_0 describes the periodic oscillations of u_ε while u is just the average of u_0. However, let us emphasize that the weak two-scale limit captures only the oscillations which are in resonance with those of the test functions $\psi(x, x/\varepsilon)$ as illustrated in the following

◇ **Exercise 5.2.** If $\psi \in L^2(\Omega; \mathcal{C}^0_{\#}(Y))$ then prove that the weak limit of $\psi(x, x/\varepsilon^2)$ in $L^2(\Omega)$ as well as its weak two-scale limit coincide with $\mathcal{M}_Y(\psi(x, \cdot))$. Observe that since the two-scale limit depends only on x, no oscillations are captured. ∎

◇ **Exercise 5.3.** Show that if u_ε converges (strongly) to u in $L^2(\Omega)$ then $u_0(x,y)$ is independent of y; in fact $u_0(x,y) = u(x)$. Show by means of examples that the converse is false. ∎

Before concluding the basic results of this theory, we will prove a very useful corollary of Theorem 5.1 which applies when one also has bounds on the gradient ∇u_ε of the sequence u_ε. According to Theorem 5.1, in this case ∇u_ε will also converge weakly in the sense of two-scales to a limit which can be written as $\nabla u(x) + \phi(x,y)$, where $u(x)$ is the weak limit of u_ε and $\mathcal{M}_Y(\phi(x,y)) = 0$. The main content of this corollary is that ϕ is a gradient of a function in $H^1_\#(Y)$. More precisely, we have

Corollary 5.4. *If $\{u_\varepsilon\}$ is a sequence converging weakly towards u in $H^1(\Omega)$ then $\{u_\varepsilon\}$ converges weakly to u in the sense of two-scales and there exists $u_1 \in L^2(\Omega; H^1_\#(Y))$ such that a subsequence of $\{\nabla u_\varepsilon\}$ converges weakly in the sense of two-scales towards $[\nabla u(x) + \nabla_y u_1(x,y)]$.* ∎

Proof of Corollary 5.4. Since u_ε (respectively ∇u_ε) is bounded in $L^2(\Omega)$ (respectively $L^2(\Omega)^N$), up to a subsequence, $u_\varepsilon \rightharpoonup u_0(x,y) \in L^2(\Omega \times Y)$ (respectively $\nabla u_\varepsilon \rightharpoonup \chi_0(x,y) \in L^2(\Omega \times Y)^N$) weakly in the sense of two-scales. By integration by parts, we have for $\Psi \in \mathcal{D}(\Omega; C^\infty_\#(Y))^N$

$$\varepsilon \int_\Omega \nabla u_\varepsilon(x) \cdot \bar\Psi(x, \frac{x}{\varepsilon})dx = -\int_\Omega u_\varepsilon \left[\mathrm{div}_y \bar\Psi(x, \frac{x}{\varepsilon}) + \varepsilon \mathrm{div}_x \bar\Psi(x, \frac{x}{\varepsilon}) \right] dx.$$

Observe that we can pass to the limit in this relation using the fact that $u_\varepsilon \rightharpoonup u_0$ weakly in the sense of two-scales and obtain

$$0 = -\frac{1}{|Y|} \int_\Omega \int_Y u_0(x,y) \mathrm{div}_y \bar\Psi(x,y) dy dx.$$

This, in particular, implies that $u_0(x,y)$ does not depend on y and so u_0 coincides with the weak $L^2(\Omega)$ limit u. This proves the first part of the corollary.

To prove the next, we choose a test function Ψ such that $\mathrm{div}_y \Psi(x,y) = 0$. By integration by parts we have

$$\int_\Omega u_\varepsilon(x) \mathrm{div}_x \bar\Psi(x, \frac{x}{\varepsilon}) dx = -\int_\Omega \nabla u_\varepsilon(x) \cdot \bar\Psi(x, \frac{x}{\varepsilon}) dx.$$

Passing to the limit using the weak two-scale convergence of u_ε and ∇u_ε, we obtain

$$\frac{1}{|Y|} \int_\Omega \int_Y u(x) \mathrm{div}_x \bar\Psi(x,y) dy dx = \frac{-1}{|Y|} \int_\Omega \int_Y \chi_0(x,y) \cdot \bar\Psi(x,y) dy dx.$$

Thus, for any function $\Psi(x,y) \in \mathcal{D}(\Omega; \mathcal{C}_\#^\infty(Y))^N$ with $\operatorname{div}_y \Psi(x,y) = 0$, we have

$$\int_\Omega \int_Y [\chi_0(x,y) - \nabla u(x)] \cdot \bar{\Psi}(x,y) dy dx = 0.$$

Recall that the orthogonal of divergence-free functions is exactly the gradients (see, for example TARTAR [1978]). In the present context, we can arrive at the desired result by means of Fourier series in y. Thus we conclude that there exists a unique function $u_1(x,y) \in L^2(\Omega; \dot{H}_\#^1(Y))$ such that

$$\chi_0(x,y) = \nabla u(x) + \nabla_y u_1(x,y).$$

This completes the proof of Corollary 5.4. ∎

5.2. Application to a Classical Example

We are now in a position to explain how the two-scale convergence method works in the classical example alluded to in the preceding paragraph. We, therefore, place ourselves in the context of the model problem of homogenization. Let $A(x) = [a_{ij}(x)] \in L_\#^\infty(Y)^{N \times N}$ be a matrix such that

$$\mathfrak{Re}\big(a_{ij}(x)\big)\xi_i\bar{\xi}_j \geq \alpha|\xi|^2$$

where $\alpha > 0$ is a positive constant independent of $x \in Y$ and $\xi \in \mathbb{C}^N$. We consider the associated second order operator in divergence form:

$$A_\varepsilon u \overset{\text{def}}{=} -\frac{\partial}{\partial x_i}\left(a_{ij}(\frac{x}{\varepsilon})\frac{\partial u}{\partial x_j}\right) = -\operatorname{div}\left(A(\frac{x}{\varepsilon})\nabla u\right).$$

Let $f \in L^2(\Omega)$ be given and u_ε be the unique weak solution of the following Dirichlet problem:

(5.13)
$$\begin{cases} A_\varepsilon u_\varepsilon = f & \text{in} \quad \Omega, \\ u_\varepsilon \in H_0^1(\Omega). \end{cases}$$

Recall that our aim is to pass to the limit in (5.13) as $\varepsilon \to 0$. As the first step, we deduce the following estimate on u_ε which is easily obtained by multiplying (5.13) by u_ε:

$$\|u_\varepsilon\|_{1,\Omega} \leq c,$$

where c is a constant independent of ε. Therefore, we can extract a subsequence (again denoted by ε) such that $u_\varepsilon \rightharpoonup u$ in $H_0^1(\Omega)$ weak. At this juncture, we apply Corollary 5.4 and derive the existence of a function $u_1 \in L^2(\Omega; H_\#^1(Y))$ such that, up to a subsequence (which we still denote by ε), ∇u_ε converges weakly in the sense of two-scales towards $[\nabla u(x) + \nabla_y u_1(x,y)]$.

The alternative method of homogenization which uses the two-scale convergence consists of multiplying (5.13) by test functions of the type $\psi(x, x/\varepsilon)$ where $\psi \in \mathcal{D}(\Omega; C^\infty_\#(Y))$. After integration by parts, we can pass to the limit in the resulting relation by using Corollary 5.4 above. As we shall see below, this procedure yields simultaneously the homogenized problem (5.2) as well as the convergence result. We now pass on to the details. In fact, multiplying (5.13) by $\bar\psi(x) + \varepsilon\bar\psi_1(x, x/\varepsilon)$ where $\psi \in \mathcal{D}(\Omega)$ and $\psi_1 \in \mathcal{D}(\Omega; C^\infty_\#(Y))$ and integrating by parts, we get

$$\int_\Omega A(\frac{x}{\varepsilon})\nabla u_\varepsilon(x) \cdot \left[\nabla\bar\psi(x) + \nabla_y\bar\psi_1(x, \frac{x}{\varepsilon}) + \varepsilon\nabla_x\bar\psi_1(x, \frac{x}{\varepsilon})\right] dx =$$
$$= \int_\Omega f(x)\left[\bar\psi(x) + \varepsilon\bar\psi_1(x, \frac{x}{\varepsilon})\right] dx.$$

Using Corollary 5.4, we can pass to the limit in the above relation along the extracted subsequence and obtain

$$(5.14) \quad \begin{cases} \dfrac{1}{|Y|}\displaystyle\int_\Omega\int_Y A(y)[\nabla u(x) + \nabla_y u_1(x,y)] \cdot [\nabla\bar\psi(x) + \nabla_y\bar\psi_1(x,y)]\,dy\,dx = \\[2mm] \qquad\qquad\qquad\qquad = \displaystyle\int_\Omega f(x)\bar\psi(x)\,dx. \end{cases}$$

This can be interpreted as the following system of differential equations for u and u_1:

$$(5.15) \quad \begin{cases} -\operatorname{div}_y\left[A(y)(\nabla u(x) + \nabla_y u_1(x,y))\right] = 0 \quad \text{in} \quad \Omega \times Y, \\[2mm] -\operatorname{div}_x\left[\dfrac{1}{|Y|}\displaystyle\int_Y A(y)(\nabla u(x) + \nabla_y u_1(x,y))dy\right] = f \quad \text{in} \quad \Omega, \\[2mm] u = 0 \quad \text{on} \quad \partial\Omega, \\[2mm] u_1(x,y) \quad Y\text{-periodic in } y. \end{cases}$$

The relations in (5.15) are referred to as the *two-scale homogenized system*.

$\diamond$ **Exercise 5.4.** Use Lax-Milgram Lemma to show that problem (5.15) has one and only one solution $(u, u_1) \in H^1_0(\Omega) \times L^2(\Omega; \dot H^1_\#(Y))$. $\blacksquare$

Grouping together all the results above, we arrive at the following homogenization theorem in the two-scale convergence analysis:

Theorem 5.5. *The sequence $\{u_\varepsilon\}$ of solutions of* (5.13) *converges weakly to u in $H_0^1(\Omega)$ and $\{\nabla u_\varepsilon\}$ converges weakly to $[\nabla u(x) + \nabla_y u_1(x, y)]$ in the sense of two-scales.* ∎

In the two-scale homogenized system, the two space variables x and y (i.e., the macroscopic and the microscopic scales) are apparently mixed. In the classical homogenization procedure, this is not the case. However, as observed below, the variables x and y can be decoupled in this particular example. Indeed, it can be easily checked that system (5.15) is equivalent to the following systems of equations:

$$
(5.16) \qquad \begin{cases} -\operatorname{div}\left(A^*\nabla u(x)\right) = f \quad \text{in} \quad \Omega, \\ u = 0 \quad \text{on} \quad \partial\Omega, \end{cases}
$$

where A^* is a matrix whose entries are given by

$$
(5.17) \quad A_{ij}^* = \frac{1}{|Y|}\int_Y A(y)\left[\nabla_y w_i(y) + \mathbf{e}_i\right]\cdot\left[\nabla_y \bar{w}_j(y) + \mathbf{e}_j\right]dy \quad \forall i,j = 1,...,N
$$

and w_j, $1 \leq j \leq N$ is the solution of the so-called cell problem

$$
(5.18) \qquad \begin{cases} -\operatorname{div}_y\left(A(y)[\nabla_y w_j(y) + \mathbf{e}_j]\right) = 0 \quad \text{in} \quad Y, \\ w_j(y) \quad Y\text{-periodic}. \end{cases}
$$

The second component u_1 is related to the first through

$$
(5.19) \qquad u_1(x, y) = \sum_{j=1}^N \frac{\partial u}{\partial x_j}(x)w_j(y).
$$

We remark that A_{ij}^*, $1 \leq i, j \leq N$ are the usual *homogenized coefficients* and (5.16) is the so-called *homogenized problem*.

Remark. For many other type of problems, the decoupling of the variables x and y exhibited above is not possible. In these cases, the form of the two-scale homogenized system is more complicated and it may involve integro-differential operators. For several interesting examples, the reader can refer to ALLAIRE [1991], [1992a,b] and NGUETSENG [1990]. ∎

Let us now pass to prove a corrector result to Theorem 5.5.

Theorem 5.6. *Assume that $\nabla_y u_1 \in L^2(\Omega; C_\#^0(Y))^N$. Then the sequence $[\nabla u_\varepsilon(x) - \nabla u(x) - \nabla_y u_1(x, x/\varepsilon)]$ converges strongly to zero in $L^2(\Omega)^N$. In particular, if $u_1 \in L^2(\Omega; C_\#^0(Y))$ and $\nabla_x u_1, \nabla_y u_1$ are elements of $L^2(\Omega; C_\#^0(Y))^N$ then*

$$
\left[u_\varepsilon(x) - u(x) - \varepsilon u_1\left(x, \frac{x}{\varepsilon}\right)\right] \longrightarrow 0 \quad \text{in} \quad H^1(\Omega) \ \text{strongly}.
$$

Proof. Using the regularity assumption on u_1, we can write

$$\int_\Omega A(\frac{x}{\varepsilon})|\nabla u_\varepsilon(x) - \nabla u(x) - \nabla_y u_1(x, \frac{x}{\varepsilon})|^2 dx = \int_\Omega f(x)\bar{u}_\varepsilon(x)dx +$$

$$+ \int_\Omega A(\frac{x}{\varepsilon})|\nabla u(x) + \nabla_y u_1(x, \frac{x}{\varepsilon})|^2 dx - \int_\Omega A(\frac{x}{\varepsilon})\nabla u_\varepsilon(x) \cdot \left[\nabla \bar{u}(x) + \nabla_y \bar{u}_1(x, \frac{x}{\varepsilon})\right]$$

$$- \int_\Omega A(\frac{x}{\varepsilon})\nabla \bar{u}_\varepsilon(x) \cdot \left[\nabla u(x) + \nabla_y u_1(x, \frac{x}{\varepsilon})\right] dx.$$

Observe that one can pass to the limit in the second term on the right side of the previous relation by applying Theorem 5.2 to $\nabla_y u_1(x, x/\varepsilon)$ or simply using Exercise 5.1. In the other terms on the right side, we merely use the weak two-scale convergence of ∇u_ε. Taking the real parts and using the coercivity of the matrix $A(x)$, we get

$$\alpha \lim_{\varepsilon \to 0} \left\| \nabla u_\varepsilon(x) - \nabla u(x) - \nabla_y u_1(x, \frac{x}{\varepsilon}) \right\|_{0,\Omega}^2 \leq \Re e \left(\int_\Omega f(x)\bar{u}(x)dx \right) -$$

$$- \frac{1}{|Y|} \Re e \left(\int_\Omega \int_Y A(y)|\nabla u(x) + \nabla_y u_1(x,y)|^2 dy dx \right).$$

Let us next observe that, by density, the weak formulation (5.14) is valid for all $\psi \in H_0^1(\Omega)$ and $\psi_1 \in L^2(\Omega; H_\#^1(Y))$. Taking $\psi = u$ and $\psi_1 = u_1$ in (5.14), we see that the right hand side of the above inequality is zero and hence the theorem. ∎

Two-scale convergence can also handle homogenization problems in perforated domains. What we need in this case is a variant of Corollary 5.4. In order to state this, let us define a sequence Ω_ε of periodically perforated domains contained in a fixed domain Ω of $\mathbb{R}^N$. The period of Ω_ε is εY^* where $Y^* = Y \backslash T$ is the subset of the basic reference cell Y introduced in Paragraph 3.1. Y^* represents the complement of the perforation T in Y. Let $\chi_{Y^*}(y)$ be the characteristic function of Y^*. Thus Ω_ε can also be characterized as

$$\Omega_\varepsilon = \left\{ x \in \Omega \mid \chi_{Y^*}(\frac{x}{\varepsilon}) = 1 \right\}.$$

The full details of the following result can be found in ALLAIRE [1992]:

Corollary 5.7. *Let $\{u_\varepsilon\}$ be a sequence such that $u_\varepsilon \in H^1(\Omega_\varepsilon)$ and $\|u_\varepsilon\|_{1,\Omega_\varepsilon} \leq c$. Denote by $\tilde{\ }$ the extension by zero inside $(\Omega \backslash \Omega_\varepsilon)$. Then there exists a subsequence (which we denote again by ε) and a pair (u_0, u_1) in $H^1(\Omega; \dot{H}_\#^1(Y^*))$ such that $\tilde{u}_\varepsilon$ and $\widetilde{\nabla u_\varepsilon}$ converge weakly in the sense of two scales towards $u_0(x)\chi_{Y^*}(y)$ and $[\nabla u_0(x) + \nabla_y u_1(x,y)]\chi_{Y^*}(y)$ respectively.* ∎

Remark. (Connection between Bloch eigenvalues and homogenized coefficients) Let us consider the operator $\bar{A}$ defined by the coefficients $(a_{pq}(y))$ where $a_{pq} = a_{qp}$:

$$\bar{A} \overset{\text{def}}{=} -\frac{\partial}{\partial y_p}\left(a_{pq}(y)\frac{\partial}{\partial y_q}\right).$$

Associated with it are the Bloch eigenvalues $\{\lambda_n(\zeta)\}$ and the corresponding eigenfunctions $\{\phi_n(\zeta,\cdot)\}$ (see Theorem 2.3). They satisfy

$$\bar{A}(\zeta)\phi(\zeta,\cdot) = \lambda(\zeta)\phi(\zeta,\cdot) \quad \zeta \in Y',$$

where $\bar{A}(\zeta)$ is the *shifted operator* (see (2.8)):

$$\bar{A}(\zeta) = -\left(\frac{\partial}{\partial y_p} + i\zeta_p\right)\left[a_{pq}(y)\left(\frac{\partial}{\partial y_q} + i\zeta_q\right)\right].$$

On the other hand, the coefficients (A^*_{pq}) defined by (5.17) appear naturally when we study homogenization problems associated with $\bar{A}$. There is a remarkable relation between A^*_{pq} and the second derivatives of the first eigenvalue $\lambda_1(\zeta)$ and it is as follows:

$$A^*_{pq} = \frac{1}{2}\frac{\partial^2 \lambda_1}{\partial \zeta_p \partial \zeta_q}(0) \quad \forall p, q = 1, ..., N.$$

This can be established, in an informal manner, by differentiating the eigenvalue equation twice with respect to ζ and evaluating the resulting relations at $\zeta = 0$. For details of this algebra, we refer to BENSOUSSAN, LIONS and PAPANICOLAOU [1978] pp. 633–638. We merely present here some important intermediate results of the calculations. Using the simplicity of the first eigenvalue at $\zeta = 0$, we can describe its first and second order variations near 0 as follows:

$$\frac{\partial \lambda_1(\zeta)}{\partial \zeta_p} = \left(\frac{\partial \bar{A}(\zeta)}{\partial \zeta_p}\phi_1(\zeta), \phi_1(\zeta)\right),$$

$$\frac{1}{2}\frac{\partial^2 \lambda_1(\zeta)}{\partial \zeta_p \partial \zeta_q} = (a_{pq}\phi_1(\zeta), \phi_1(\zeta)) + \left(\left(\frac{\partial \bar{A}(\zeta)}{\partial \zeta_p} - \frac{\partial \lambda_1(\zeta)}{\partial \zeta_p}\right)\frac{\partial \phi_1(\zeta)}{\partial \zeta_q}, \phi_1(\zeta)\right) +$$
$$\left(\left(\frac{\partial \bar{A}(\zeta)}{\partial \zeta_q} - \frac{\partial \lambda_1(\zeta)}{\partial \zeta_q}\right)\frac{\partial \phi_1(\zeta)}{\partial \zeta_p}, \phi_1(\zeta)\right).$$

Moreover, the first derivative of $\phi_1(\zeta)$ satisfies

$$\left(\bar{A}(\zeta) - \lambda_1(\zeta)\right)\frac{\partial \phi_1(\zeta)}{\partial \zeta_p} + \left(\frac{\partial \bar{A}(\zeta)}{\partial \zeta_p} - \frac{\partial \lambda_1(\zeta)}{\partial \zeta_p}\right)\phi_1(\zeta) = 0.$$

We know that $\lambda_1(0) = 0$ and $\phi_1(0) = 1$. If we use this information in the above formulae evaluated at $\zeta = 0$, we get

$$\frac{\partial \lambda_1}{\partial \zeta_p}(0) = 0 \quad \forall p = 1, ..., N,$$

$$\frac{\partial \phi_1}{\partial \zeta_p}(0) = iw_p \quad \forall p = 1, ..., N,$$

where w_p are solutions defined by (5.18). Inserting these into the above expression for the second derivative of λ_1, we arrive at the desired relation between A_{pq}^* and the second derivative of λ_1. ∎

5.3. Homogenization of the Laplace Model

With the insight gained in the two-scale convergence analysis, we now study, following the standard homogenization procedure, the asymptotic behaviour of the Laplace model when the number of tubes goes to infinity. To this end, we consider problem (3.30) and rescale it in such a way that the container of the fluid is independent of n and the distance between the immersed tubes and the size of the tubes goes to zero when $n \to \infty$. We introduce therefore the following change of variables:

(5.20)
$$\begin{cases} \text{Put } \varepsilon = 1/n, \quad \text{replace } x \text{ by } x/n, \quad \Omega_n \text{ by } \Omega_\varepsilon \overset{\text{def}}{=} \varepsilon \Omega_n, \\[4pt] \Gamma_n \text{ by } \Gamma_0 \overset{\text{def}}{=} \frac{1}{n}\Gamma_n, \quad Q_n \text{ by } Q_\varepsilon \overset{\text{def}}{=} Q_n \dagger, \\[4pt] Y_j \; (j \in Q_n) \text{ by } Y_j^\varepsilon \overset{\text{def}}{=} \varepsilon Y_j \; (j \in Q_\varepsilon), \Gamma_j \text{ by } \Gamma_j^\varepsilon \overset{\text{def}}{=} \varepsilon \Gamma_j, Y_j^* \text{ by } \varepsilon Y_j^*, \\[4pt] \phi^{(n)} \text{ by } \phi_\varepsilon(x) \overset{\text{def}}{=} \phi^{(n)}(nx), \quad \lambda^{(n)} \text{ by } \lambda_\varepsilon \overset{\text{def}}{=} \lambda^{(n)} \quad \text{and} \\[4pt] K_n \text{ by } K_\varepsilon \overset{\text{def}}{=} 4n^2 = 4\varepsilon^{-2}. \end{cases}$$

Observe that with the above rescaling, we see that

$$\Omega_\varepsilon \subset \Omega \overset{\text{def}}{=}]-2\pi, 2\pi[^2 \quad \forall \varepsilon > 0$$

and Γ_0 is exactly the boundary of Ω. From (3.30), it can easily be verified that $(\lambda_\varepsilon, \phi_\varepsilon)$ are solutions of the following problem in Ω_ε: Find $\lambda_\varepsilon \in \mathbb{C}$ for which there is a non-constant function ϕ_ε such that

(5.21)
$$\begin{cases} \Delta \phi_\varepsilon = 0 \quad \text{in} \quad \Omega_\varepsilon, \\[4pt] \dfrac{\partial \phi_\varepsilon}{\partial n} = \varepsilon^{-2} \lambda_\varepsilon \Big(\displaystyle\int_{\Gamma_j^\varepsilon} \phi_\varepsilon \mathbf{n} d\gamma \Big) \cdot \mathbf{n} \quad \text{on} \quad \Gamma_j^\varepsilon, \; j \in Q_\varepsilon, \\[4pt] \dfrac{\partial \phi_\varepsilon}{\partial n} = 0 \quad \text{on} \quad \Gamma_0. \end{cases}$$

---

† This means that the holes in Ω_ε will be numbered in the same fashion as in Ω_n.

To simplify the situation a little bit, we take Dirichlet condition on Γ instead of Neumann. Recall that such a change does not drastically affect the asymptotic behaviour as mentioned several times in Paragraph 2.4 and in §3. Therefore, in the sequel, we concentrate our attention on the following system satisfied by $(\lambda_\varepsilon, \phi_\varepsilon)$:

$$(5.22) \qquad \begin{cases} \Delta\phi_\varepsilon = 0 \quad \text{in} \quad \Omega_\varepsilon, \\ \dfrac{\partial\phi_\varepsilon}{\partial n} = \varepsilon^{-2}\lambda_\varepsilon\left(\displaystyle\int_{\Gamma_j^\varepsilon}\phi_\varepsilon \mathbf{n}d\gamma\right)\cdot\mathbf{n} \quad \text{on} \quad \Gamma_j^\varepsilon, \ j \in Q_\varepsilon, \\ \phi_\varepsilon = 0 \quad \text{on} \quad \Gamma_0 \end{cases}$$

As usual, to solve (5.22), we introduce the Green's operator S_ε acting on $\mathbb{C}^{2K_\varepsilon}$ defined by the following rule:

$$(5.23) \qquad S_\varepsilon\mathbf{a} = \left\{\varepsilon^{-2}\int_{\Gamma_j^\varepsilon}\varphi_\varepsilon \mathbf{n}d\gamma\right\}_{j\in Q_\varepsilon} \qquad \forall\mathbf{a} \in \mathbb{C}^{2K_\varepsilon},$$

where φ_ε is the weak solution of the following boundary-value problem:

$$(5.24) \qquad \begin{cases} \Delta\varphi_\varepsilon = 0 \quad \text{in} \quad \Omega_\varepsilon, \\ \dfrac{\partial\varphi_\varepsilon}{\partial n} = \mathbf{a}_j\cdot\mathbf{n} \quad \text{on} \quad \Gamma_j^\varepsilon, \ j \in Q_\varepsilon, \\ \varphi_\varepsilon = 0 \quad \text{on} \quad \Gamma_0. \end{cases}$$

The operator S_ε is self-adjoint and positive definite and further its characteristic values are nothing but the eigenvalues of problem (5.22). We arrange the eigenvalue of S_ε as follows:

$$(5.25) \qquad 0 < \mu_1^{(\varepsilon)} \leq \cdots \leq \mu_{2K_\varepsilon}^{(\varepsilon)}.$$

Recall that we are interested in the behaviour of problem (5.22) as $\varepsilon \to 0$. To this end, we shall use the two-scale convergence analysis and prove the strong convergence of S_ε and then apply Theorems 4.7 and 4.9 of Chapter I. Since S_ε is not defined on the same fixed space, we need to define an extension $\tilde{S}_\varepsilon$ of S_ε acting on the fixed space $L^2(\Omega)^2$. To do this, we first introduce two linear operators $U_\varepsilon \in \mathcal{L}(L^2(\Omega)^2, \mathbb{C}^{2K_\varepsilon})$ and $V_\varepsilon \in \mathcal{L}(\mathbb{C}^{2K_\varepsilon}, L^2(\Omega)^2)$:

$$(5.26) \qquad U_\varepsilon\mathbf{a} = \left\{\frac{1}{|Y_j^\varepsilon|}\int_{Y_j^\varepsilon}\mathbf{a}(x)dx\right\}_{j\in Q_\varepsilon} \qquad \forall\mathbf{a} \in L^2(\Omega)^2,$$

$$(5.27) \qquad V_\varepsilon\{\mathbf{a}_j\}_{j\in Q_\varepsilon} = \sum_{j\in Q_\varepsilon}\mathbf{a}_j\chi_{Y_j^\varepsilon} \qquad \forall\{\mathbf{a}_j\}_{j\in Q_\varepsilon} \in \mathbb{C}^{2K_\varepsilon},$$

where $|Y_j^\varepsilon| = \varepsilon^2 |Y|$ is the measure of Y_j^ε. The consideration of these two operators is very natural and is motivated from the fact that the motion of each tube is rigid and so it is normal to approximate $\mathbf{a}$ by functions which are piece-wise constant in each cell Y_j^ε. We note, in passing, that the composition $V_\varepsilon U_\varepsilon$ is nothing but the standard projection operator from $L^2(\Omega)^2$ onto the subspace of piece-wise constant functions in each cell Y_j^ε, $j \in Q_\varepsilon$. Moreover, a brief calculation shows that the adjoint $V_\varepsilon^* = \varepsilon^N U_\varepsilon$. Since the size of the cells in Ω_ε goes to zero as $\varepsilon \to 0$, it is classical that the following result holds:

Proposition 5.8. *For each* $\mathbf{a} \in L^2(\Omega)^2$,

$$V_\varepsilon U_\varepsilon \mathbf{a} \longrightarrow \mathbf{a} \quad in \quad L^2(\Omega)^2. \ \blacksquare$$

Using these operators, we can extend S_ε to $\tilde{S}_\varepsilon : L^2(\Omega)^2 \longrightarrow L^2(\Omega)^2$ as follows:

$$(5.28) \qquad\qquad \tilde{S}_\varepsilon \mathbf{a} = V_\varepsilon S_\varepsilon U_\varepsilon \mathbf{a}.$$

This new operator is also self-adjoint and has finite rank. Its spectrum coincides with $\sigma(S_\varepsilon)$ except for the eigenvalue zero which is of infinite multiplicity. The study of the spectral family of S_ε is reduced to that of the spectral family of $\tilde{S}_\varepsilon$ via the identification made in (5.27).

To identify the weak limit of $\tilde{S}_\varepsilon$, let $\mathbf{a}, \mathbf{b}$ be given in $L^2(\Omega)^2$ and let $\varphi_\varepsilon, \psi_\varepsilon$ be the solutions of (5.24) associated with $U_\varepsilon \mathbf{a}, U_\varepsilon \mathbf{b}$ respectively, i.e., ψ_ε satisfies

$$(5.29) \qquad \begin{cases} \Delta \psi_\varepsilon = 0 \quad \text{in} \quad \Omega_\varepsilon, \\[2mm] \dfrac{\partial \psi_\varepsilon}{\partial n} = (U_\varepsilon \mathbf{b})_j \cdot \mathbf{n} \quad \text{on} \quad \Gamma_j, \ j \in Q_\varepsilon, \\[2mm] \psi_\varepsilon = 0 \quad \text{on} \quad \Gamma_0 \end{cases}$$

and φ_ε is the solution of an analogous system. By the very definition of $\tilde{S}_\varepsilon$, we have

$$\int_\Omega \tilde{S}_\varepsilon \mathbf{a} \cdot \bar{\mathbf{b}} dx = \sum_{j \in Q_\varepsilon} (\varepsilon^{-2} \int_{\Gamma_j^\varepsilon} \varphi_\varepsilon \mathbf{n} d\gamma) \cdot (\int_{Y_j^\varepsilon} \bar{\mathbf{b}} dx).$$

On the other hand, multiplying (5.29) by φ_ε and integrating by parts, we get

$$\int_{\Omega_\varepsilon} \nabla \varphi_\varepsilon \cdot \nabla \bar{\psi}_\varepsilon dx = \sum_{j \in Q_\varepsilon} (\int_{\Gamma_j^\varepsilon} \varphi_\varepsilon \mathbf{n} d\gamma) \cdot (U_\varepsilon \bar{\mathbf{b}})_j.$$

Comparing both relations obtained above, we conclude that

$$(5.30) \qquad\qquad \int_\Omega \tilde{S}_\varepsilon \mathbf{a} \cdot \bar{\mathbf{b}} dx = |Y| \int_{\Omega_\varepsilon} \nabla \varphi_\varepsilon \cdot \nabla \bar{\psi}_\varepsilon dx.$$

To study the asymptotic behaviour as $\varepsilon \to 0$, we consider the operator S which is nothing but the weak limit of $\tilde{S}_\varepsilon$. In order to describe S, we need to pass to the limit on the right side of (5.30) which amounts to homogenization of problems of the type (5.29). For this purpose, we use the basic results of the two-scale convergence. More precisely, we have

Theorem 5.9. *Let $\{\psi_\varepsilon\}$ be the sequence of solutions of (5.29) and $\sim$ denote the extension by zero inside $(\Omega \backslash \Omega_\varepsilon)$. Then $\{\tilde{\psi}_\varepsilon\}$ and $\{\widetilde{\nabla \psi_\varepsilon}\}$ converge weakly in the sense of two-scales towards $\psi(x)\chi_{Y^*}(y)$ and $[\nabla \psi(x) + \nabla_y \psi_1(x,y)]\chi_{Y^*}(y)$ respectively, where the pair (ψ, ψ_1) is the unique solution in $H_0^1(\Omega) \times L^2(\Omega; \dot{H}_\#^1(Y^*))$ of the following two-scale homogenized system:*

$$
(5.31) \quad
\begin{cases}
-\operatorname{div}_x \displaystyle\int_{Y^*} [\nabla_x \psi(x) + \nabla_y \psi_1(x,y)]\, dy = |T| \operatorname{div}_x \mathbf{b} & in \quad \Omega, \\[2mm]
-\operatorname{div}_y [\nabla_x \psi(x) + \nabla_y \psi_1(x,y)] = 0 & in \quad \Omega \times Y^*, \\[2mm]
[\nabla_x \psi(x) + \nabla_y \psi_1(x,y) - \mathbf{b}] \cdot \mathbf{n} = 0 & on \quad \Omega \times \Gamma, \\[2mm]
\psi = 0 & on \quad \Gamma_0, \\[2mm]
\psi_1(x,y) \quad Y\text{-periodic in } y \text{ for a.e. } x \in \Omega.
\end{cases}
$$

Furthermore, $\widetilde{\nabla \psi_\varepsilon}$ converges strongly in the sense of two-scales to its limit, i.e.,

$$
(5.32) \quad \lim_{\varepsilon \to 0} \left\| \widetilde{\nabla \psi_\varepsilon}(x) - \left[\nabla \psi(x) + \nabla_y \psi_1(x, \tfrac{x}{\varepsilon})\right] \chi_{Y^*}(\tfrac{x}{\varepsilon}) \right\|_{0,\Omega} = 0
$$

provided $\nabla_y \psi_1 \in L^2(\Omega; \mathcal{C}_\#^0(Y^*))$.

Proof. *First Step:* Here we obtain estimates on the solution ψ_ε of problem (5.29). Indeed, we multiply (5.29) by $\tilde{\psi}_\varepsilon$ and integrate by parts; we get

$$
(5.33) \quad \int_{\Omega_\varepsilon} |\nabla \psi_\varepsilon|^2 dx = \sum_{j \in Q_\varepsilon} \left(\frac{1}{|Y_j^\varepsilon|} \int_{Y_j^\varepsilon} \mathbf{b}(x) dx \right) \cdot \left(\int_{\Gamma_j^\varepsilon} \tilde{\psi}_\varepsilon \mathbf{n} d\gamma \right).
$$

To estimate the second factor on the right side, we use the idea of Bramble-Hilbert Inequality (3.35). Since we are interested in the dependence of the constants on ε, we go through the details. In fact, we have

$$
\left| \int_{\Gamma_j^\varepsilon} \tilde{\psi}_\varepsilon \mathbf{n} d\gamma \right| = \left| \int_{\Gamma_j^\varepsilon} [\tilde{\psi}_\varepsilon - \mathcal{M}_{\varepsilon Y_j^*}(\tilde{\psi}_\varepsilon)] \mathbf{n} d\gamma \right| \leq c\varepsilon^{1/2} \left\| \psi_\varepsilon - \mathcal{M}_{\varepsilon Y_j^*}(\psi_\varepsilon) \right\|_{0,\Gamma_j^\varepsilon},
$$

where we have used Cauchy-Schwarz's Inequality. On the other hand, by Poincaré-Wirtinger Inequality (1.11) of Chapter I, we know that there is a constant $c > 0$ such that

$$
\|\psi - \mathcal{M}_{Y^*}(\psi)\|_{0,\Gamma} \leq c\|\nabla_y \psi\|_{0,Y^*}.
$$

Applying this inequality to $\psi(y) = \psi_\varepsilon(x)$ with $y = x/\varepsilon$, we deduce that

$$\|\psi_\varepsilon - \mathcal{M}_{\varepsilon Y_j^*}(\psi_\varepsilon)\|_{0,\Gamma_j^\varepsilon} \le c\varepsilon^{1/2}\|\nabla\psi_\varepsilon\|_{0,\varepsilon Y_j^*}.$$

Combining this with the above inequality, we conclude that

$$|\int_{\Gamma_j^\varepsilon} \bar{\psi}_\varepsilon \mathbf{n}d\gamma| \le c\varepsilon\|\nabla\psi_\varepsilon\|_{0,\varepsilon Y_j^*}.$$

Returning to (5.33), we obtain then

$$\|\nabla\psi_\varepsilon\|_{0,\Omega_\varepsilon}^2 \le c \sum_{j \in Q_\varepsilon} \|\mathbf{b}\|_{0,Y_j^\varepsilon}\|\nabla\psi_\varepsilon\|_{0,\varepsilon Y_j^*}$$

because of the obvious inequality

$$|\int_{Y_j^\varepsilon} \mathbf{b}(x)dx| \le c\varepsilon\|\mathbf{b}\|_{0,Y_j^\varepsilon}.$$

Thus we reach the desired apriori estimate, namely that

$$\|\nabla\psi_\varepsilon\|_{0,\Omega_\varepsilon} \le c\|\mathbf{b}\|_{0,\Omega}.$$

To deduce the estimate on ψ_ε itself, one can use Poincaré Inequality in Ω_ε (see CIORANESCU and SAINT JEAN PAULIN [1979]) to get

$$\|\psi_\varepsilon\|_{0,\Omega_\varepsilon} \le c\|\nabla\psi_\varepsilon\|_{0,\Omega_\varepsilon}$$

and so we conclude that

(5.34) $\|\psi_\varepsilon\|_{1,\Omega_\varepsilon} \le c\|\mathbf{b}\|_{0,\Omega}.$

Second Step: As a consequence of the first step and Corollary 5.7, we obtain a pair $(\psi,\psi_1) \in H_0^1(\Omega) \times L^2(\Omega; \dot{H}_\#^1(Y^*))$ such that, up to a subsequence, $\tilde{\psi}_\varepsilon$ and $\widetilde{\nabla\psi_\varepsilon}$ converge weakly in the sense of two-scales towards $\psi(x)\chi_{Y^*}(y)$ and $[\nabla\psi(x) + \nabla_y\psi_1(x,y)]\chi_{Y^*}(y)$ respectively. Our task now is to identify the limit (ψ,ψ_1) as the solution of (5.31). Indeed, we multiply (5.29) by $\bar{\theta}(x) + \varepsilon\bar{\theta}_1(x,x/\varepsilon)$ where $\theta \in \mathcal{D}(\Omega)$ and $\theta_1 \in \mathcal{D}(\Omega; C_\#^\infty(Y))$. Integrating by parts in Ω_ε, we get

(5.35)
$$\begin{cases} \int_{\Omega_\varepsilon} \nabla\psi_\varepsilon \cdot \left[\nabla\bar{\theta}(x) + \nabla_y\bar{\theta}_1(x,\frac{x}{\varepsilon}) + \varepsilon\nabla_x\bar{\theta}_1(x,\frac{x}{\varepsilon})\right]dx = \\ \qquad\qquad = \sum_{j \in Q_\varepsilon} (U_\varepsilon\mathbf{b})_j \cdot \int_{\Gamma_j^\varepsilon} \left[\bar{\theta}(x) + \varepsilon\bar{\theta}_1(x,\frac{x}{\varepsilon})\right]\mathbf{n}d\gamma \end{cases}$$

Integration by parts in the boundary integral on the right side yields

$$\int_{\Gamma_j^\epsilon} \left[\bar\theta(x) + \epsilon \bar\theta_1(x, \frac{x}{\epsilon}) \right] \mathbf{n} d\gamma = -\int_{T_j^\epsilon} \left[\nabla \bar\theta(x) + \nabla_y \bar\theta_1(x, \frac{x}{\epsilon}) + \epsilon \nabla_x \bar\theta_1(x, \frac{x}{\epsilon}) \right] dx.$$

Using then the operators U_ϵ and V_ϵ, we can rewrite (5.35) as follows:

$$\int_\Omega \widetilde{\nabla \psi}_\epsilon \cdot \left[\nabla \bar\theta(x) + \nabla_y \bar\theta_1(x, \frac{x}{\epsilon}) + \epsilon \nabla_x \bar\theta_1(x, \frac{x}{\epsilon}) \right] dx =$$

$$= -\int_\Omega \chi_T(\frac{x}{\epsilon}) V_\epsilon U_\epsilon \mathbf{b} \cdot \left[\nabla \bar\theta(x) + \nabla_y \bar\theta_1(x, \frac{x}{\epsilon}) + \epsilon \nabla_x \bar\theta_1(x, \frac{x}{\epsilon}) \right] dx.$$

By definition of two-scale convergence, the left side of the above relation converges to

$$\frac{1}{|Y|} \int_\Omega \int_{Y^*} [\nabla \psi(x) + \nabla_y \psi_1(x, y)] \cdot [\nabla \bar\theta(x) + \nabla_y \bar\theta_1(x, y)] \, dy dx.$$

On the other hand, since $V_\epsilon U_\epsilon \mathbf{b} \longrightarrow \mathbf{b}$ in $L^2(\Omega)$, the right side converges to

$$\frac{1}{|Y|} \int_\Omega \int_T \mathbf{b}(x) \cdot [\nabla \bar\theta(x) + \nabla_y \bar\theta_1(x, y)] \, dy dx.$$

Combining these two results and integrating by parts, we get

$$(5.36) \quad \begin{cases} \displaystyle\int_\Omega \int_{Y^*} [\nabla \psi(x) + \nabla_y \psi_1(x, y)] \cdot [\nabla \bar\theta(x) + \nabla_y \bar\theta_1(x, y)] \, dy dx = \\ \displaystyle\qquad = \int_\Omega \mathbf{b}(x) \cdot \int_\Gamma n(y) \bar\theta_1(x, y) d\gamma dx + |T| \int_\Omega \bar\theta(x) \operatorname{div} \mathbf{b}(x) dx \end{cases}$$

which is none other than the variational formulation of problem (5.31). This proves that (ψ, ψ_1) is the weak solution of (5.31). Since (5.31) admits a unique solution by Lax-Milgram Lemma, the whole sequences $\{\tilde\psi_\epsilon\}$ and $\{\widetilde{\nabla \psi}_\epsilon\}$ converge weakly towards their corresponding limits.

Third Step: We show here the strong convergence of $\widetilde{\nabla \psi}_\epsilon$ in the sense of two-scales. Applying Allaire's result (Theorem 5.2), we see that all we need is to prove the convergence of energy, i.e., we have to prove that

$$(5.37) \quad \lim_{\epsilon \to 0} \|\nabla \psi_\epsilon\|_{0,\Omega_\epsilon}^2 = \frac{1}{|Y|} \|\nabla \psi(x) + \nabla_y \psi_1(x, y)\|_{0,\Omega \times Y^*}^2.$$

To this end, we reconsider (5.33) and rewrite the right side as follows by doing a simple integration by parts:

$$(5.38) \qquad -\int_\Omega \chi_T(\tfrac{x}{\varepsilon}) V_\varepsilon U_\varepsilon \mathbf{b} \cdot \nabla \bar{\psi}_\varepsilon dx.$$

Here we have also denoted by ψ_ε an extension of ψ_ε inside the holes such that

$$(5.39) \qquad \psi_\varepsilon \in H^1(\Omega), \quad \|\psi_\varepsilon\|_{1,\Omega} \le c,$$

where c is a constant independent of ε; it depends only on the right side of (5.34).

Now, $\chi_T(\tfrac{x}{\varepsilon})\nabla\bar{\psi}_\varepsilon$ remains bounded in $L^2(\Omega)^2$. Since $\widetilde{\nabla\psi}_\varepsilon$ converges weakly in the sense of two-scales to $[\nabla\psi(x) + \nabla_y\psi_1(x,y)]\chi_{Y^*}(y)$, applying standard arguments, it is straightforward to check that $\chi_T(\tfrac{x}{\varepsilon})\nabla\bar{\psi}_\varepsilon$ also has a two-scale weak limit and it is none other than $\chi_T(y)[\nabla\bar{\psi}(x) + \nabla_y\bar{\psi}_1(x,y)]$. It is now easy to pass to the limit in (5.38) and obtain

$$\lim_{\varepsilon\to 0} \|\nabla\psi_\varepsilon\|_{0,\Omega_\varepsilon}^2 = -\frac{1}{|Y|}\int_\Omega\int_T \mathbf{b}(x) \cdot [\nabla\bar{\psi}(x) + \nabla_y\bar{\psi}_1(x,y)]\,dydx.$$

The choice $\theta = \psi$ and $\theta_1 = \psi_1$ in (5.36) shows that the right side of the above relation is nothing but that of (5.37). This completes the proof. ∎

Using the standard notations of the classical process of homogenization, problem (5.31) can be recast as follows: Let $\tilde{A} = (\tilde{A}_{k\ell})$ be the usual matrix whose entries are the homogenized coefficients obtained in the homogenization of the operator $(-\Delta)$ in a periodically perforated domain Ω_ε with Neumann condition on the boundary of the holes. We know, from CIORANESCU and SAINT JEAN PAULIN [1979], that $\tilde{A}_{k\ell}$ are given by

$$(5.40) \quad \tilde{A}_{k\ell} = \frac{1}{|Y|}\int_{Y^*}[\nabla v_k + \mathbf{e}_k]\cdot[\nabla v_\ell + \mathbf{e}_\ell]dy = \frac{1}{|Y|}\int_{Y^*}[\nabla v_k + \mathbf{e}_k]\cdot \mathbf{e}_\ell dy,$$

where $v_\ell = v_\ell(y)$, $\ell = 1,2$, are solutions of the following local boundary-value problems in Y^*:

$$(5.41) \qquad \begin{cases} -\text{div}_y[\nabla_y v_\ell + \mathbf{e}_\ell] = 0 & \text{in } Y^*, \\ [\nabla_y v_\ell + \mathbf{e}_\ell]\cdot\mathbf{n} = 0 & \text{on } \Gamma, \\ v_\ell \text{ } Y\text{-periodic.} \end{cases}$$

On using the functions v_ℓ, it is not difficult to check that (5.31) decouples in (x,y) variables. Indeed, ψ_1 is expressed in terms of ψ through the relation

$$(5.42) \qquad \psi_1(x,y) = \sum_{\ell=1}^2 v_\ell(y)\left[\frac{\partial\psi}{\partial x_\ell}(x) - b_\ell(x)\right]$$

and $\psi(x)$ is the unique solution of the following homogenized problem in Ω:

(5.43)
$$\begin{cases} \operatorname{div}(\tilde{A}\nabla\psi) = \operatorname{div}(\tilde{A} - I)\mathbf{b} & \text{in } \Omega, \\ \psi = 0 & \text{on } \Gamma_0. \end{cases}$$

◇ **Exercise 5.5.** Show that $\tilde{A}$ and $(I - \tilde{A})$ are both self-adjoint and positive-definite. ∎

We are now in a position to pass to the limit in (5.30). Let us denote by $\varphi(x)\chi_{Y^*}(y)$ and $[\nabla\varphi(x) + \nabla_y\varphi_1(x,y)]\chi_{Y^*}(y)$ the two-scale weak limits of $\tilde{\varphi}_\varepsilon$ and $\overline{\nabla\varphi}_\varepsilon$ respectively provided by Theorem 5.9. Then we have the convergence of the right side of (5.30). More precisely, we have

(5.44)
$$\begin{cases} \displaystyle\lim_{\varepsilon\to 0}\int_{\Omega_\varepsilon}\nabla\varphi_\varepsilon \cdot \nabla\bar{\psi}_\varepsilon dx = \\ \displaystyle= \frac{1}{|Y|}\int_\Omega\int_{Y^*}[\nabla\varphi(x) + \nabla_y\varphi_1(x,y)]\cdot[\nabla\bar{\psi}(x) + \nabla_y\bar{\psi}_1(x,y)]dy\,dx. \end{cases}$$

Indeed, the left side of (5.44) can be written as

$$\int_\Omega \widetilde{\nabla\varphi}_\varepsilon \cdot \widetilde{\nabla\bar{\psi}}_\varepsilon dx = \int_\Omega \widetilde{\nabla\varphi}_\varepsilon \cdot \left[\widetilde{\nabla\bar{\psi}}_\varepsilon - \left[\nabla\bar{\psi}(x) + \nabla_y\bar{\psi}_1(x,\tfrac{x}{\varepsilon})\right]\chi_{Y^*}(\tfrac{x}{\varepsilon})\right]dx +$$
$$+ \int_\Omega \widetilde{\nabla\varphi}_\varepsilon \cdot \left[\nabla\bar{\psi}(x) + \nabla_y\bar{\psi}_1(x,\tfrac{x}{\varepsilon})\right]\chi_{Y^*}(\tfrac{x}{\varepsilon})dx$$

and it is easy to pass to the limit on the right side using Theorem 5.9 and this yields (5.44). A simple integration by parts in the two-scale homogenized system (5.31) shows that the right side of (5.44) is indeed equal to

(5.45)
$$\frac{1}{|Y|}\int_\Omega \bar{\mathbf{b}}(x)\cdot\left[-|T|\nabla\varphi(x) + \int_\Gamma \varphi_1(x,y)\mathbf{n}d\gamma\right]dx.$$

Combining (5.30), (5.44) and (5.45), we see that the weak limit S of $\tilde{S}_\varepsilon$ acts in $L^2(\Omega)^2$ into itself and it is defined by the following rule:

(5.46)
$$Sa(x) = -|T|\nabla\varphi(x) + \int_\Gamma \varphi_1(x,y)\mathbf{n}d\gamma \quad \forall a \in L^2(\Omega)^2.$$

We can rewrite S by eliminating φ_1 altogether. For this purpose, let us remark that the pair (φ, φ_1) is the weak solution in $H_0^1(\Omega) \times L^2(\Omega; \dot{H}_\#^1(Y^*))$

of the two-scale homogenized system (5.31) associated with **a**. Also an expression analogous to (5.42) holds, i.e.,

$$(5.47) \qquad \varphi_1(x, y) = \sum_{\ell=1}^{2} v_\ell(y) \left[\frac{\partial \varphi}{\partial x_\ell}(x) - a_\ell(x) \right],$$

where φ is characterized as the solution of the homogenized problem:

$$(5.48) \qquad \begin{cases} \operatorname{div}(\tilde{A}\nabla\varphi) = \operatorname{div}(\tilde{A} - I)\mathbf{a} \quad \text{in} \quad \Omega, \\ \varphi = 0 \quad \text{on} \quad \Gamma_0. \end{cases}$$

Replacing φ_1 in (5.46) by the expression (5.47) and using (5.41), we obtain the following alternative expression for S

$$S\mathbf{a} = |Y|(\tilde{A} - I)(\nabla\varphi - \mathbf{a}) - |T|\mathbf{a} \quad \forall \mathbf{a} \in L^2(\Omega)^2,$$

where φ is the solution of (5.48). Following the terminology introduced in Paragraph 4.1 of Chapter II, we can call the operator S the *homogenized added mass matrix*.

◇ **Exercise 5.6.** Show that S is self-adjoint, positive-definite and non-compact. ∎

By the very definition of S, for every $\mathbf{a} \in L^2(\Omega)^2$ we have that

$$\tilde{S}_\varepsilon \mathbf{a} \rightharpoonup S\mathbf{a} \quad \text{in} \quad L^2(\Omega)^2 \text{ weak.}$$

The purpose of the following result is to show that the convergence is indeed strong:

Theorem 5.10. *For all* $\mathbf{a} \in L^2(\Omega)^2$, $\tilde{S}_\varepsilon \mathbf{a}$ *converges to* $S\mathbf{a}$ *in* $L^2(\Omega)^2$.

Proof. Let $\{\mathbf{a}_\varepsilon\}$ by a sequence in $L^2(\Omega)^2$ which converges weakly to a limit function $\mathbf{a} \in L^2(\Omega)^2$. We have to prove that, for all $\mathbf{b} \in L^2(\Omega)^2$

$$(5.49) \qquad \lim_{\varepsilon \to 0} \int_\Omega \tilde{S}_\varepsilon \mathbf{b} \cdot \mathbf{a}_\varepsilon \, dx = \int_\Omega S\mathbf{b} \cdot \mathbf{a} \, dx.$$

Let η_ε be the solution of (5.24) associated with $U_\varepsilon \mathbf{a}_\varepsilon$, i.e., η_ε satisfies

$$(5.50) \qquad \begin{cases} \Delta\eta_\varepsilon = 0 \quad \text{in} \quad \Omega_\varepsilon, \\ \dfrac{\partial\eta_\varepsilon}{\partial n} = (U_\varepsilon \mathbf{a}_\varepsilon)_j \cdot \mathbf{n} \quad \text{on} \quad \Gamma_j, \ j \in Q_\varepsilon, \\ \eta_\varepsilon = 0 \quad \text{on} \quad \Gamma_0. \end{cases}$$

Multiplying (5.50) by ψ_ε (unique solution of (5.29)) and integrating by parts as was done in the proof of Proposition 5.8, we conclude that

$$\int_\Omega \tilde{S}_\varepsilon \mathbf{b} \cdot \mathbf{a}_\varepsilon dx = |Y| \int_{\Omega_\varepsilon} \nabla \psi_\varepsilon \cdot \nabla \bar{\eta}_\varepsilon dx.$$

We will prove (5.49) by passing to the limit in the right hand side of this identity. But to do this, we need first to homogenize problem (5.50). Since the application of the two-scale method to problem (5.50) follows step by step the proof of Theorem 5.9, we shall not go into details and simply underline the main differences.

From the first step of Theorem 5.9, we deduce that there exists c such that

$$\|\eta_\varepsilon\|_{1,\Omega_\varepsilon} \leq c\|U_\varepsilon \mathbf{a}_\varepsilon\|_{0,\Omega}.$$

Thus $\|\eta_\varepsilon\|_{1,\Omega_\varepsilon}$ remains bounded as $\varepsilon \to 0$. Applying Corollary 5.7 we obtain a pair (η, η_1) in $H_0^1(\Omega) \times L^2(\Omega; \dot{H}_\#^1(Y^*))$ such that, except by a subsequence, $\tilde{\eta}_\varepsilon$ and $\widetilde{\nabla \eta_\varepsilon}$ converge weakly in the sense of two-scales to $\eta(x)\chi_{Y^*}(y)$ and $[\nabla \eta(x) + \nabla_y \eta_1(x,y)]\chi_{Y^*}(y)$ respectively.

Next, to identify the limit we follow the second step of Theorem 5.9. Everything works analogously, the only difference comes from the argument to be used in order to pass to the limit in the right hand side of (5.35). In the case of problem (5.50), this term becomes

$$\sum_{j \in Q_\varepsilon} (U_\varepsilon \mathbf{a}_\varepsilon)_j \cdot \int_{\Gamma_j^\varepsilon} \left[\bar{\theta}(x) + \varepsilon \bar{\theta}_1\left(x, \frac{x}{\varepsilon}\right) \right] \mathbf{n} d\gamma =$$

$$= - \int_\Omega \chi_T\left(\frac{x}{\varepsilon}\right) V_\varepsilon U_\varepsilon \mathbf{a}_\varepsilon \cdot \left[\nabla \bar{\theta}(x) + \nabla_y \bar{\theta}_1\left(x, \frac{x}{\varepsilon}\right) + \varepsilon \nabla_x \bar{\theta}_1\left(x, \frac{x}{\varepsilon}\right) \right] dx =$$

$$= - \int_\Omega \mathbf{a}_\varepsilon \cdot V_\varepsilon U_\varepsilon \chi_T\left(\frac{x}{\varepsilon}\right) \left[\nabla \bar{\theta}(x) + \nabla_y \bar{\theta}_1\left(x, \frac{x}{\varepsilon}\right) + \varepsilon \nabla_x \bar{\theta}_1\left(x, \frac{x}{\varepsilon}\right) \right] dx.$$

It is not difficult to check that $V_\varepsilon U_\varepsilon \chi_T(x/\varepsilon)$ converges strongly to $\int_Y \chi_T(y)dy = |T|$ in $L^2(\Omega)$. With this property, we can pass to the limit in this term and easily verified that (η, η_1) is the unique solution of the two-scale homogenized system (5.31) where $\mathbf{b}$ is changed by $\mathbf{a}$. Now, to pass to the limit in (5.49) we just need to use the strong convergence result (5.32); we get

$$\begin{cases} \lim_{\varepsilon \to 0} \int_\Omega \tilde{S}_\varepsilon \mathbf{b} \cdot \mathbf{a}_\varepsilon dx = \lim_{\varepsilon \to 0} |Y| \int_{\Omega_\varepsilon} \nabla \psi_\varepsilon \cdot \nabla \bar{\eta}_\varepsilon dx = \\ \qquad\qquad = -\frac{1}{|Y|} \int_\Omega \int_{Y^*} [\nabla \psi(x) + \nabla_y \psi_1(x,y)] \cdot [\nabla \bar{\eta}(x) + \nabla_y \bar{\eta}_1(x,y)] dy dx \end{cases}$$

which gives the desired result up to an integration by parts in (5.31). Theorem 5.10 is therefore proved. ∎

The convergence of the spectral family $\{E_\tau(\tilde{S}_\varepsilon)\}_\tau$ towards $\{E_\tau(S)\}_\tau$ is now an easy corollary of Theorem 4.7 of Chapter I. Indeed, we have

Theorem 5.11. *For all $\tau \in \mathbb{R}$ such that τ is not an eigenvalue of S, the following convergence holds: For each $\mathbf{a} \in L^2(\Omega)^2$,*

$$E_\tau(\tilde{S}_\varepsilon)\mathbf{a} \longrightarrow E_\tau(S)\mathbf{a} \quad in \quad L^2(\Omega)^2. \quad ∎$$

On the other hand, if we use Theorem 4.9 of Chapter I, we can always ensure the distributional convergence of the spectral families:

Theorem 5.12. *The spectral family of $\tilde{S}_\varepsilon$ converges to that of S in the following sense: For every $\mathbf{a}, \mathbf{b} \in L^2(\Omega)^2$, we have*

$$\left(E_\tau(\tilde{S}_\varepsilon)\mathbf{a}, \mathbf{b}\right) \rightharpoonup \left(E_\tau(S)\mathbf{a}, \mathbf{b}\right)$$

in the space of tempered distributions on $\mathbb{R}$. ∎

In general, we do not have much information on the nature of $\sigma(S)$. It can be proved (see ALLAIRE and CONCA [1995]) that the spectrum of S is purely essential (i.e., $\sigma(S) = \sigma_{ess}(S)$, but we have very little information about existence of eigenvalues for S. However, if the hole is sufficiently symmetrically placed inside the reference cell Y, we have the following result:

Theorem 5.13. *If the homogenized matrix $\tilde{A}$ is the scalar $\alpha_1 I$, then*

$$\sigma(S) = \sigma_p(S) = \{\nu, \mu\},$$

where ν and μ are defined by

$$\begin{cases} \nu = -|T| + (1 - \alpha_1)|Y|, \\ \mu = -|T| + \dfrac{1 - \alpha_1}{\alpha_1}|Y|. \end{cases}$$

The corresponding eigenspaces are given by

$$\begin{aligned} F_\nu &= \left\{\mathbf{b} \in L^2(\Omega)^2 \mid \operatorname{div}\mathbf{b} = 0 \quad in \quad \Omega\right\}, \\ F_\mu &= \left\{\mathbf{b} \in L^2(\Omega)^2 \mid \exists q \in H_0^1(\Omega), \ \mathbf{b} = \nabla q\right\}. \end{aligned}$$

Proof. It is well-known that F_ν and F_μ are closed subspaces of $L^2(\Omega)^2$ and they provide an orthogonal decomposition of $L^2(\Omega)^2$ (see GIRAULT and RAVIART [1986] or TEMAM [1977]). Thus, given $\mathbf{a} \in L^2(\Omega)^2$ arbitrary, there exists a unique pair $(q, \mathbf{b}) \in H_0^1(\Omega) \times F_\nu$ such that $\mathbf{a} = \nabla q + \mathbf{b}$. Moreover, q is the unique solution of

$$\begin{cases} \Delta q = \operatorname{div} \mathbf{a} & \text{in} \quad \Omega, \\ q = 0 & \text{on} \quad \Gamma_0. \end{cases}$$

Comparing this with the homogenized problem (5.48) it follows that

$$\varphi = [(\alpha_1 - 1)/\alpha_1]q.$$

Therefore, using the definition of S, we conclude that

$$S\mathbf{a} = \mu\nabla q + \nu\mathbf{b}$$

where ν, μ are defined as in the theorem. From this decomposition of S, all assertions made in Theorem 5.13 follow. ∎

Remark. From a physical point of view, a scalar homogenized matrix means that the resulting *effective material* associated with the homogenization procedure is *isotropic*. In this case, from Theorem 5.13, we see that there are two kind of macroscopic displacements of the tubes. Those in the eigenspace F_ν which represent incompressible displacements (corresponding to a zero fluid velocity), and those in F_μ which are gradient-type displacements yielding a non-zero fluid velocity. In the *anisotropic* case (i.e. the homogenized matrix is not scalar), a decomposition that simple is not easily available. ∎

We are now in a position to complete our discussion begun in Paragraph 3.4, Comment 2, concerning the limiting spectrum of problem (3.30) (i.e., Laplace model in a sequence of periodically perforated domains). Recall that the non-standard homogenization technique provided us with a homogenized spectrum for the sequence S_n of Green operators associated with (3.30) (see Theorem 3.15 and Theorem 3.17). Let us denote this limiting set by σ_{Bloch}. Recall also that σ_{Bloch} has a band structure, that it is purely essential, and that we were not entirely satisfied with the strong convergence of the operators S_n because it only led to a very weak convergence of the corresponding spectra. As we already discussed, it can happen that σ_{Bloch} does not capture all the possible cluster points of sequences in $\sigma(S_n)$, i.e, σ_{Bloch} may not recover all of the limit set σ_∞.

The main goal in this paragraph was to homogenize a suitable renormalized version of problem (3.30), namely, we homogenized problem (5.21)†.

---

† This is not strictly true because we slightly modified the external boundary condition in (5.21). Recall that we finally take Dirichlet condition on Γ instead of Neumann and this led us to consider problem (5.22). However, for the purposes of the present discussion, let us assume that the homogenization process reaches in both problems to qualitatively very similar conclusions.

This task was achieved by using the two-scale convergence method. The result of this limiting process was a new homogenized spectrum which we denoted by $\sigma(S)$ (see Theorems 5.10–5.13). In this alternative approach the same difficulty encountered above with σ_{Bloch} is still present, since the "poor" convergence of the sequence $\tilde{S}_\varepsilon$ given by Theorem 5.10 implies merely that $\sigma(S) \subset \sigma_\infty$, and the inclusion is generally strict.

Grouping together both approaches, there are several questions which arise naturally. Since there are no obvious reasons to think that $\sigma(S)$ coincides with σ_{Bloch}, it is therefore very interesting to understand any discrepancy between the "Bloch" spectrum and what we can call the (classical) homogenized spectrum $\sigma(S)$. Another interesting point consists in trying to find a unified procedure of homogenizing problems (3.30) or (5.21) so that both limiting spectra could be recovered. Furthermore, can one homogenize (3.30) in such a way that the whole set σ_∞ is recovered at the limit? These points and other related questions is the leitmotif of the article ALLAIRE and CONCA [1995].

5.4. Homogenization of the Helmholtz Model

This paragraph is devoted to the study of the asymptotic behaviour of the Helmholtz model as the number of tubes goes to infinity by the two-scale convergence method. Since this method has been presented in the case of the Laplace model with full details in Paragraph 5.3 and its application to the Helmholtz model is very similar, we shall simply concentrate our efforts to underline the main differences.

We consider thus the sequence of problems (4.1) and rescale them according to the prescription (5.20) where, of course, $\omega^2(n)$ is replaced by ω_ε^2 and $\mathbf{s}^{(n)}$ is replaced by $\mathbf{s}^\varepsilon$. As in the Laplace model, the Neumann boundary condition on Γ_0 is replaced by Dirichlet condition for simplification. We are thus led to the following spectral problem in Ω_ε: Find $\omega_\varepsilon^2 \in \mathbb{C}$ for which there exist a non-zero function ϕ_ε and $\mathbf{s}_\varepsilon \in \mathbb{C}^{2K_\varepsilon}$ such that

(5.51)
$$\begin{cases} \Delta\phi_\varepsilon + \omega_\varepsilon^2 \phi_\varepsilon = 0 & \text{in} \quad \Omega_\varepsilon, \\[2mm] \dfrac{\partial\phi_\varepsilon}{\partial n} = \varepsilon^{-1}\mathbf{s}_j^\varepsilon \cdot \mathbf{n} & \text{on} \quad \Gamma_j^\varepsilon, \ j \in Q_\varepsilon, \\[2mm] (k - m\omega_\varepsilon^2)\mathbf{s}_j^\varepsilon = \rho_0\varepsilon^{-1}\omega_\varepsilon^2 \displaystyle\int_{\Gamma_j^\varepsilon} \phi_\varepsilon \mathbf{n} d\gamma, \\[4mm] \phi_\varepsilon = 0 & \text{on} \quad \Gamma_0. \end{cases}$$

As usual, to solve (5.51), we introduce the Green's operator R_ε in $\mathcal{L}(L^2(\Omega_\varepsilon) \times \mathbb{C}^{2K_\varepsilon})$ defined by the following rule (see (2.20) of Chapter II):

(5.52)
$$R_\varepsilon(\phi, \mathbf{a}) = \left(\vartheta_\varepsilon, \rho_0 k^{-1}\left\{\varepsilon^{-2}\int_{\Gamma_j^\varepsilon} \vartheta_\varepsilon \mathbf{n} d\gamma\right\}_{j \in Q_\varepsilon}\right) + mk^{-1}(0, \mathbf{a})$$

for all $(\phi, \mathbf{a}) \in L^2(\Omega_\varepsilon) \times \mathbf{C}^{2K_\varepsilon}$, where ϑ_ε is the unique weak solution of the following boundary-value problem:

(5.53)
$$\begin{cases} \Delta \vartheta_\varepsilon + \phi = 0 & \text{in} \quad \Omega_\varepsilon, \\ \dfrac{\partial \vartheta_\varepsilon}{\partial n} = \mathbf{a}_j \cdot \mathbf{n} & \text{on} \quad \Gamma_j^\varepsilon, \ j \in Q_\varepsilon, \\ \vartheta_\varepsilon = 0 & \text{on} \quad \Gamma_0. \end{cases}$$

Of course, the characteristic values of R_ε coincide with the frequencies ω_ε^2 of (5.51). We are therefore interested in studying the behaviour of R_ε as $\varepsilon \to 0$. Since the domains of R_ε vary with ε, our first task, as in the Laplace model, is to modify the operator so that we have a fixed space. To do this, we shall use the linear operators U_ε and V_ε introduced in (5.26) and (5.27) respectively as well as the usual operator which is extension by zero inside the holes. More precisely, let $W_\varepsilon \in \mathcal{L}(L^2(\Omega_\varepsilon), L^2(\Omega))$ be defined by $W_\varepsilon \phi = \tilde{\phi}$ for every $\phi \in L^2(\Omega_\varepsilon)$.

We are now in a position to modify R_ε and define a new operator $\tilde{R}_\varepsilon \colon L^2(\Omega) \times L^2(\Omega)^2 \longrightarrow L^2(\Omega) \times L^2(\Omega)^2$ by

(5.54)
$$\tilde{R}_\varepsilon = \begin{bmatrix} W_\varepsilon & 0 \\ 0 & V_\varepsilon \end{bmatrix} R_\varepsilon \begin{bmatrix} i_\varepsilon & 0 \\ 0 & U_\varepsilon \end{bmatrix},$$

where i_ε is just the restriction to Ω_ε of functions on Ω. When no confusion arises, we suppress i_ε in the writing. As is usual in the case of Helmholtz model, we modify the inner product by inserting weights; more precisely, we introduce

(5.55)
$$\langle (\phi, \mathbf{a}), (\psi, \mathbf{b}) \rangle = \int_\Omega \phi \bar{\psi} dx + \rho_0^{-1} k \int_\Omega \mathbf{a}(x) \cdot \bar{\mathbf{b}}(x) dx.$$

$\diamond$ **Exercise 5.7.** Show that $\tilde{R}_\varepsilon$ is self-adjoint and positive in $L^2(\Omega) \times L^2(\Omega)^2$ equipped with the inner product (5.55). ∎

We are interested in describing the asymptotic behaviour of $\tilde{R}_\varepsilon$ as $\varepsilon \to 0$. To this end, let $(\phi, \mathbf{a})$ and $(\psi, \mathbf{b})$ be arbitrary elements of $L^2(\Omega) \times L^2(\Omega)^2$. One can easily compute the inner product:

(5.56)
$$\begin{cases} \langle \tilde{R}_\varepsilon(\phi, \mathbf{a}), (\psi, \mathbf{b}) \rangle = \\ |Y| \displaystyle\int_{\Omega_\varepsilon} \nabla \vartheta_\varepsilon \cdot \nabla \bar{\eta}_\varepsilon dx + \rho_0^{-1} m \int_\Omega V_\varepsilon U_\varepsilon \mathbf{a}(x) \cdot \bar{\mathbf{b}}(x) dx + (1 - |Y|) \int_\Omega \tilde{\vartheta}_\varepsilon \bar{\psi} dx, \end{cases}$$

where ϑ_ε and η_ε are the solutions of (5.53) associated with $(\phi, U_\varepsilon \mathbf{a})$ and $(\psi, U_\varepsilon \mathbf{b})$ respectively; for example, η_ε is the unique solution of

(5.57)
$$\begin{cases} \Delta \eta_\varepsilon + \psi = 0 & \text{in} \quad \Omega_\varepsilon, \\ \dfrac{\partial \eta_\varepsilon}{\partial n} = (U_\varepsilon \mathbf{b})_j \cdot \mathbf{n} & \text{on} \quad \Gamma_j^\varepsilon, \ j \in Q_\varepsilon, \\ \eta_\varepsilon = 0 & \text{on} \quad \Gamma_0. \end{cases}$$

We are thus led to pass to the limit in the right hand side of (5.56). As before, we shall use the two-scale convergence analysis for this purpose. First, let us describe the behaviour of η_ε as $\varepsilon \to 0$.

Theorem 5.14. *Let* $\{\eta_\varepsilon\}$ *be the sequence of solutions of* (5.57). *Then* $\{\widetilde{\eta}_\varepsilon\}$ *and* $\{\widetilde{\nabla \eta}_\varepsilon\}$ *converge weakly in the sense of two-scales towards* $\eta(x)\chi_{Y^*}(y)$ *and* $[\nabla \eta(x) + \nabla_y \eta_1(x,y)]\chi_{Y^*}(y)$ *respectively, where the pair* (η, η_1) *is the unique solution in* $H_0^1(\Omega) \times L^2(\Omega; \dot{H}_\#^1(Y^*))$ *of the following two-scale homogenized system:*

$$(5.58) \quad \begin{cases} -\operatorname{div}_x \displaystyle\int_{Y^*} [\nabla_x \eta(x) + \nabla_y \eta_1(x,y)]\,dy = |T|\operatorname{div}_x \mathbf{b} + |Y^*|\psi \quad in \quad \Omega, \\[2mm] -\operatorname{div}_y [\nabla_x \eta(x) + \nabla_y \eta_1(x,y)] = 0 \quad in \quad \Omega \times Y^*, \\[2mm] [\nabla_x \eta(x) + \nabla_y \eta_1(x,y) - \mathbf{b}(x)] \cdot \mathbf{n} = 0 \quad on \quad \Omega \times \Gamma, \\[2mm] \eta = 0 \quad on \quad \Gamma_0, \\[2mm] \eta_1(x,y) \quad Y\text{-periodic in } y \text{ for a.e. } x \in \Omega. \end{cases}$$

Furthermore, $\widetilde{\nabla \eta}_\varepsilon$ *converges strongly in the sense of two-scales to its limit, i.e.,*

$$(5.59) \qquad \lim_{\varepsilon \to 0} \left\| \widetilde{\nabla \eta}_\varepsilon(x) - \left[\nabla \eta(x) + \nabla_y \eta_1\left(x, \frac{x}{\varepsilon}\right)\right]\chi_{Y^*}\left(\frac{x}{\varepsilon}\right) \right\|_{0,\Omega} = 0$$

provided $\nabla_y \eta_1 \in L^2(\Omega; C_\#^0(Y^*))$.

Proof. *First Step:* Following the method in the first step of Theorem 5.9, we derive the estimate on η_ε: there exists a constant $c > 0$ independent of ε such that

$$(5.60) \qquad \|\eta_\varepsilon\|_{1,\Omega_\varepsilon} \le c\big[\|\psi\|_{0,\Omega_\varepsilon} + \|\mathbf{b}\|_{0,\Omega_\varepsilon}\big].$$

Second Step: Using now Corollary 5.7, we deduce the existence of (η, η_1) in $H_0^1(\Omega) \times L^2(\Omega; \dot{H}_\#^1(Y^*))$ such that for a subsequence $\widetilde{\eta}_\varepsilon$ and $\widetilde{\nabla \eta}_\varepsilon$ converge weakly in the sense of two-scales to $\eta(x)\chi_{Y^*}(y)$ and $[\nabla \eta(x) + \nabla_y \eta_1(x,y)]\chi_{Y^*}(y)$ respectively. Next, we multiply (5.57) by $\bar{\theta}(x) + \varepsilon\bar{\theta}_1(x, x/\varepsilon)$ where $\theta \in \mathcal{D}(\Omega)$ and $\theta_1 \in \mathcal{D}(\Omega; C_\#^\infty(Y))$. Integrating by parts in Ω_ε, we get

$$(5.61) \quad \begin{cases} \displaystyle\int_{\Omega_\varepsilon} \nabla \eta_\varepsilon \cdot \left[\nabla\bar{\theta}(x) + \nabla_y \bar{\theta}_1\left(x, \frac{x}{\varepsilon}\right) + \varepsilon\nabla_x \bar{\theta}_1\left(x, \frac{x}{\varepsilon}\right)\right] dx - \\[4mm] -\displaystyle\int_{\Omega_\varepsilon} \psi(x)\left[\bar{\theta}(x) + \varepsilon\bar{\theta}_1\left(x, \frac{x}{\varepsilon}\right)\right] = \sum_{j \in Q_\varepsilon}(U_\varepsilon \mathbf{b})_j \cdot \displaystyle\int_{\Gamma_j^\varepsilon}\left[\bar{\theta}(x) + \varepsilon\bar{\theta}_1\left(x, \frac{x}{\varepsilon}\right)\right] \mathbf{n}\,d\gamma. \end{cases}$$

By definition of two-scale convergence, the left side of the above relation converges to

$$\frac{1}{|Y|}\int_\Omega\int_{Y^*}[\nabla\eta(x)+\nabla_y\eta_1(x,y)]\cdot[\nabla\bar\theta(x)+\nabla_y\bar\theta_1(x,y)]\,dy\,dx-\frac{1}{|Y|}\int_\Omega\int_{Y^*}\psi(x)\bar\theta(x)\,dy\,dx$$

To pass to the limit in the right side of (5.61), we act exactly in the same way as in the Laplace model and obtain its limit to be equal to

$$\frac{1}{|Y|}\int_\Omega\int_T \mathbf{b}(x)\cdot[\nabla\bar\theta(x)+\nabla_y\bar\theta_1(x,y)]\,dy\,dx.$$

Combining these two results and integrating by parts, we get

$$(5.62)\quad\begin{cases}\displaystyle\int_\Omega\int_{Y^*}[\nabla\eta(x)+\nabla_y\eta_1(x,y)]\cdot[\nabla\bar\theta(x)+\nabla_y\bar\theta_1(x,y)]\,dy\,dx-\\[2mm]\displaystyle-|Y^*|\int_\Omega\psi(x)\bar\theta(x)=\int_\Omega\mathbf{b}(x)\cdot\int_\Gamma\bar\theta_1(x,y)\,d\gamma\,dx+|T|\int_\Omega\bar\theta(x)\operatorname{div}\mathbf{b}(x)\,dx\end{cases}$$

which is none other than the variational formulation of (5.58). Since (5.58) can be easily shown to admit a unique solution via Lax-Milgram Lemma, the entire sequences $\{\tilde\eta_\varepsilon\}$ and $\{\widetilde{\nabla\eta_\varepsilon}\}$ converge weakly towards their corresponding limits.

Third Step: The strong convergence of $\widetilde{\nabla\eta_\varepsilon}$ is proved exactly in the same way as in the Laplace model. More precisely, it is easily shown that we have

$$(5.63)\qquad\lim_{\varepsilon\to0}\|\nabla\eta_\varepsilon\|_{0,\Omega_\varepsilon}^2=\frac{1}{|Y|}\|\nabla\eta(x)+\nabla_y\eta_1(x,y)\|_{0,\Omega\times Y^*}^2$$

which, according to Theorem 5.2, is equivalent to the strong convergence. ∎

As was done in the Laplace model, one can recast problem (5.58) by separating the x,y variables. The first component η is characterized as the unique solution of the homogenized problem:

$$(5.64)\quad\begin{cases}\operatorname{div}(\tilde A\nabla\eta)=\operatorname{div}(\tilde A-I)\mathbf{b}-\dfrac{|Y^*|}{|Y|}\psi&\text{in }\Omega,\\[3mm]\eta=0&\text{on }\Gamma_0,\end{cases}$$

where $\tilde A$ is the same as the matrix (5.40) appearing in the Laplace model. On the other hand, the second component η_1 is expressed as

$$(5.65)\qquad\eta_1(x,y)=\sum_{\ell=1}^2 v_\ell(y)\Big[\frac{\partial\eta}{\partial x_\ell}(x)-b_\ell(x)\Big].$$

We are now in a position to pass to the limit in (5.56). Let us denote by $\vartheta(x)\chi_{Y^*}(y)$ and $[\nabla\vartheta(x) + \nabla_y\vartheta_1(x,y)]\chi_{Y^*}(y)$ the two-scale weak limits of $\tilde{\vartheta}_\varepsilon$ and $\widehat{\nabla\vartheta}_\varepsilon$ respectively provided by Theorem 5.14. Passing to the limit on the right side of (5.56), we have

$$\lim_{\varepsilon\to 0}\langle\tilde{R}_\varepsilon(\phi,\mathbf{a}),(\psi,\mathbf{b})\rangle = \int_\Omega\int_{Y^*}[\nabla\vartheta(x)+\nabla_y\vartheta_1(x,y)]\cdot[\nabla\bar{\eta}(x)+\nabla_y\bar{\eta}_1(x,y)]dydx$$

$$+\rho_0^{-1}m\int_\Omega\mathbf{a}(x)\cdot\bar{\mathbf{b}}(x)dx + \frac{|Y^*|}{|Y|}(1-|Y|)\int_\Omega\vartheta(x)\bar{\psi}(x)dx.$$

By taking $\theta = \vartheta$ and $\theta_1 = \vartheta_1$ in (5.62), the right hand side can be simplified and we get

$$\frac{|Y^*|}{|Y|}\int_\Omega\vartheta\bar{\psi}dx + \int_\Omega\bar{\mathbf{b}}\cdot\left[\int_\Gamma\vartheta_1\mathbf{n}d\gamma - |T|\nabla\vartheta + \rho_0^{-1}m\mathbf{a}\right]dx$$

Thus, we see that the weak limit R of $\tilde{R}_\varepsilon$ acts in $L^2(\Omega)\times L^2(\Omega)^2$ into itself and is given by

$$(5.66)\qquad R(\phi,\mathbf{a}) = (\frac{|Y^*|}{|Y|}\vartheta, \rho_0k^{-1}\left[\int_\Gamma\vartheta_1\mathbf{n}d\gamma - |T|\nabla\vartheta\right] + mk^{-1}\mathbf{a}).$$

We can rewrite R by eliminating ϑ_1 altogether. In fact, the expression analogous to (5.65) for ϑ_1 is the following:

$$(5.67)\qquad \vartheta_1(x,y) = \sum_{\ell=1}^{2}v_\ell(y)\left[\frac{\partial\vartheta}{\partial x_\ell}(x) - a_\ell(x)\right].$$

Substituting this into (5.66) and using the definition (5.40) of the homogenized matrix $\tilde{A}$, we get

$$(5.68)\ R(\phi,\mathbf{a}) = (\frac{|Y^*|}{|Y|}\vartheta, \rho_0k^{-1}|Y|\left[(\tilde{A}-I)\nabla\vartheta - (\tilde{A}-\frac{|Y^*|}{|Y|}I)\mathbf{a}\right] + mk^{-1}\mathbf{a}),$$

where ϑ is the solution of the homogenized problem:

$$(5.69)\qquad\begin{cases}\operatorname{div}(\tilde{A}\nabla\vartheta) = \operatorname{div}(\tilde{A}-I)\mathbf{a} - \dfrac{|Y^*|}{|Y|}\phi & \text{in}\quad\Omega,\\[2mm]\vartheta = 0 & \text{on}\quad\Gamma_0.\end{cases}$$

◇ **Exercise 5.8.** Show that the operator R is self-adjoint and positive-definite with respect to the inner product (5.55). Further, it is non-compact. ∎

Theorem 5.15. *For all* $(\phi, \mathbf{a}) \in L^2(\Omega) \times L^2(\Omega)^2$, $\tilde{R}_\varepsilon(\phi, \mathbf{a})$ *converges to* $R(\phi, \mathbf{a})$ *in* $L^2(\Omega) \times L^2(\Omega)^2$.

Proof. It follows exactly the same method as the proof of Theorem 5.10 which provides a similar result for the Laplace model. We will therefore omit it. ∎

The convergence of the spectral family $\{E_\tau(\tilde{R}_\varepsilon)\}_\tau$ towards $\{E_\tau(R)\}_\tau$ is now an easy corollary of Theorems 4.7 and 4.9 of Chapter I. We record them for the sake of completeness.

Theorem 5.16. *For* $\tau \in \mathbb{R}$ *that is not an eigenvalue of* R, *the following convergence holds: For each* $(\phi, \mathbf{a}) \in L^2(\Omega) \times L^2(\Omega)^2$,

$$E_\tau(\tilde{R}_\varepsilon)(\phi, \mathbf{a}) \longrightarrow E_\tau(R)(\phi, \mathbf{a}) \quad in \quad L^2(\Omega) \times L^2(\Omega)^2. \quad \blacksquare$$

Theorem 5.17. *The spectral family of* $\tilde{R}_\varepsilon$ *converges to that of* $\tilde{R}$ *in the following sense: For every* $(\phi, \mathbf{a}), (\psi, \mathbf{b}) \in L^2(\Omega) \times L^2(\Omega)^2$, *we have*

$$\langle E_\tau(\tilde{R}_\varepsilon)(\phi, \mathbf{b}), (\psi, \mathbf{b}) \rangle \rightarrow \langle E_\tau(R)(\phi, \mathbf{a}), (\psi, \mathbf{b}) \rangle$$

in the space of tempered distributions on $\mathbb{R}$. ∎

In general, we do not have much knowledge on the nature of $\sigma(R)$. However if the homogenized matrix $\tilde{A}$ is scalar, we can assert a few things.

◇ **Exercise 5.9.** Suppose that $\tilde{A} = \alpha_1 I$. Show then that R can be expressed as follows:

$$R(\phi, \mathbf{a}) = \Big(\frac{|Y^*|}{|Y|} \vartheta, \lambda_0 \mathbf{b} + \lambda_1 \nabla q + \lambda_2 \nabla L(\phi) \Big),$$

where $\mathbf{a}$ is decomposed uniquely as $\mathbf{a} = \nabla q + \mathbf{b}$ with $q \in H^1_0(\Omega)$ and div $\mathbf{b} = 0$, $L(\phi)$ is the unique solution of $\Delta L(\phi) = \phi$ in Ω, $\phi = 0$ on Γ_0 and ϑ is the solution of the homogenized problem:

$$\begin{cases} \alpha_1 \Delta \vartheta = (\alpha_1 - 1)\text{div } \mathbf{a} - \dfrac{|Y^*|}{|Y|} \phi & \text{in} \quad \Omega, \\ \vartheta = 0 & \text{on} \quad \Gamma_0. \end{cases}$$

The constants λ_0, λ_1 and λ_2 are given by

$$\begin{cases} \lambda_0 = \rho_0 k^{-1}|Y|\Big[1 - \alpha_1 - \dfrac{|T|}{|Y|} + \dfrac{m\rho_0^{-1}}{|Y|}\Big], \\ \lambda_1 = \rho_0 k^{-1}|Y|\Big[\dfrac{1-\alpha_1}{\alpha_1} - \dfrac{|T|}{|Y|} + \dfrac{m\rho_0^{-1}}{|Y|}\Big], \\ \lambda_2 = \rho_0 k^{-1}|Y|\dfrac{1-\alpha_1}{\alpha_1}\Big[1 - \dfrac{|T|}{|Y|}\Big]. \end{cases}$$

Deduce that λ_0 is an eigenvalue of infinite multiplicity for R. Does R have other eigenvalues? ∎

§6. Asymptotic Expansions in Fluid-Solid Structures

As mentioned in §3, in industrial problems, the number of tubes immersed in the fluid is usually very large; for example heat exchangers contain several thousands of them; in the core of nuclear reactors, there is again a large number of tubes through which combustion takes place. In such situations, the number of unknowns in the problems that we study is incredibly great; on the other hand, the discretization of such problems poses numerous difficulties in numerical computations because of geometry. Thus we feel that it is not easy to resolve directly such problems even with the help of present day computers. This is one of the principal reasons that led us in §3 and §4 to study theoretically the asymptotic behaviour of the solutions as the number of tubes increases to infinity. In these sections, we did this by following the *non-standard homogenization method* introduced in Paragraph 2.4. In the present section and the previous one, we adapt two techniques from *standard homogenization* to overcome the above difficulties. While §5 was devoted to applying *two-scale convergence analysis*, we follow, in this section, the classical two-scale asymptotic expansion method introduced in the book by BENSOUSSAN, LIONS and PAPANICOLAOU [1978].

6.1. Two-Scale Expansion for the Laplace Model

With the notations introduced in §5, we reconsider the renormalized version (5.22) of the Laplace model, i.e., find $(\lambda_\varepsilon, \phi_\varepsilon)$ such that ϕ_ε is not identically zero and

$$(6.1) \quad \begin{cases} \Delta\phi_\varepsilon = 0 \quad \text{in} \quad \Omega_\varepsilon, \\[2mm] \dfrac{\partial\phi_\varepsilon}{\partial n} = \varepsilon^{-2}\lambda_\varepsilon(\int_{\Gamma_j^\varepsilon} \phi_\varepsilon \mathbf{n} d\gamma) \cdot \mathbf{n} \quad \text{on} \quad \Gamma_j^\varepsilon, \; j \in Q_\varepsilon, \\[3mm] \phi_\varepsilon = 0 \quad \text{on} \quad \Gamma_0. \end{cases}$$

The analysis of this problem is, by Green's operator technique, reduced to the homogenization of (5.29), i.e.,

$$(6.2) \quad \begin{cases} \Delta\psi_\varepsilon = 0 \quad \text{in} \quad \Omega_\varepsilon, \\[2mm] \dfrac{\partial\psi_\varepsilon}{\partial n} = (U_\varepsilon \mathbf{b})_j \cdot \mathbf{n} \quad \text{on} \quad \Gamma_j^\varepsilon, \; j \in Q_\varepsilon, \\[2mm] \psi_\varepsilon = 0 \quad \text{on} \quad \Gamma_0, \end{cases}$$

where $\mathbf{b} \in L^2(\Omega)^2$ and $U_\varepsilon \mathbf{b} \in \mathbb{C}^{2K_\varepsilon}$ is defined by (5.26):

$$(U_\varepsilon \mathbf{b})_j = \frac{1}{|Y_j^\varepsilon|} \int_{Y_j^\varepsilon} \mathbf{b}(x)dx \quad \forall j \in Q_\varepsilon.$$

Recall that $(U_\varepsilon \mathbf{b})_j$ represents the speed of the jth tube and ψ_ε is the velocity potential of the fluid.

It is not clear how to suggest an expansion scheme for (6.2). However, we will be able to do so in a slight modification of (6.2). The modified problem can be written as follows:

(6.3)
$$\begin{cases} \Delta \psi_\varepsilon = 0 \quad \text{in} \quad \Omega_\varepsilon, \\ \dfrac{\partial \psi_\varepsilon}{\partial n} = \mathbf{b}(x_j^\varepsilon) \cdot \mathbf{n} \quad \text{on} \quad \Gamma_j^\varepsilon, \; j \in Q_\varepsilon, \\ \psi_\varepsilon = 0 \quad \text{on} \quad \Gamma_0, \end{cases}$$

where x_j^ε is the point in Y_j^ε which corresponds to the origin (see Figure 6.1). Note that the error committed in passing from (6.2) to (6.3) is of order $O(\varepsilon)$ in $H^1(\Omega_\varepsilon)$ norm. This is quite clear from the proof of the first step in Theorem 5.9 if we use the fact that $|U_\varepsilon \mathbf{b} - \mathbf{b}(x_j^\varepsilon)| \leq c\varepsilon$. In particular, this shows that the limits of problems (6.2) and (6.3) are same.

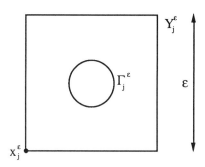

Figure 6.1.

In the sequel, we consider (6.3) and construct an asymptotic expansion for its solution as $\varepsilon \to 0$. As one can see, there are two variations involved in (6.3): the periodic variation of the domain and the non-periodic variation of $\mathbf{b}$. The idea of the two-scale expansion procedure, which takes care of periodic variations, consists in proposing the following ansatz for the solution ψ_ε of (6.3):

(6.4) $$\psi_\varepsilon(x) = \psi_0(x,y) + \varepsilon \psi_1(x,y) + \varepsilon^2 \psi_2(x,y) + \cdots |_{y=\frac{x}{\varepsilon}},$$

where $\psi_0, \psi_1, \ldots$ are defined for $x \in \Omega, y \in Y^*$ and they are assumed to be Y^*-periodic in y for all $x \in \Omega$. In this formal method, the solution ψ_ε and the functions $\psi_0, \psi_1, \ldots$ are all assumed to be as smooth as we please.

Since the variation $\mathbf{b}$ is not periodic, the idea is to use a suitable Taylor expansion for $\mathbf{b}$ and this is the new element of modification we add to the classical procedure. To this end, we suppose that $\mathbf{b}$ is smooth and expand

$$(6.5) \qquad \mathbf{b}_\ell(x_j^\varepsilon) = b_\ell(x) - \varepsilon \nabla_x b_\ell(x) \cdot y + O(\varepsilon^2) \quad \forall \ell = 1, 2, \ x \in \varepsilon Y_j^*,$$

where $y = \varepsilon^{-1}(x - x_j^\varepsilon)$.

In this context, x is called the macroscopic variable and y is the microscopic variable. Following the general principle of two-scale expansion procedure, we treat them as two independent variables. The operator $\partial/\partial x_k$ applied to a function $u(x, x/\varepsilon)$ becomes then $(\partial/\partial x_k + \varepsilon^{-1}\partial/\partial y_k)$ and the Laplacian operator becomes $(\varepsilon^{-2}\Delta_y + 2\varepsilon^{-1}\Delta_{xy} + \Delta_x)$ where

$$(6.6) \qquad \Delta_{xy} = \sum_{k=1}^{2} \frac{\partial^2}{\partial x_k \partial y_k}.$$

Substituting (6.4) into the first equation in (6.2) and identifying the different powers of ε, we get the following equations in $\Omega \times Y^*$:

$$(6.7) \qquad \begin{cases} \Delta_y \psi_0 = 0, \\ \Delta_y \psi_1 + 2\Delta_{xy}\psi_0 = 0, \\ \Delta_y \psi_2 + 2\Delta_{xy}\psi_1 + \Delta_x\psi_0 = 0, \ldots \end{cases}$$

On the other hand, using (6.4) and (6.6) into the boundary condition on Γ_j^ε, we get the following equations on $\Omega \times \Gamma$:

$$(6.8) \qquad \begin{cases} \nabla_y \psi_0 \cdot \mathbf{n} = 0, \\ \nabla_y \psi_1 \cdot \mathbf{n} = -\nabla_x \psi_0 \cdot \mathbf{n} + \mathbf{b}(x) \cdot \mathbf{n}, \\ \nabla_y \psi_2 \cdot \mathbf{n} = -\nabla_x \psi_1 \cdot \mathbf{n} - \nabla_x b_\ell(x) \cdot y_j n_\ell, \ldots \end{cases}$$

To resolve the various equations obtained above, we consider the following problem in the basic reference cell Y^*: Given $F \in L_\#^2(Y^*)$ and $G \in L_\#^2(\Gamma)$, find $\vartheta = \vartheta(y) \in \dot{H}_\#^1(Y^*)$ such that

$$(6.9) \qquad \begin{cases} \Delta_y \vartheta = F \quad \text{in} \quad Y^*, \\ \nabla_y \vartheta \cdot \mathbf{n} = G \quad \text{on} \quad \Gamma, \\ \vartheta \quad Y\text{-periodic.} \end{cases}$$

It is classically known that the existence and uniqueness of a solution for (6.9) modulo constants is guaranteed iff the following condition (Fredholm's Alternative) holds:

$$(6.10) \qquad \int_{Y^*} F \, dy - \int_{\Gamma} G \, d\gamma = 0.$$

It follows from this that ψ_0 is independent of y:

(6.11)
$$\psi_0(x, y) = \psi(x),$$

where ψ is to be determined. Using this information in (6.7), (6.8), we see, for each $x \in \Omega$, that ψ_1 satisfies

(6.12)
$$\begin{cases} \Delta_y \psi_1 = 0 & \text{in} \quad Y^*, \\ \nabla_y \psi_1 \cdot \mathbf{n} = -\nabla_x \psi \cdot \mathbf{n} + \mathbf{b}(x) \cdot \mathbf{n} & \text{on} \quad \Gamma \end{cases}$$

which is a problem of type (6.9). Therefore it has a unique solution (up to an additive function of x) iff

$$\nabla_x \psi(x) \cdot \int_\Gamma \mathbf{n} d\gamma = \mathbf{b}(x) \cdot \int_\Gamma \mathbf{n} d\gamma.$$

This obviously holds and each side is in fact equal to zero. Therefore (6.12) is well-posed and its solution can be expressed as

(6.13)
$$\psi_1(x, y) = \sum_{\ell=1}^{2} v_\ell(y) \left[\frac{\partial \psi}{\partial x_\ell}(x) - b_\ell(x) \right],$$

where v_ℓ, $\ell = 1, 2$ are the solutions of the local boundary value problems (5.41).

Now, let us write down the system satisfied by ψ_2 using (6.11):

(6.14)
$$\begin{cases} \Delta_y \psi_2 = -\Delta_x \psi - 2\Delta_{xy} \psi_1 & \text{in} \quad Y^*, \\ \nabla_y \psi_2 \cdot \mathbf{n} = -\nabla_x \psi_1 \cdot \mathbf{n} - \nabla_x b_\ell(x) \cdot y_j n_\ell & \text{on} \quad \Gamma. \end{cases}$$

Since ψ_1 is Y-periodic in y, it follows from Green's formula that

$$\int_{Y^*} \Delta_{xy} \psi_1 \, dy - \int_\Gamma \nabla_x \psi_1 \cdot \mathbf{n} d\gamma = 0$$

and so the Fredholm's Alternative for (6.14) is reduced to

(6.15)
$$-|Y^*| \Delta_x \psi(x) - \int_{Y^*} \Delta_{xy} \psi_1 \, dy + \int_\Gamma \nabla_x b_\ell(x) \cdot y_j n_\ell d\gamma = 0.$$

First, we calculate the third term on the left side of the above relation. A simple integration by parts in Y^* shows that

(6.16)
$$\nabla_x b_\ell(x) \cdot \int_\Gamma y n_\ell d\gamma = -|T| \nabla_x b_\ell(x) \cdot \mathbf{e}_\ell = -|T| \text{div} \, \mathbf{b}(x).$$

The second term is computed using the expression (6.13) for ψ_1. Indeed, we have

$$-\int_{Y^*} \Delta_{xy}\psi_1 \, dy = -\text{div}_x \int_{Y^*} \nabla_y v_\ell \Big[\frac{\partial \psi}{\partial x_\ell} - b_\ell\Big] dy =$$

$$= -\text{div}_x \int_{Y^*} (\nabla_y v_\ell + \mathbf{e}_\ell - \mathbf{e}_\ell) \Big[\frac{\partial \psi}{\partial x_\ell} - b_\ell\Big] dy =$$

$$= -\text{div}_x (|Y|\tilde{A} - |Y^*|I)[\nabla_x \psi - \mathbf{b}],$$

where we have used the definition (5.40) of the homogenized coefficients. The first term in (6.15) is written as $-\text{div}_x(|Y^*|\nabla_x \psi)$. Putting all these together in (6.15), we see that ψ satisfies

(6.17)
$$\begin{cases} \text{div}\,(\tilde{A}\nabla\psi) = \text{div}\,(\tilde{A} - I)\mathbf{b} \quad \text{in} \quad \Omega, \\ \psi = 0 \quad \text{on} \quad \Gamma_0 \end{cases}$$

which is nothing but the homogenized problem (5.43). Moreover (6.13) coincides with (5.42). Thus this method gives exactly the same result as the two-scale convergence analysis done in §5.

Remark. The manipulations heading to (6.17) using the expansion (6.6) are not classical. It is remarkable that we could arrive at the homogenized system (6.17) directly by using the two-scale convergence without encountering any of the modifications introduced in the two-scale expansion. ∎

After providing the two-scale expansion for problem (6.2), we turn our attention to the corresponding eigenvalue problem (6.1). We cannot hope to write down an expansion for the eigenvalues as in classical cases treated by KESAVAN [1979] and VANNINATHAN [1981]. This is essentially because the nature of the spectrum in (6.1) changes as $\varepsilon \to 0$. To overcome this difficulty, what one can do is to come back to the original wave equation associated with (6.1) (see Paragraph 2.1 of Chapter I) and suggest a two-scale expansion for the solution. This is indeed possible and it has been carried out using just the same kind of expansions presented in this paragraph. The full details can be found in CONCA, PLANCHARD, THOMAS and VANNINATHAN [1994].

6.2. Two-Scale Expansion for the Helmholtz Model

Application of the two-scale expansion to the Helmholtz model is very similar to that of the Laplace model and for this reason, we will be very brief. We will simply underline the main differences. Let us recall from §5, that the analysis of the spectrum in the Helmholtz model was reduced to the homogenization of (5.57). Introducing the same modification as for the

Laplace model, we are led to consider the following variant of (5.57): Given $(\psi, \mathbf{b}) \in L^2(\Omega) \times L^2(\Omega)^2$, find η_ε satisfying

(6.18)
$$\begin{cases} \Delta \eta_\varepsilon + \psi = 0 & \text{in} \quad \Omega_\varepsilon, \\ \dfrac{\partial \psi_\varepsilon}{\partial n} = \mathbf{b}(x_j^\varepsilon) \cdot \mathbf{n} & \text{on} \quad \Gamma_j^\varepsilon, \ j \in Q_\varepsilon, \\ \eta_\varepsilon = 0 & \text{on} \quad \Gamma_0, \end{cases}$$

where x_j^ε is the same point as in (6.3). Following the general procedure outlined in Paragraph 6.1, we consider the expansions

(6.19)
$$\begin{cases} \eta_\varepsilon(x) = \eta_0(x, y) + \varepsilon \eta_1(x, y) + \varepsilon^2 \eta_2(x, y) + \cdots |_{y=\frac{x}{\varepsilon}}, \\ \mathbf{b}_\ell(x_j^\varepsilon) = b_\ell(x) - \varepsilon \nabla_x b_\ell(x) \cdot y + O(\varepsilon^2), \quad x \in \varepsilon Y_j^*, \end{cases}$$

where $y = \varepsilon^{-1}(x - x_j^\varepsilon)$. Substitution of these expansions into (6.18) and the identification of various powers of ε will yield the following sets of equations which correspond to (6.7) and (6.8) respectively:

(6.20)
$$\begin{cases} \Delta_y \eta_0 = 0, \\ \Delta_y \eta_1 + 2\Delta_{xy} \eta_0 = 0, \\ \Delta_y \eta_2 + 2\Delta_{xy} \eta_1 + \Delta_x \eta_0 + \psi = 0, \dots \end{cases}$$

and

(6.21)
$$\begin{cases} \nabla_y \eta_0 \cdot \mathbf{n} = 0, \\ \nabla_y \eta_1 \cdot \mathbf{n} = -\nabla_x \eta_0 \cdot \mathbf{n} + \mathbf{b}(x) \cdot \mathbf{n}, \\ \nabla_y \eta_2 \cdot \mathbf{n} = -\nabla_x \eta_1 \cdot \mathbf{n} - \nabla_x b_\ell(x) \cdot y_j n_\ell, \dots \end{cases}$$

Comparing (6.20) with (6.7), we note that the only difference is the presence of the additional term involving ψ in the third equation. Therefore, we draw the following conclusions which are analogues of (6.11), (6.13) and (6.17):

(6.22)
$$\begin{cases} \eta_0(x, y) = \eta(x), \\ \eta_1(x, y) = \displaystyle\sum_{\ell=1}^{2} v_\ell(y) \left[\dfrac{\partial \eta}{\partial x_\ell}(x) - b_\ell(x) \right] \end{cases}$$

and η satisfies the homogenized equation

(6.23)
$$\begin{cases} \operatorname{div}(\tilde{A} \nabla \eta) = \operatorname{div}(\tilde{A} - I)\mathbf{b} - \dfrac{|Y^*|}{|Y|} \psi & \text{in} \quad \Omega, \\ \eta = 0 & \text{on} \quad \Gamma_0. \end{cases}$$

This problem is the same as (5.64) which was obtained by the two-scale convergence analysis in Paragraph 5.4.

Here we have applied the two-scale expansion procedure to problem (6.18) associated with the Helmholtz model. The same procedure can be extended to the corresponding wave equation and the reader is referred to CONCA, PLANCHARD, THOMAS and VANNINATHAN [1994] Chapter III.

Open Questions

The contents of this book are outcome of a long collaboration of the authors. During this period, several questions arose out of our discussions. We present below a list of those which seem to us the most interesting and relevant and for which we do not have complete answers. They have been arranged indicating the corresponding chapter and section with which they are concerned.

Chapter II

§3 and §4

Question 1. (Compressible Stokes model) Existence and bounds for the compressible Stokes model (1.26) are to be established (see Exercise 2.2).

Question 2. (Navier-Stokes models) Can one extend Stokes' models (incompressible and compressible) to the case when the fluid obeys Navier-Stokes equations? If so, how can existence of the spectrum be proved and bounds be obtained? What is the behaviour if the viscosity $\nu \to 0$?

Question 3. (Moving tubes) All the models treated in this book supposed that the domains do not vary with time. How can movement of tubes be incorporated into the Laplace, Helmholtz and Stokes' models?

Chapter III

§1

Question 4. Is there any direct proof of Theorem 1.5 without using condition (P) at least in the case of periodically perforated domains?

§2

Question 5. (Measurable selection of Bloch waves) Can one simplify the work of WILCOX [1978] to get just a measurable selection of Bloch waves for the classical example studied in §2?

§3

Question 6. (Refined properties of $\sigma(S)$) Can one find suitable geometrical conditions on the tubes which ensure that
- the limit operator S does not have eigenvalues?
- the bands of its spectrum are disjoint/overlapping?
- the spectrum of S is absolutely continuous?

§4

Question 7. (Measurable selection of Bloch waves) Can one extend the method employed in the Laplace model to obtain a measurable selection of Bloch waves in the Helmholtz model?

Question 8. Same as Question 6 above but for the Helmholtz model.

§4 and §5

Question 9. (Relation between standard and non-standard homogenization) Is there any relation between the limit operator obtained for the Helmholtz model via the non-standard homogenization and the homogenized operator provided by the two-scale convergence technique?

§5

Question 10. (Refined properties of $\sigma(S)$ and $\sigma(R)$) What can one say about the nature of the spectra of the homogenized operators S and R when $\tilde{A}$ is not a scalar matrix? In particular, what can one say about existence of eigenvalues?

§2, §3 and §4

Question 11. (Non-standard homogenization of the Stokes' models) How can one generalize the non-standard homogenization procedure to cover Stokes' models?

§5 and §6

Question 12. (Standard homogenization of the Stokes' models) How can one generalize the standard homogenization method to cover Stokes' models?

References

ADAMS R.A. [1975] *Sobolev Spaces*, Academic Press, New York.

AGUIRRE F. and CONCA C. [1988] Eigenfrequencies of a tube bundle immersed in a fluid, *Appl. Math. Optim.* **18**, 1–38.

ALLAIRE G. [1991] Homogénéisation et convergence à deux échelles. Application à un problème de convection diffusion, *C.R. Acad. Sci. Paris Série I* t.**312**, 581–586.

ALLAIRE G. [1992a] Homogenization and two-scale convergence, *SIAM J. Math. Anal.* **23**, 1482–1518.

ALLAIRE G. [1992b] Homogenization of the unsteady Stokes equations in porous media, in *Progress in Partial Differential Equations*: *Calculus of Variations, Applications*, C. BANDLE et al. eds., Pitman Research Notes in Mathematics Series **267**, 109–123, Longman Scientific & Technical, Harlow.

ALLAIRE G. and CONCA C. [1995] Bloch wave homogenization for a spectral problem in fluid-solid structures, *Arch. Rational Mech. Anal.* (in press).

BAI D. [1982] Flow-induced vibration of multi-span tube bundles of large condensers: experimental studies on full scale models in steam cross flow, in *Proceedings UKAEA/BNES Third Keswick International Conference Vibrations in Nuclear Plants*, Paper 2.3, Keswick.

BARBU V. [1976] *Nonlinear Semigroups and Differential Equations in Banach Spaces*, Editura Academiei & Noordhoff International Publishing, Bucharest-Leyden.

BENSOUSSAN A., LIONS J.L. and PAPANICOLAOU G. [1978] *Asymptotic Analysis in Periodic Structures*, North-Holland, Amsterdam.

BERGER M.S. [1977] *Nonlinearity and Functional Analysis*, Academic Press, New York.

BERGMANN S. and SCHIFFER M. [1953] *Kernel Functions and Elliptic Differential Equations in Mathematical Physics*, Academic Press, New York.

BLEVINS R.D. [1977] *Flow-Induced Vibrations*, Van Nostrand Reinhold, New York.

BLOCH F. [1928] Über die Quantenmechanik der Electronen in Kristallgittern, *Z. Phys.* **52**, 555–600.

BOURBAKI N. [1965] *Eléments de Mathématiques. Vol. XII: Integration*, Hermann, Paris.

BREZIS H. [1983] *Analyse Fonctionnelle. Théorie et Applications*, Masson, Paris.

BRILLOUIN L. [1953] *Propagation of Waves in Periodic Structures*, Dover, New York.

CASEAU P. [1972] Electricité de France (private communication).

CHEN S.S. [1975] Vibrations of nuclear fuel-bundles, *Nuclear Engineering Design* **35**, 399–422.

CHEN S.S. [1987] *Flow-Induced Vibration of Circular Cylindrical Structures*, Hemisphere Publishing Corp., Washington.

CIARLET P.G. [1978] *The Finite Element Method for Elliptic Problems*, North-Holland, Amsterdam.

CIORANESCU D. and SAINT JEAN PAULIN J. [1979] Homogenization in open sets with holes, *J. Math. Anal. Appl.* **71**, 590–607.

CONCA C. and DURÁN M. [1995] A numerical study of a spectral problem in solid-fluid type structures, *Numer. Methods Partial Differential Equations* (in press).

CONCA C., DURÁN M. and PLANCHARD J. [1992a] A quadratic eigenvalue problem involving Stokes equations, *Comput. Methods Appl. Mech. Engrg.* **100**, 295–313.

CONCA C., DURÁN M. and PLANCHARD J. [1992b] A bound for the number of nonreal solutions of a quadratic eigenvalue problem, *Adv. Math. Sci. Appl.* **1**, 229–249.

CONCA C., PLANCHARD J. and VANNINATHAN M. [1989a] Un problème de fréquences propres en couplage fluide-structure, *Comput. Methods Appl. Mech. Engrg.* **75**, 27–37.

CONCA C., PLANCHARD J. and VANNINATHAN M. [1989b] Existence and location of eigenvalues for fluid-solid structures, *Comput. Methods Appl. Mech. Engrg.* **77**, 253–291.

CONCA C., PLANCHARD J. and VANNINATHAN M. [1990] Limits of the resonance spectrum of tube arrays immersed in a fluid, *J. Fluids Structures* **4**, 541–558.

CONCA C., PLANCHARD J. and VANNINATHAN M. [1993] Limiting behaviour of a spectral problem in fluid-solid structures, *Asymptotic Analysis* **6**, 365–389.

CONCA C., PLANCHARD J., THOMAS B. and VANNINATHAN M. [1994] *Problèmes Mathématiques en Couplage Fluide-Structure. Applications aux Faisceaux Tubulaires*, Collection de la Direction des Etudes et Recherches d'Electricité de France **85**, Editions Eyrolles, Paris.

CONCA C. and PUSCHMANN H. [1992] A note on real eigenvalues in Hermitian quadratic problems of restricted rank, *Appl. Math Lett.* **6**(6), 9–13.

CONCA C. and VANNINATHAN M. [1988] A spectral problem arising in fluid-solid structures, *Comput. Methods Appl. Mech. Engrg.* **69**, 215–242.

CONNORS H.J. [1970] Fluid elastic vibrations of tube arrays excited by cross-flow, in *Flow Induced Vibrations in Heat Exchangers*, R.D. REIFF ed., A.S.M.E., New York.

COURANT R. and HILBERT D. [1953] *Methods of Mathematical Physics. Vol. I*, Interscience Publishers Inc., New York.

CRACKNELL A.P. and WONG K.C. [1973] *The Fermi Surface*, Clarendon Press, Oxford.

DAUTRAY R. and LIONS J.L. [1988-90] *Mathematical Analysis and Numerical Methods for Science and Technology. I. Physical Modelling and Potential Theory, II. Functional and Variational Methods, III. Spectral Theory and Applications, IV Integral Equations and Numerical Methods, V. Evolution Problems I, VI. Evolution Problems II*, Springer-Verlag, Berlin.

DENY J. and LIONS J.L. [1953-54] Les espaces du type de Beppo-Levi, *Ann. Inst. Fourier* **5**, 305–370.

DIESTEL J. and UHL J.J.JR. [1977] *Vector Measures*, Mathematical Surveys **15**, American Mathematical Society, Providence.

DUNFORD N. and SCHWARTZ J.T. [1964] *Linear Operators. Part II: Spectral Theory*, Wiley-Interscience, New York.

DURÁN M. [1991] *Estudio de las Frecuencias y Modos Propios de Vibración de una Estructura Metálica Inmersa en un Fluido Viscoso Incompresible*, Thesis, University of Chile, Santiago.

EASTHAM M.S.P. [1973] *The Spectral Theory of Periodic Differential Equations*, Scottish Academic Press, Edinburgh.

EKELAND I. and TEMAM R. [1973] *Analyse Convexe et Problèmes Variationnels*, Dunod and Gauthier-Villars, Paris.

FLEURY F. and SÁNCHEZ-PALENCIA E. [1986] Asymptotics and spectral properties of the acoustic vibration of a body perforated by narrow channels, *Bull. Sci. Math.* **110**, 149–176.

GELFAND I.M. [1950] Expansion in series of eigenfunctions of an equation with periodic coefficients, *Dokl. Akad. Nauk SSSR* **73**, 1117–1120.

GIBERT R.J. [1988] *Vibrations des Structures – Interactions avec les Fluides – Sources d'Excitations Aléatoires* (Ecole d'Été d'Analyse Numérique, CEA-EDF-INRIA), Collection de la Direction des Etudes et Recherches d'Electricité de France **69**, Editions Eyrolles, Paris.

GIRAULT V. and RAVIART P.A. [1986] *Finite Element Methods for Navier-Stokes Equations. Theory and Algorithms*, SCM **5**, Springer-Verlag, Berlin.

GRÉGOIRE J.P. [1973] Propagation des ondes acoustiques dans un fluide, Electricité de France, Direction des Etudes et Recherches, Internal Report HI 1051-02.

GRÉGOIRE J.P., NÉDELEC J.C. and PLANCHARD J. [1975] A method for computing the eigenfrequencies of an acoustic resonator, in *Applications of Methods of Functional Analysis to Problems of Mechanics*, Lecture Notes in Mathematics **503**, 343–353.

GRÉGOIRE J.P., NÉDELEC J.C. and PLANCHARD J. [1976] A method of finding the eigenvalues and eigenfunctions of self-adjoint operators, *Comput. Methods Appl. Mech. Engrg.* **8**, 201–214.

GRISVARD P. [1985] *Elliptic Problems in Nonsmooth Domains*, Pitman, London.

GRISVARD P. [1989] Contrôlabilité exacte des solutions de l'équation des ondes en présence de singularités, *J. Math. Pures Appl.* **68**, 215–259.

HUET D. [1976] *Décomposition Spectrale et Opérateurs*, Presses Universitaires de France, Paris.

HUTSON V. and PYM J.S. [1980] *Applications of Functional Analysis and Operator Theory*, Academic Press, London.

IBNOU-ZAHIR M. [1984] *Problèmes de Valeurs Propres avec des Conditions aux Limites Non Locales*, Thesis, University Pierre and Marie Curie, Paris.

IBNOU-ZAHIR M. and PLANCHARD J. [1983a] Natural frequencies of a tube bundle in an incompressible fluid, *Comput. Methods Appl. Mech. Engrg.* **41**, 47–68.

IBNOU-ZAHIR M. and PLANCHARD J. [1983b] Fréquences et modes propres d'un faisceau de 100 tubes rigides supportés élastiquement et placés dans un fluide incompressible, Electricité de France, Direction des Etudes et Recherches, Internal Report HI 4419-07.

KATO T. [1966] *Perturbation Theory for Linear Operators*, Springer-Verlag, Berlin.

KESAVAN S. [1979] Homogenization of elliptic eigenvalue problems. Part I, *Appl. Math. Optim.* **5**, 153–167; Part II, *Appl. Math. Optim.* **5**, 197–216.

KRANTZ S.G. [1982] *Function Theory of Several Complex Variables*, Wiley-Interscience, New York.

KRANTZ S.G. and PARKS H.R. [1992] *A Primer of Real Analytic Functions*, Birkhäuser Verlag, Basel.

KUCHMENT P. [1993] *Floquet Theory for Partial Differential Equations*, Birkhäuser Verlag, Basel.

LANDAU L. and LIFCHITZ E. [1989] *Mécanique des Fluides*, 2nd edition, revised and completed, Editions Mir & Editions Librairie du Globe, Moscow (translated from Russian).

LANZA DE CRISTOFORIS M. and ANTMAN S.S. [1991] The large deformation of nonlinearly elastic tubes in two-dimensional flows, *SIAM J. Math. Anal.* **22**, 1193–1221.

LIONS J.L. [1968] *Sur le Contrôle Optimal de Systèmes Gouvernés par des Equations aux Dérivées Partielles*, Dunod, Paris.

LIONS J.L. [1969] *Quelques Méthodes de Résolution des Problèmes aux Limites Non Linéaires*, Dunod, Paris.

LIONS J.L. [1988] *Contrôlabilité Exacte, Perturbations et Stabilisation de Systèmes Distribués. Tome 1: Contrôlabilité Exacte*, Masson, Paris.

LIONS J.L. and MAGENES E. [1968] *Problèmes aux Limites Non Homogènes et Applications. Vol. 1*, Dunod, Paris.

LOJACIEWICZ S. [1991] *Complex Analytic Geometry*, Birkhäuser, Boston.

MAZ'JA V.G. [1985] *Sobolev Spaces*, Springer-Verlag, Berlin.

MORAND H. and OHAYON R. [1992] *Interactions Fluides-Structures*, RMA **23**, Masson, Paris.

MÜLLER C. [1966] *Spherical Harmonics*, Lectures Notes in Mathematics **17**, Springer-Verlag, Berlin.

MURAT F. [1977-78] *H-Convergence*, Séminaire d'Analyse Fonctionnelle et Numérique de l'Université d'Alger, mimeographed notes. English translation [1994], in *Topics in the Mathematical Modeling of Composite Materials*, R. KOHN ed., Series Progress in Nonlinear Differential Equations and their Applications, Birkhäuser Verlag, Boston.

NEČAS J. [1967] *Les Méthodes Directes en Théorie des Equations Elliptiques*, Masson, Paris.

NGUETSENG G. [1989] A general convergence result for a functional related to the theory of homogenization, *SIAM J. Math. Anal.* **20**, 608–623.

NGUETSENG G. [1990] Asymptotic analysis for a stiff variational problem arising in mechanics, *SIAM J. Math. Anal.* **21**, 1394–1414.

NIKODYM [1933] Sur une classe de fonctions considerées dans le problème de Dirichlet, *Fund. Math.* **21**, 129–150.

ODEH F. and KELLER J.B. [1964] Partial differential equations with periodic coefficients and Bloch waves in crystals, *J. Math. Phys.* **5**, 1499–1504.

PAIDOUSSIS M.P. [1966] Dynamics of flexible cylinders in axial flow theory. Parts I and II, *J. Fluid Mech.* **26**, 717–751.

PALIS J. and DE MELO W. [1982] *Geometric Theory of Dynamical Systems*, Springer-Verlag, New York.

PLANCHARD J. [1980] Computation of the acoustic eigenfrequencies of cavities containing a tube bundle, *Comput. Methods Appl. Mech. Engrg.* **24**, 125–135.

PLANCHARD J. [1982] Eigenfrequencies of a tube-bundle placed in a confined fluid, *Comput. Methods Appl. Mech. Engrg.* **30**, 75–93.

PLANCHARD J. [1985] Vibrations of nuclear fuel assemblies: a simplified model, *Nuclear Engineering and Design* **86**, 383–391.

PLANCHARD J. [1987] Global behaviour of large elastic tube bundles immersed in a fluid, *Comput. Mech.* **2**, 105–118.

PLANCHARD J. [1988] Comportement des faisceaux de tubes immergés, in *Aspects Théoriques et Numériques de la Dynamique des Structures* (Ecole d'Eté d'Analyse Numérique, CEA-EDF-INRIA), Y. BAMBERGER et al., Collection de la Direction des Etudes et Recherches d'Electricité de France **70**, 165–242, Editions Eyrolles, Paris.

PLANCHARD J. and THOMAS B. [1993] On the dynamical stability of cylinders placed in cross-flow, *J. Fluids Structures* **7**, 321–339.

RAVIART P.A. and THOMAS J.M. [1983] *Introduction à l'Analyse Numérique des Equations aux Dérivées Partielles*, Masson, Paris.

RAYLEIGH L. (Sir John William Strutt) [1877] *The Theory of Sound*, 2nd edition, revised [1945], Dover Publications Inc., New York.

REED M. and SIMON B. [1972-78] *Methods of Modern Mathematical Physics. I. Functional Analysis, II. Fourier Analysis and Self-Adjointness, III. Scattering Theory, IV. Analysis of Operators*, Academic Press, New York.

RÉMY F.N. [1982] Flow-induced vibrations of tube bundles in two phase cross flow, in *Proceedings UKAEA/BNES Third Keswick International Conference Vibrations in Nuclear Plants*, Paper 1.9, Keswick.

RIESZ F. and NAGY B.Sz. [1955] *Leçons d'Analyse Fonctionnelle*, 3rd edition, Gauthier-Villars & Akadémiai Kiadó, Paris-Budapest, reprinted version [1990], Editions Jacques-Gabay, Paris.

ROSEAU M. [1966] *Vibrations Nonlinéaires*, Springer-Verlag, Berlin.

ROSEAU M. [1987] *Vibrations in Mechanical Systems*, Springer-Verlag, Berlin.

RUDIN W. [1979] *Functional Analysis*, 2nd edition, Tata Mc Graw-Hill, New Delhi.

SÁNCHEZ-HUBERT J. and SÁNCHEZ-PALENCIA E. [1989] *Vibration and Coupling of Continuous Systems. Asymptotic Methods*, Springer-Verlag, Berlin.

SÁNCHEZ-HUBERT J. and TURBÉ N. [1986] Ondes élastiques dans une bande périodique, *RAIRO Modél. Math. Anal. Numér.* **20**, 539–561.

SÁNCHEZ-PALENCIA E. [1980] *Non-Homogeneous Media and Vibration Theory*, Lecture Notes in Physics **127**, Springer-Verlag, Berlin.

SCHWARTZ L. [1957] Théorie des distributions à valeurs vectorielles (I), *Ann. Inst. Fourier* **7**, 1–142.

SCHWARTZ L. [1965] *Méthodes Mathématiques pour les Sciences Physiques*, Hermann, Paris.

SCHWARTZ L. [1966] *Théorie des Distributions*, 2nd edition [1978], revised and completed, Hermann, Paris.

SOBOLEV S.L. [1964] *Partial Differential Equations of Mathematical Physics*, Addison-Wesley, New York.

STOKER J.J. [1950] *Nonlinear Vibrations in Mechanical and Electrical Systems*, Interscience Publishers Inc., New York.

TARTAR L. [1977] *Problèmes d'Homogénéisation dans les Equations aux Dérivées Partielles*, Cours Peccot au Collège de France, partially written in MURAT [1977-78].

TARTAR L. [1978] *Topics in Nonlinear Analysis*, Publications Mathématiques d'Orsay **78.13**, University of Paris-Sud, Paris.

TARTAR L. [1980] *Incompressible Fluid Flow in a Porous Medium. Convergence of the Homogenization Process*, Appendix in the book by SÁNCHEZ-PALENCIA [1980].

TEMAM R. [1977] *Navier-Stokes Equations*, North-Holland, Amsterdam.

THOMAS L.E. [1973] Time dependent approach to scattering from impurities in a crystal, *Comm. Math. Phys.* **3**, 335–343.

THOMPSON R.C. [1976] The behaviour of eigenvalues and singular values under perturbations of restricted rank, *Linear Algebra Appl.* **13**, 69–78.

VAN DER WAERDEN B.L. [1953] *Modern Algebra. Vol. I*, revised English edition, Frederick Ungar Publishing Co., New York.

VANNINATHAN M. [1981] Homogenization of eigenvalue problems in perforated domains, *Proc. Indian Acad. Sci. Math. Sci.* **90**, 239–271.

WEAVER D.S. and KOROYANNAKIS R. [1982] The cross-flow response of a tube array in water; a comparison with the same in air, *ASME J. Pressure Vessel Tech.* **104**, 139–146.

WILCOX C. [1978] Theory of Bloch waves, *J. Anal. Math.* **33**, 146–167.

WILKINSON J. [1965] *The Algebraic-Eigenvalue Problem*, Clarendon Press, Oxford.

YOSIDA K. [1980] *Functional Analysis*, 6th edition, Springer-Verlag, Berlin.

ZIMAN J.M. [1972] *Principles of the Theory of Solids*, 2nd edition, Cambridge University Press, London.

Subject Index

A

absolutely continuous spectrum, 68, 69, 204
acoustic resonator, 21, 164
added mass matrix, 152, 302
adjoint operator, 34
Allaire Theorem, 283
Analytic Implicit Function Theorem, 13
anisotropic case, 305
Arzelà-Ascoli Theorem, 12, 40
Atkinson's Theorem, 202

B

Banach space-valued function, 12
Banach-Steinhaus Theorem, 84, 87
band structure, 192, 237, 273, 305
Beppo-Levi space(s), 174, 175
Bloch spectrum, 306
Bloch Wave(s), 63, 185, 187, 233, 271
— coefficients, 192, 195, 235, 272
— — method, 184, 209, 210, 260
Continuous — — Decomposition Theorem, 192, 233, 272
Discrete — — Decomposition Theorem, 215
Bochner integral, 10
bounded domain, 3
— operator, 31
Bramble-Hilbert Lemma, 177
Brillouin zone
first — —, 187

C

Carathéodory class, 284
Cauchy problem, 275
cell
— problem, 291

reciprocal —, 187, 188
reference —, 186, 188
closed operator, 31
Closed Graph Theorem, 25, 29
complex-analytic
— — function, 12
— — of type (A), 199
compressible
— case model, 97
— Stokes model, 100, 319
condition (P), 180
conjugate-gradient method, 246
continuity equation, 96, 99
continuous spectrum, 33
convergence
— of spectral families, 84, 86, 87, 240, 245, 273, 304, 311
point-wise —, 54, 90
strong —, 9, 84, 86
strong two-scale —, 284
weak —, 9, 87
weak two-scale —, 283
corrector result, 284, 291
Courant-Fischer Theorem, 47

D

De Rham's Theorem, 115
Dini's Theorem, 54
Dirac mass, 27, 176
Dirichlet
— boundary condition, 15, 19, 62, 73, 156, 229, 295
— problem, 8, 289
discrete spectrum, 67
distribution(s), 2
— derivative, 10, 61
tempered —; see tempered distributions
Hilbert space-valued —, 9
divergence-free, 99
domain
bounded —, 3

SNEL S.A.
Rue Saint-Vincent 12 - 4020 Liège
août 1995